Susceptibility to Infectious Diseases

In the past ten years, substantial progress has been made in identifying why some people are particularly susceptible to specific infectious diseases. Extensive evidence has now accumulated that host genes are important determinants of the outcome of infection for many common pathogens. This book summarises the advances that have been made in understanding the complexity of host genetic susceptibility. The diseases covered include those of great public health importance, such as malaria and HIV, and those of current topical interest, such as Creutzfeldt–Jakob disease. Many different techniques have been used to identify host genes involved in infectious disease susceptibility. Each chapter describes how these discoveries were made, and the book is therefore useful to anyone planning genetic studies on a multifactorial disease, regardless of whether it has an infectious etiology.

RICHARD BELLAMY is an infectious diseases physician with research interests in host genetics, tuberculosis, HIV, and public health in developing countries. He is currently Director of the Obaapavita Trial, a study on the effects of vitamin A supplements on maternal mortality in rural Ghana.

Published titles

1. *Bacterial Adhesion to Host Tissues*. Edited by Michael Wilson 0521801079
2. *Bacterial Evasion of Host Immune Responses*. Edited by Brian Henderson and Petra Oyston 0521801737
3. *Dormancy and Low-Growth States in Microbial Disease*. Edited by Anthony R. M. Coates 0521809401

Forthcoming titles in the series

Bacterial Invasion of Host Cells. Edited by Richard Lamont 0521809541
Mammalian Host Defence Peptides. Edited by Deirdre Devine and Robert Hancock 0521822203
The Dynamic Bacterial Genome. Edited by Peter Mullany 0521821576
Bacterial Protein Toxins. Edited by Alistair Lax 052182091X
The Influence of Bacterial Communities on Host Biology. Edited by Margaret McFall Ngai, Brian Henderson, and Edward Ruby 0521834651
The Yeast Cell Cycle. Edited by Jeremy Hyams 0521835569
Salmonella Infections. Edited by Pietro Mastroeni and Duncan Maskell 0521835046

Over the past decade, the rapid development of an array of techniques in the fields of cellular and molecular biology has transformed whole areas of research across the biological sciences. Microbiology has perhaps been influenced most of all. Our understanding of microbial diversity and evolutionary biology, and of how pathogenic bacteria and viruses interact with their animal and plant hosts at the molecular level, for example, have been revolutionized. Perhaps the most exciting recent advance in microbiology has been the development of the interface discipline of cellular microbiology, a fusion of classical microbiology, microbial molecular biology, and eukaryotic cellular and molecular biology. Cellular microbiology is revealing how pathogenic bacteria interact with host cells in what is turning out to be a complex evolutionary battle of competing gene products. Molecular and cellular biology are no longer discrete subject areas but vital tools and an integrated part of current microbiological research. As part of this revolution in molecular biology, the genomes of a growing number of pathogenic and model bacteria have been fully sequenced, with immense implications for our future understanding of microorganisms at the molecular level.

Advances in Molecular and Cellular Microbiology is a series edited by researchers active in these exciting and rapidly expanding fields. Each volume will focus on a particular aspect of cellular or molecular microbiology and will provide an overview of the area, as well as examine current research. This series will enable graduate students and researchers to keep up with the rapidly diversifying literature in current microbiological research.

ADVANCES IN MOLECULAR AND CELLULAR MICROBIOLOGY

AMCM

Contents

Contributors

Laurent Abel
Human Genetics of Infectious Diseases
INSERM U550
Necker Medical School
University René Descartes
156 rue de Vaugirard
75015 Paris, France

Alexandre Alcaïs
Human Genetics of Infectious Diseases
INSERM U550
Necker Medical School
University René Descartes
156 rue de Vaugirard
75015 Paris, France

Abdulhamid M. Alkout
Department of Medical Microbiology
The Medical School
University of Edinburgh
Teviot Place
Edinburgh, Scotland, United Kingdom

Osama M. Almadani
Department of Medical Microbiology and Forensic Medicine Unit
The Medical School
University of Edinburgh
Teviot Place
Edinburgh, Scotland, United Kingdom

Laurent Argiro
Immunologie et Génétique des Maladies Parasitaires
Institut National de la Santé et de la Recherche Médicale
INSERM U399
Laboratoire de Parasitologie–Mycologie
Faculté de Médicine
Marseille, France

Richard Bellamy
Kintampo Health Research Centre
P.O. Box 200
Kintampo, Ghana

Matthew Bishop
Departments of Pathology and Clinical Neurosciences
CJD Surveillance Unit
University of Edinburgh
Western General Hospital
Crewe Road
Edinburgh, Scotland, United Kingdom

C. Caroline Blackwell
Discipline of Immunology and Microbiology and Hunter Immunology Unit
University of Newcastle
Newcastle, Australia
and
Institute for Scientific Evaluation of Naturopathy
University of Cologne
Cologne, Germany

Marion Bonnet
Laboratory of Human Genetics of Infectious Diseases
Université René Descartes
INSERM U550
Necker Medical School
156 rue de Vaugirard
75730 Paris, Cedex 15, France

J. Matthias Braun
Department of Medical Microbiology
The Medical School

University of Edinburgh
Teviot Place
Edinburgh, Scotland, United Kingdom
and
Institute for Scientific Evaluation of Naturopathy
University of Cologne
Cologne, Germany

Anthony Busuttil
Forensic Medicine Unit
The Medical School
University of Edinburgh
Teviot Place
Edinburgh, Scotland, United Kingdom

Jean-Laurent Casanova
Laboratory of Human Genetics of Infectious Diseases
Université René Descartes
INSERM U550
Necker Medical School
156 rue de Vaugirard
75730 Paris, Cedex 15, France

Christophe Chevillard
Immunologie et Génétique des Maladies Parasitaires
Institut National de la Santé et de la Recherche Médicale
INSERM U399
Laboratoire de Parasitologie–Mycologie
Faculté de Médicine
Marseille, France

Alan W. Cuthbert
Department of Medicine
University of Cambridge
Addenbrooke's Hospital
Level 5 (box 157)
Hills Road
Cambridge CB2 2QQ, United Kingdom

Alain J. Dessein
Immunologie et Génétique des Maladies Parasitaires
Institut National de la Santé et de la Recherche Médicale

INSERM U399
Laboratoire de Parasitologie–Mycologie
Faculté de Médecine
Marseille, France

Hélia Dessein
Immunologie et Génétique des Maladies Parasitaires
Institut National de la Santé et de la Recherche Médicale
INSERM U399
Laboratoire de Parasitologie–Mycologie
Faculté de Médecine
Marseille, France

Omar R. El Ahmer
Department of Medical Microbiology
The Medical School
University of Edinburgh
Teviot Place
Edinburgh, Scotland, United Kingdom

Nasureldin El Wali
Institute of Nuclear Medicine and Molecular Biology
University of Gezira
Wad Medani, Sudan

Philippe Gros
Department of Biochemistry
Centre for the Study of Host Resistance
McGill Cancer Centre
McGill University
3655 Drummond
Montreal, Quebec, Canada H3G 176

Tyler Harris
Nuffield Department of Clinical Laboratory Sciences and National Blood
Service – Oxford Centre
John Radcliffe Hospital
Headington, Oxford OX3 9DU, United Kingdom

Sandrine Henri
Immunologie et Génétique des Maladies Parasitaires
Institut National de la Santé et de la Recherche Médicale

INSERM U399
Laboratoire de Parasitologie–Mycologie
Faculté de Médecine
Marseille, France

Dominique Hillaire
Immunologie et Génétique des Maladies Parasitaires
Institut National de la Santé et de la Recherche Médicale
INSERM U399
Laboratoire de Parasitologie–Mycologie
Faculté de Médecine
Marseille, France

J. W. Ironside
Departments of Pathology and Clinical Neurosciences
CJD Surveillance Unit
University of Edinburgh
Western General Hospital
Crewe Road
Edinburgh, Scotland, United Kingdom

Dominic L. Jack
Institute of Child Health
University College London
30 Guilford Street
London WC1N 1EH
United Kingdom and
Division of Genomic Medicine
University of Sheffield Medical School
Beech Hill Road
Sheffield S10 2RX, United Kingdom

Valerie S. James
Department of Medical Microbiology
The Medical School
University of Edinburgh
Teviot Place
Edinburgh, Scotland, United Kingdom

Richard A. Kaslow
Department of Epidemiology and International Health
School of Public Health

University of Alabama at Birmingham
Birmingham, Alabama 35294, USA

Gachuhi Kimani
Kenya Medical Research Institute
Biomedical Sciences Research Centre
Nairobi, Kenya

Nigel Klein
Institute of Child Health
University College London
30 Guilford Street
London WC1N 1EH, United Kingdom

Doris A. C. Mackenzie
Department of Medical Microbiology
The Medical School
University of Edinburgh
Teviot Place
Edinburgh, Scotland, United Kingdom

Mubarak Magzoub
Institute of Nuclear Medicine and Molecular Biology
University of Gezira
Wad Medani, Sudan

Sandrine Marquet
Immunologie et Génétique des Maladies Parasitaires
Institut National de la Santé et de la Recherche Médicale
INSERM U399
Laboratoire de Parasitologie–Mycologie
Faculté de Médecine
Marseille, France

Carole Eboumbou Moukoko
Immunologie et Génétique des Maladies Parasitaires
Institut National de la Santé et de la Recherche Médicale
INSERM U399
Laboratoire de Parasitologie–Mycologie
Faculté de Médecine
Marseille, France

Aluizio Prata
Faculty of Medicine do Triangulo Mineiro
Ubéraba, Brazil

David J. Roberts
Nuffield Department of Clinical Laboratory Sciences and National Blood
Service – Oxford Centre
John Radcliffe Hospital
Headington, Oxford OX3 9DU, United Kingdom

Virmondes Rodriques, Jr.
Faculty of Medicine do Triangulo Mineiro
Ubéraba, Brazil

Erwin Schurr
Centre for the Study of Host Resistance
Departments of Medicine and Human Genetics
McGill University
3655 Drummond
Montreal, Quebec, Canada H3G 176

Claire Soudais
Laboratory of Human Genetics of Infectious Diseases
Université René Descartes
INSERM U550
Necker Medical School
156 rue de Vaugirard
75730 Paris, Cedex 15, France

Jianming Tang
Division of Geographic Medicine
Department of Medicine
School of Medicine
University of Alabama at Birmingham
Birmingham, Alabama 35294, USA

Mark R. Thursz
Imperial College
Faculty of Medicine at St. Mary's Hospital
London W2 1NY, United Kingdom

Malcolm W. Turner
Institute of Child Health
University College London
30 Guilford Street
London WC1N 1EH, United Kingdom

Mauno Vihinen
Institute of Medical Technology
FIN-33014 University of Tampere Finland and
Tampere University Hospital
FIN-33520 Tampere, Finland

Donald M. Weir
Department of Medical Microbiology
The Medical School
University of Edinburgh
Teviot Place
Edinburgh, Scotland, United Kingdom

Thomas Williams
Wellcome Trust–KEMRI Centre for Geographic Medicine
P.O. Box 230
Kilifi, Kenya
and
Department of Paediatrics
Faculty of Medicine
Imperial College of Science Technology and Medicine
Exhibition Road
London SW7 2 AZ, United Kingdom

Leland J. Yee
Imperial College
Faculty of Medicine at St. Mary's Hospital
London W2 1NY, United Kingdom
and
London School of Hygiene and Tropical Medicine
London WC1E 7HT, United Kingdom

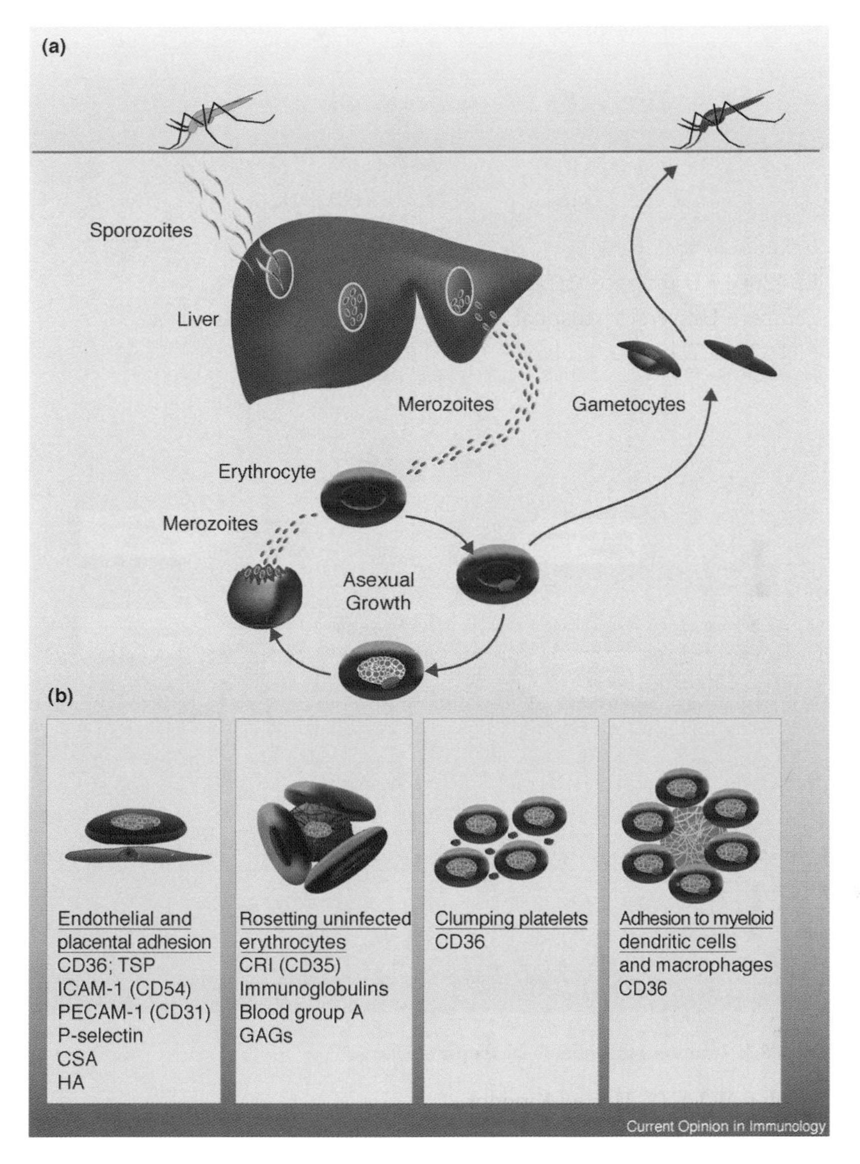

Figure 6.1. The life cycle of the malaria parasite.

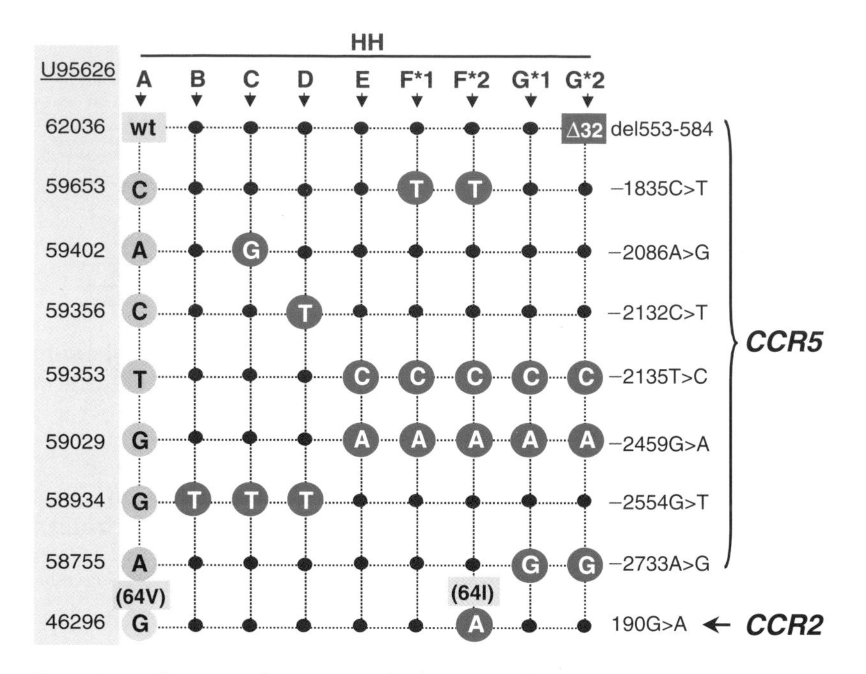

Figure 7.3. Delineation of *CCR2-CCR5* haplotypes on chromosome 3.

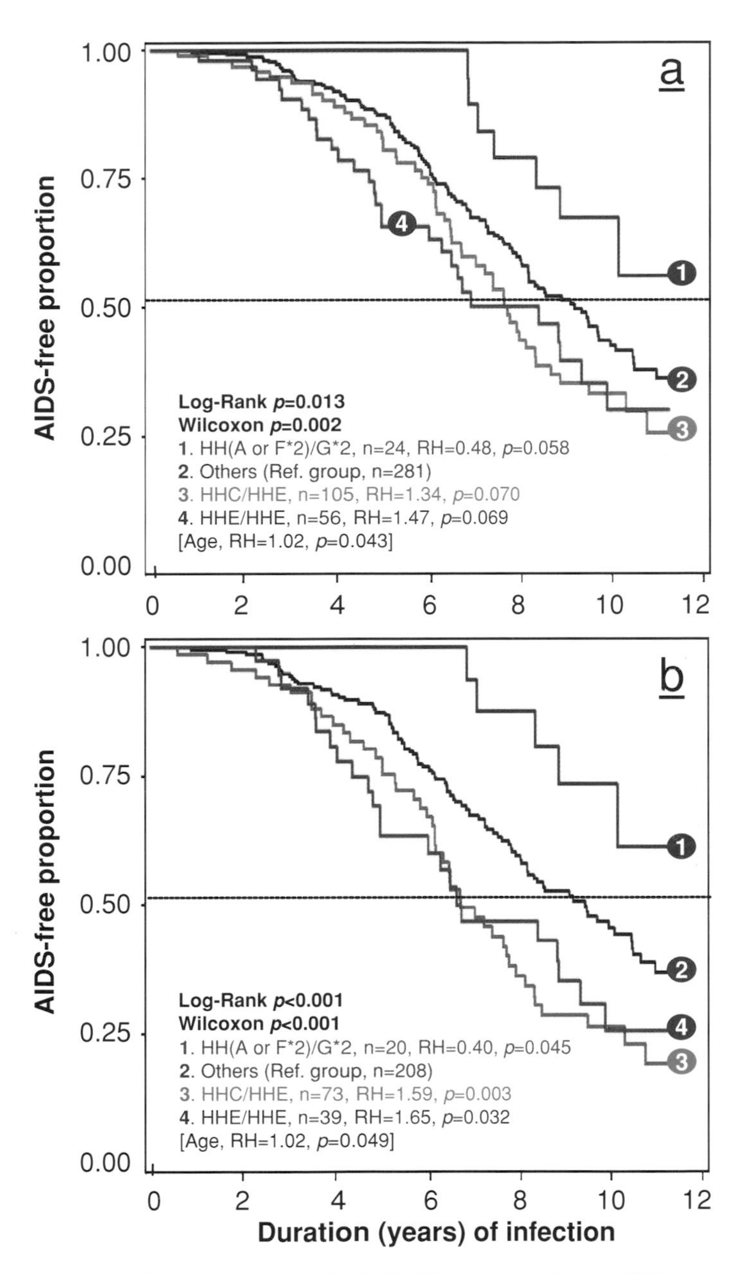

Figure 7.4. Rates of progression to AIDS (CDC 1987) during 11.5 years of follow-up among 470 (a) and 340 (b) HIV-1 seroconverted Caucasians not receiving antiretroviral therapy.

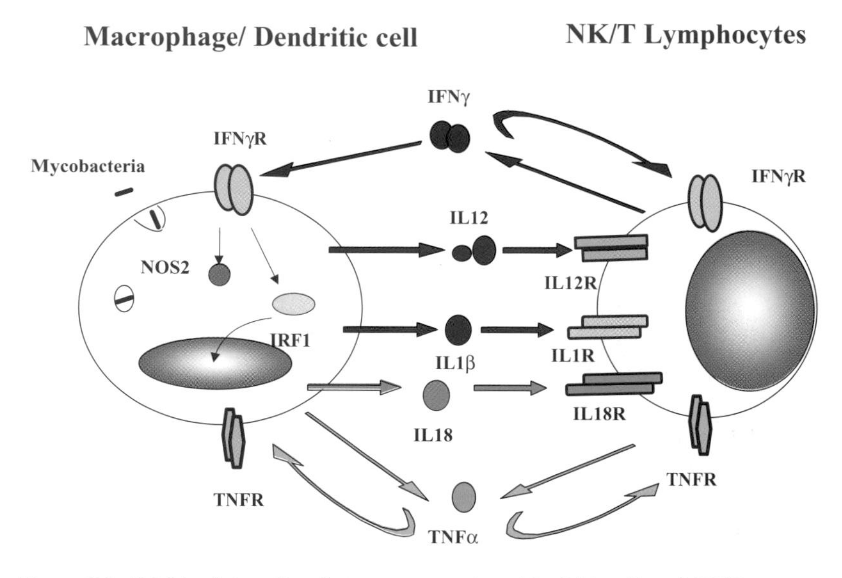

Figure 9.1. Cytokine interactions between macrophage/dendritic cells and NK/T lymphocytes.

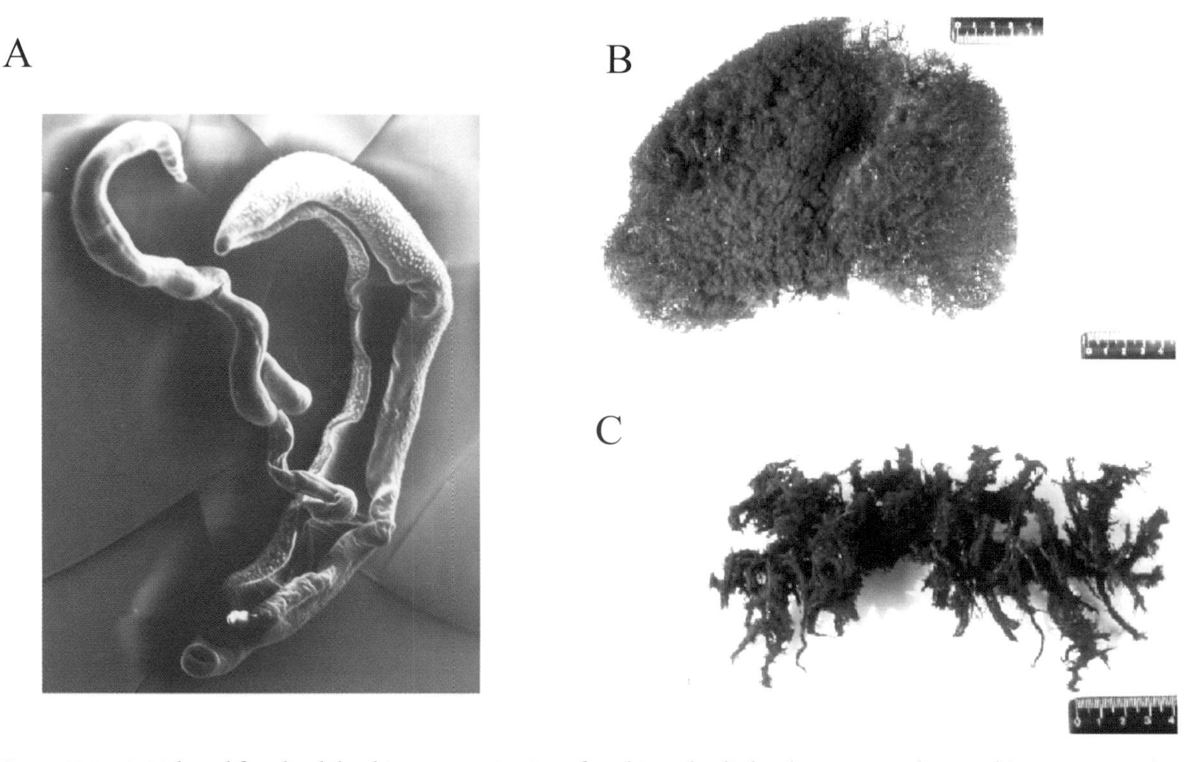

Figure 12.1. (A) Male and female adult schistosomes. (B) Liver of a subject who died with severe PPF due to *Schistosoma mansoni* infection. (C) The hepatic vascular tree injected with acrylic blue liquid resin.

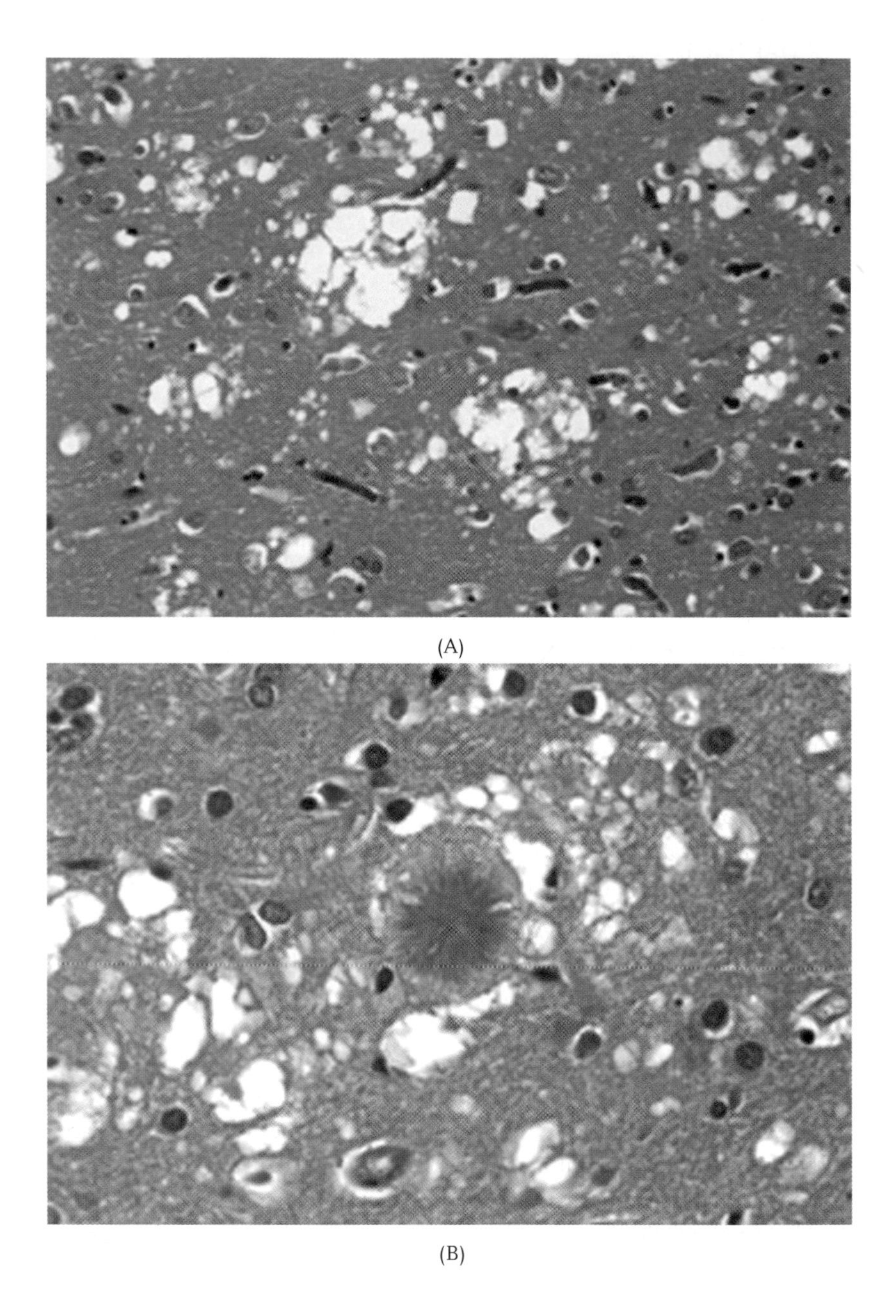

(A)

(B)

Figure 13.2. (A) Confluent spongiform degeneration in the neuropil of sporadic CJD (haematoxylin and eosin). (B) Florid plaque within the cerebral cortex of a vCJD case (haematoxylin and eosin).

CHAPTER 1

Introduction

Richard Bellamy
Kintampo Health Research Centre, Ghana

Patients suffering from a serious illness frequently ask "Why did this happen to me?" When the disease is cancer or cardiovascular disease, patients recognise the risk of inheriting "bad genes" from parents as readily as the risks from smoking and diet. It is all too clear that if both our parents suffered myocardial infarcts at an early age we must be at increased risk of the same thing happening to ourselves. However, when asked why someone developed a serious infection, we generally blame lack of acquired immunity, environmental factors, or bad luck. Increasingly it appears that "bad luck" really means the genes we have inherited.

It is a common misapprehension that our genes are not important in determining our ability to fight off infectious diseases. In fact a study of almost 1,000 adoptees in Denmark found that the host genetic component of susceptibility to premature death from infection is greater than for cancer and cardiovascular disease (Sorensen et al., 1988). This is not unexpected as common diseases which cause high mortality exert the greatest evolutionary effects on the human genome. Prior to this century infectious diseases were the major cause of death in the western world and still are in many developing countries. From this we can surmise that microorganisms have been the major selective force in recent human evolution. In other words the interaction between the genes of our ancestors and those of human pathogens have resulted in what makes each of us genetically unique today.

When a population is exposed to an environmental factor for many generations, evolution results in adaptation to it. This is most apparent in the differences in skin and eye colour which occur between populations exposed to different amounts of sunlight. Similarly, the longer a population has been exposed to an infectious disease, the more resistant we should expect the current members of that population to be to it. After several generations of exposure the more genetically susceptible individuals are killed off and the frequency of disease-resistance genes in the population increases. This can most clearly be seen for malaria because it is restricted to certain geographic regions and exerts high mortality. Comparisons of gene frequencies in different populations have enabled the identification of several genetic variants conferring malaria resistance, including sickle cell haemoglobin, glucose-6-phosphate dehydrogenase deficiency, α-thalassaemia and the Duffy-negative erythrocyte phenotype. Conversely, when a population is first exposed to an infectious disease, we should expect them to be highly susceptible to it. This was strikingly observed when the population of the Qu'Appelle Indian Reservation, Saskatchewan, first came into contact with tuberculosis in the 1890s. Initially the annual tuberculosis-related death rate was almost 10% of the population. After 40 years, when two generations had passed, more than half of the families were eliminated and the annual tuberculosis death rate had fallen to only 0.2%. This fall in mortality rate is believed to be because of "weaning out" of tuberculosis-susceptibility genetic factors (Motulsky, 1960). Massive death rates from measles, smallpox, and other infections, which occurred when the conquistadors first visited the Americas, are also likely to have been due to a combination of genetic susceptibility and lack of acquired immunity.

The following chapter, by Alcaïs and Abel, provides an overview of the approaches which can be used to identify the host genes involved in susceptibility to infectious diseases. It is clear that no single method could be used to identify all of these genes. A wide range of methods must be used to dissect out the complex genetic factors underlying the multifactorial aetiology of susceptibility to specific pathogens. The task is not easy and the results of different studies have sometimes been contradictory. However, the subsequent chapters of the book show that difficulties have been overcome and substantial progress has already been made in understanding genetic susceptibility to many different pathogens. A wide range of approaches has been used, including population linkage and association studies, extrapolations from mouse-models of disease, and *in vitro* studies of immune function. Each chapter illustrates a different scientific approach providing insight into the uses and limitations of each method.

In Chapter 3, Vihinen provides a summary of the large number of rare, monogenic immunodeficiency syndromes which have now been identified. In many cases the molecular basis underlying the condition has been identified and catalogues and databases now provide ready access to the current state of knowledge. Yee and Thursz, in Chapter 4, describe the extreme polymorphism of the major histocompatibility complex and how this has evolved in response to pressure from microorganisms. The present significance of this variability is shown by examples of how possessing particular human leukocyte antigen genotypes may increase our risk of developing serious complications following exposure to specific pathogens. Children with cystic fibrosis are more susceptible to infections with bacteria such as *Pseudomonas aeruginosa* and *Burkholderia cepacia*. In Chapter 5, Cuthbert discusses the molecular basis underlying cystic fibrosis and why it results in susceptibility to specific microorganisms. He also discusses the intriguing possibility that the common gene mutations causing cystic fibrosis have been selected for by conferring heterozygote resistance to other pathogens.

The greatest advances in our understanding of susceptibility to any single infectious disease have been with malaria. In Chapter 6, Roberts discusses how the geographical restriction of this disease and its high mortality have resulted in marked variability in the frequency of common variants in the haemoglobin and other genes. Human immunodeficiency virus (HIV) is a very new disease compared to malaria and has not yet had sufficient time to exert a major influence on the evolution of the human genome. However the recognition that some persons who had been repeatedly exposed to HIV had never become infected led to the suspicion that these subjects may have innate immunity to the condition. In Chapter 7, Tang and Kaslow describe how studies on such subjects determined that their macrophages could not be infected by HIV and led to the discovery that they lacked expression of the membrane protein, chemokine receptor 5, due to a 32-basepair deletion in this gene. In contrast, the discovery of the *Nramp1* gene was made by studying a mouse model of susceptibility to mycobacteria and other intracellular pathogens. Gros and Schurr, in Chapter 8, describe the long process of identifying this murine gene by positional cloning and the subsequent studies to ascertain its function. Large population studies have since been performed confirming that the human homologue of this gene is important in human susceptibility to tuberculosis. In Chapter 9, Casanova describes how a very different approach was used to identify how five genes in the interferon-γ signalling pathway are involved in human susceptibility to mycobacterial infections. Investigation of children who suffered from recurrent infections with atypical mycobacteria, or who developed disseminated infections following vaccination with bacille Calmette-Guerin, led to these discoveries.

Mannose binding lectin (MBL) deficiency is discussed by Jack, Klein, and Turner in Chapter 10. This defect of opsonisation was first described in a child with recurrent bacterial infections in 1968. Since then many infectious diseases have been found to be associated with MBL deficiency. The three principal gene mutations causing MBL deficiency have been found at very high frequency in most populations studied. Whether these gene variants have been selected for by conferring resistance to an intracellular pathogen is still uncertain. Polymorphism in blood groups and secretor status may also have evolved due to exposure to an infectious disease. A large number of studies have found associations between infectious diseases and blood group phenotypes and/or secretor status. In Chapter 11, Blackwell and colleagues discuss how blood group antigens may act as receptors for microorganisms, facilitating mucosal colonisation, and tissue invasion.

In Chapter 12, Dessein et al. describe how they identified that two different genes influence immunity to *Schistosoma mansoni* and subsequent development of liver disease. A gene in the cytokine cluster on chromosome 5q31–33 influences worm burden and a gene on chromosome 6q21–23, in the region of the interferon-γ receptor 1 gene, determines who will develop periportal fibrosis. In the final chapter Bishop and Ironside discuss prion diseases. The interactions between the host genotype and the different prion proteins offer valuable insight into the nature of these diseases. This has proved valuable in developing models of the likely future epidemic curve for new variant Creutzfeldt–Jakob disease.

This is an exciting time to be studying the role of genetic factors in multifactorial diseases. With the success of the human genome project and advances in molecular biology and bioinformatics, significant advances in our understanding of complex diseases should be forthcoming. For many infectious diseases it is clear that interaction between many host genes and environmental factors will be involved in determining the outcome of infection. The greater the number of genes involved, the more difficult it will be to predict who will develop a particular infection and who will die from it. For example, it is uncertain if it will ever be possible to predict exactly who will develop new variant Creutzfeldt–Jakob disease, tuberculosis, or cerebral malaria and who will not. However, identifying host disease-susceptibility genes will provide valuable insight into disease pathogenesis.

The chapters in this book discuss how the advances, which have been made in host genetics, may eventually find applications in the development of novel therapeutic and preventative strategies. Identification of MHC associations with disease may lead to development of vaccines, the cystic fibrosis transmembrane regulator may eventually be replaced by gene therapy, chemokine receptor blockers may be used in the treatment of HIV, mannose binding lectin replacement may become available, and antiadhesion therapy may be used to stop pathogens binding to host cell receptors. For those working in this field there is still much work to be done. In this era of emerging infections and antibiotic-resistant bacteria, physicians need every possible weapon to combat human pathogens. Advancing our understanding of the interaction between host and pathogen genomes should hopefully provide some new weapons to add to our arsenal.

REFERENCES

Motulsky, A. G. (1960). Metabolic polymorphisms and the role of infectious diseases in human evolution. *Hum. Biol.* **32**, 28–62.

Sorensen, T. I. A., Nielsen, G. G., Andersen, P. K., et al. (1988). Genetic and environmental influences on premature death in adult adoptees. *N. Engl. J. Med.* **318**, 727–732.

CHAPTER 2

Application of genetic epidemiology to dissecting host susceptibility/resistance to infection illustrated with the study of common mycobacterial infections

Alexandre Alcaïs and Laurent Abel

Human Genetics of Infectious Diseases, Necker Medical School, University René Descartes

1. INTRODUCTION

In the general context of genetic dissection of complex traits, genetics of human infectious diseases present the following several advantages and specificities: (1) there is a known causative agent which is absolutely required to become infected and to get the disease, but generally not sufficient stressing the importance of the host background; (2) environmental factors influencing the risk of infection are generally known and can be taken into account in the analysis when they are accurately measured; (3) there is a strong orientation in the choice of candidate genes based on the function of the gene and its known role in the response to the studied pathogen and/or on mouse–human chromosome homology exploiting the identification of murine resistance loci; and (4) the identification of major genes involved in the response to a given infectious pathogen takes advantage of the opportunity to study several complementary traits related to this pathogen. Among these traits are clinical phenotypes which are usually binary (affected/unaffected) but can take into account time to onset of the disease (e.g., time of progression to AIDS for HIV-infected patients), biological phenotypes measuring infection which can be either quantitative (e.g., infection intensities measured by fecal egg counts in schistosomiasis) or binary (HIV seropositive/seronegative), and immunological phenotypes measuring the immune response (antibody or cytokine levels, skin test response, etc.) more or less specific to a given antigen. The panel of phenotypes available for a given infectious disease allows performing complementary studies in order to investigate the genetic control of the different steps of the pathogenic process leading to the disease itself.

In humans, the role of genetic factors in infectious diseases has been suggested by several observations (Abel, 2002; Abel and Dessein, 1998). One of the most important, and probably the oldest one, is the very large variability in the response observed between individuals exposed to the same infectious agent as follows: (1) there is generally a certain fraction of subjects who are never infected; (2) among subjects who are infected there is a strong variation in infection levels (when they can be measured); (3) only some infected subjects will develop a clinical disease; and (4) among subjects presenting with the disease there is a large spectrum of clinical manifestations (more or less severe, early or late onset, etc.). A dramatic illustration of this variable response was provided by the accidental immunisation of children with a virulent strain of *Mycobacterium tuberculosis* in 1926 in Lübeck, Germany. Of 251 children who received the same dose of mycobacteria 77 died and 127 had radiological signs of disease, whereas 47 showed no detectable disease. Furthermore, this large interindividual variability often contrasts with intraethnic and intrafamilial similarities. Familial clustering has been found in most of the infectious diseases studied, raising the problem of the distinction between environmental (family shared factors increasing the risk of infection) and genetic causes of familial aggregation. Twin studies have been helpful to estimate the extent of the genetic contribution and have showed higher concordance rates in monozygotic than dizygotic twins in many infectious diseases (e.g., leprosy and tuberculosis). Subsequently, the specific statistical methods of genetic epidemiology are used to further investigate this genetic contribution and identify the main gene(s) involved in the control of infectious disease related phenotypes.

Genetic epidemiology methods (Abel and Dessein, 1998; Khoury et al., 1993; Lander and Schork, 1994) combine epidemiological and genetic information with the ultimate goal to identify the genes (and the polymorphisms of these genes) which have a significant influence on the phenotype under study and the possible interactions of these genes with relevant environmental factors. Epidemiological data concern the collection of the measured risk factors which could influence the trait under study (e.g., factors influencing the contamination by the infectious agent, age). Genetic information is represented by the knowledge of the familial relationships between study subjects (collection of families) and the typing of genetic markers. The recent establishment of the genetic map of the human genome based on highly polymorphic markers (Dib et al., 1996) and the growing availability of single nucleotide polymorphisms (SNPs) located within candidate genes (Kruglyak, 1999; Wang et al., 1998) are now fundamental tools for these genetic studies. Numerous methods have been (and are being) developed which have their respective advantages and disadvantages, which are described in the following sections. Consequently, there is no unique optimal strategy to investigate genes involved in human infectious diseases, and the choice of a design for a particular study depends on several factors related to the phenotype (nature, frequency, familial distribution, etc.): the population, the accurate measure of environmental factors, and the known genetic background among other factors.

As summarised in Fig. 2.1, two basic types of situation can be observed in the human genetics of infectious diseases. In certain rare infections, the family structure (e.g., consanguineous parents) or the familial relationships between infected subjects suggest simple Mendelian inheritance (monogenic). Although rare, a number of Mendelian syndromes of susceptibility to infectious agents have been described, notably the predisposition to infection by weakly virulent mycobacteria (Casanova and Abel, 2002). Mendelian resistance to some pathogens, such as *Plasmodium vivax* (Barnwell et al., 1989; Miller et al., 1976), has also been described, and the molecular basis of the Duffy blood group system accounting for this resistance to *P. vivax* has been elucidated (Iwamoto et al., 1995; Tournamille et al., 1995). More commonly, the genetic predisposition is expected to be more complex (polygenic). The distinction between these two categories is somewhat blurred because other genes may have a substantial impact on the clinical expression of a Mendelian predisposition and because polygenic susceptibility may primarily reflect the effect of a predominant gene often referred to as a major gene (Abel and Casanova, 2000). Therefore, some aspects of the strategies for searching for the genetic factors in these two situations are specific and others are complementary. The following sections provide an overview of the two most popular tools used in genetic epidemiology and displayed in Fig. 2.1, i.e., linkage and association studies. The use of these methods is illustrated with the main findings obtained in the study of human genes influencing susceptibility to common mycobacterial infections.

2. LINKAGE ANALYSIS

In the analysis of complex traits such as infectious diseases, linkage studies are used to locate chromosomal regions containing the gene(s) of interest by either focusing on a few candidate regions or using a genome-wide search. The main interest of the whole genome approach is to ensure that all major loci involved in the control of a phenotype are identified and to provide the opportunity to discover new major genes, and consequently physiopathological pathways that were not previously suspected of contributing to the phenotype under study. Unfortunately, unlike the analysis of simple monogenic diseases, a fine mapping of the gene(s) of interest cannot be expected from

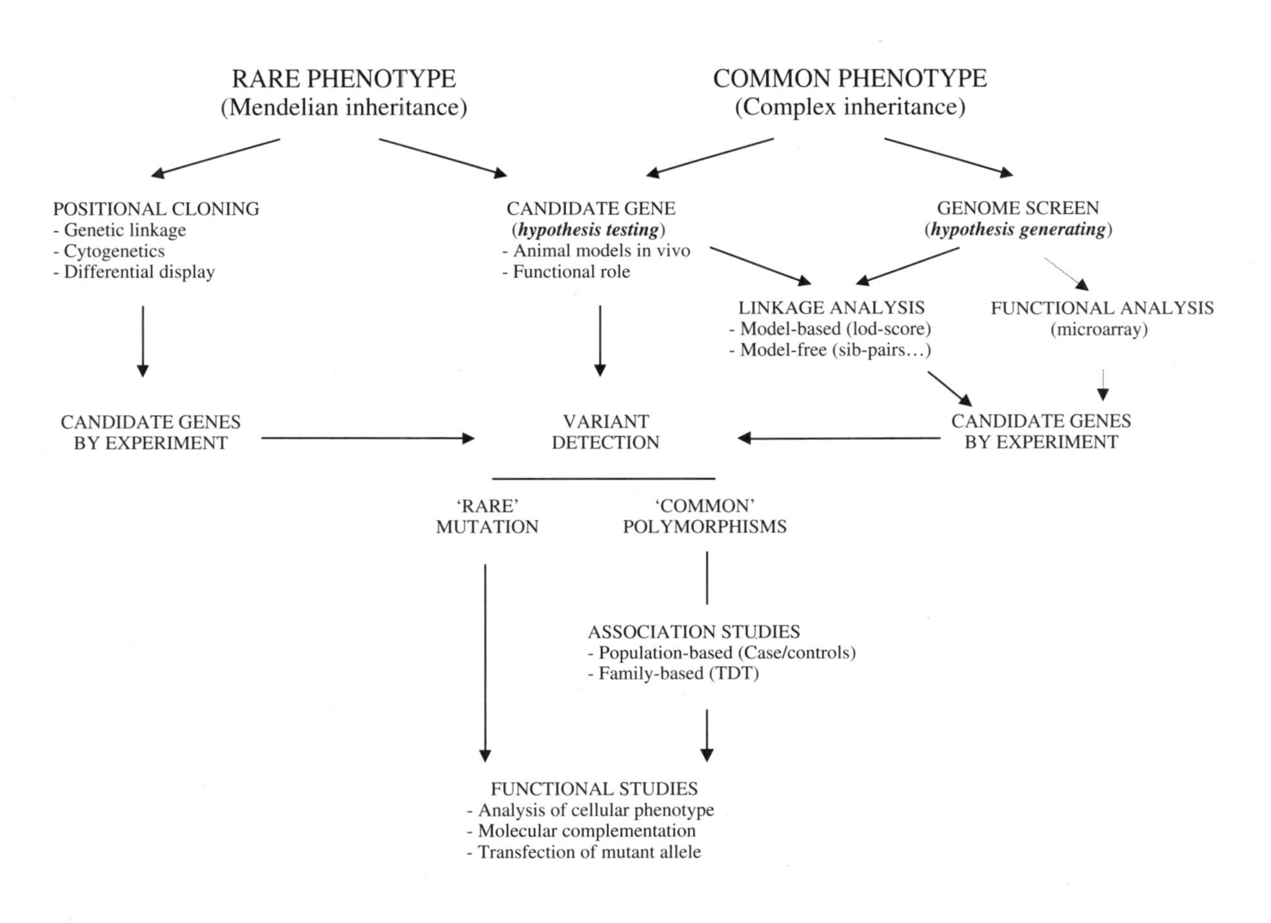

RARE PHENOTYPE
(Mendelian inheritance)
COMMON PHENOTYPE
(Complex inheritance)
POSITIONAL CLONING
- Genetic linkage
- Cytogenetics
- Differential display
CANDIDATE GENE
(hypothesis testing)
- Animal models in vivo
- Functional role
GENOME SCREEN
(hypothesis generating)
LINKAGE ANALYSIS
- Model-based (lod-score)
- Model-free (sib-pairs…)
FUNCTIONAL ANALYSIS
(microarray)
CANDIDATE GENES
BY EXPERIMENT
VARIANT
DETECTION
CANDIDATE GENES
BY EXPERIMENT
'RARE'
MUTATION
'COMMON'
POLYMORPHISMS
ASSOCIATION STUDIES
- Population-based (Case/controls)
- Family-based (TDT)
FUNCTIONAL STUDIES
- Analysis of cellular phenotype
- Molecular complementation
- Transfection of mutant allele

linkage studies of complex infectious phenotypes. When successful, linkage analyses generally identify a region of about 10–20 cM (~10,000 to 20,000 kb) which may still contain hundreds of genes. As detailed in the association studies section, the next step is to test the role of polymorphisms of candidate genes located within the identified region. The general principle of linkage analysis is to search for chromosomal regions that segregate nonrandomly with the phenotype of interest within families. According to what is known, or what one is disposed to assume, about the phenotype mode of inheritance, linkage analysis methods are usually classified as model-based or model-free. Although the terms parametric and nonparametric are sometimes used, they should be avoided since all model-free (nonparametric) approaches are actually parametric in the sense that they necessitate, more or less explicitly, the estimation of at least one parameter.

2.1 Model-Based Methods

Model-based linkage analysis by the classical lod-score method (Morton, 1955) requires one to define the model specifying the relationship between the phenotype and factors that may influence its expression, mainly a putative gene with two alleles (d, D) and other relevant risk factors, often referred to as the genetic (or phenotype/genotype) model. In the context of a binary

---

Figure 2.1. (*facing page*). Methods and strategies for identifying human mycobacterial susceptibility genes. The molecular basis of rare Mendelian predisposition to mycobacterial disease may be investigated using several strategies. Linkage analysis is usually the first step in the positional cloning approach, although the identification of visible cytogenetic abnormalities may be helpful. The candidate gene approach involves the prior selection of genes (generally based on studies of animal models or comparison with other human inherited disorders with a related clinical phenotype), which are then tested by functional assays and/or mutation detection. Another potentially fruitful strategy is based on studying the differential expression of genes in tissues from affected and healthy individuals. To determine the molecular basis of complex predisposition to common mycobacterial diseases, one may use candidate gene hypothesis testing as defined above or generate new hypotheses by means of genomewide screens. At present these screens are performed by linkage study but functional analysis (e.g., through DNA chips) is likely to be used in the near future. The role of polymorphisms within candidate genes defined a priori or identified "by experiment" (e.g., located within a linked region pinpointed by a linkage genome scan) is tested in association studies (which may be population based or family based). Statistical evidence for an association should be validated by functional studies, aimed at determining the impact of the polymorphism studied on gene function and, potentially, on the phenotype of interest.

phenotype (e.g., affected/unaffected and seronegative/seropositive), this genetic model should specify, in addition to the frequency of the deleterious allele denoted as D, the penetrance vector, i.e., the probability for an individual to be affected given his genotype (dd, Dd, or DD) and his own set of relevant covariates such as age, factors of exposure to the infectious agent, etc. In the context of a quantitative phenotype (e.g., infection levels), the complete specification of the genetic model includes, in addition to the frequency of the allele predisposing to high values of the trait denoted as D, the three genotype-specific means and variances which may also be influenced by some individual covariates; given the genotype, the distribution of the phenotype is then assumed to be normal so that the overall distribution is a mixture of three normal distributions. An elegant way to express this phenotype/genotype model is to use a regressive approach for binary (Bonney, 1986) as well as for quantitative (Bonney, 1984) traits.

The genetic model is generally provided and estimated by segregation analysis, which is the first step to determine from family data the mode of inheritance of a given phenotype. The aim of segregation analysis is to discriminate between the different factors causing familial resemblance, with the main goal to test for the existence of a single gene, called a major gene. The major gene term does not mean that it is the only gene involved in the expression of the phenotype, but that, among the set of involved genes, there is at least one gene with an effect important enough to be distinguished from the others. For a binary phenotype, this effect can be expressed in terms of relative risks, e.g., the ratio of the probability for a subject to be affected given he has a DD genotype to the same probability given he has a dd genotype. For a quantitative phenotype, this effect is measured by the proportion of the phenotypic variance explained by the major gene (heritability due to the gene). A detailed review of the pros and cons of segregation analysis can be found in Jarvik (1998).

When there is evidence for a major gene, model-based linkage analysis allows one to confirm and to locate this gene, denoted below as the phenotype locus. It tests in families whether the phenotype locus cosegregates with genetic markers of known chromosomal location and provides an estimation of the recombination rate between these two loci (Ott, 1999). Linkage with the phenotype locus can be tested marker by marker (two-point analysis) or considering a set of linked markers (multipoint analysis). In this analysis, as in segregation analysis, all the inferences for individual genotypes at the phenotype locus are made from the individual phenotypes and the specified phenotype/genotype model (e.g., according to his quantitative phenotypic value, the probability that an individual carries genotype dd, dD, or DD at the phenotype locus will be computed from the mixture of the three normal distributions described above and for which means and variances have been estimated through segregation analysis).

The lod-score approach is certainly the most powerful linkage method when the assumed genetic model is (or is close enough to) the true model. This is the case in a situation of monogenic inheritance where a simple genetic model can be assumed. A nice illustration in the syndrome of Mendelian susceptibility to mycobacterial infection was provided by the identification of *IFNGR1*, the gene encoding the interferon-γ receptor ligand-binding chain, in two kindreds (Jouanguy et al., 1996; Newport et al., 1996) by means of homozygosity mapping, a specific model-based linkage approach assuming inheritance of a rare recessive deleterious allele in consanguineous families (Lander and Botstein, 1987). However, a misspecification of the genetic model can lead to both severe loss of power to detect linkage (and therefore to false exclusion of the region containing the phenotype locus) and bias in the estimation of the recombination fraction (i.e., the genetic distance) between the phenotype locus and the marker locus (Clerget-Darpoux et al., 1986). Nevertheless, such a misspecification does not affect the robustness of the method, i.e., it does not lead to false conclusions in favor of linkage, as long as only one phenotype/genotype model is tested. When there is some knowledge about the prevalence of the disease under study and the level of familial aggregation, a common procedure to reduce the risk of misspecification is to generate a limited number of realistic genetic models to use in lod-score analysis. However, when performing the analysis under a number of different genetic models, one needs to introduce a correction for multiple testing and adjust the significance level of the lod score (MacLean et al., 1993). The same issue occurs when several markers are tested, and guidelines which have been proposed to adapt lod score thresholds to the context of the genome-wide search (Lander and Kruglyak, 1995) are detailed at the end of this section. Another source of problem arises when marker data are missing for some family members. In this case, linkage analysis also depends on marker allele frequencies and misspecification of these frequencies can affect both the power and the robustness of the method. Note that the two latter problems (multiple marker testing and misspecification of marker allele frequencies) are also common to model-free methods.

2.2 Model-Free Methods

Model-free linkage approaches (allele sharing methods such as sib-pair studies) allow locating the genetic factors influencing a phenotype without

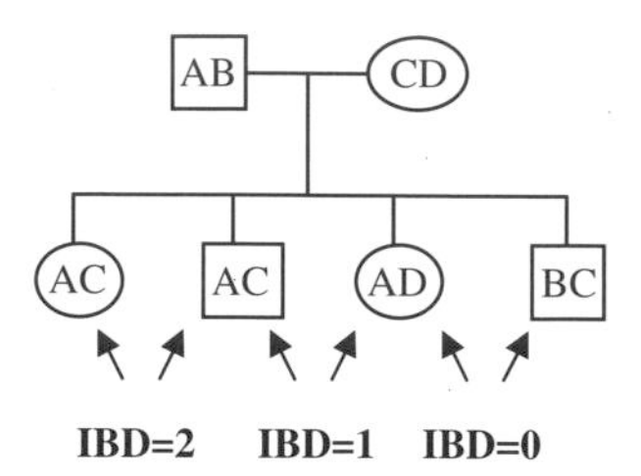

Figure 2.2. Expected distribution of Identical By Descent (IBD) alleles in a nuclear family. For simplicity the two parents are assumed to be heterozygous at a marker (AB × CD respectively). Two sibs can share 2, 1, or 0 alleles IBD. In a sample of sib-pairs, the expected IBD distribution is 0.25, 0.5, and 0.25 for IBD values of 2, 1, and 0, respectively. Overall, the mean proportion of alleles shared IBD by two sibs is 0.5. Different approaches have been developed to assess departure from the expected values that are based either on the IBD distribution or on the mean IBD value (see text).

specifying the phenotype/genotype model. Therefore, they are strongly recommended when little is known about this model (i.e., no segregation analysis has been performed or no clear major gene model can be inferred from segregation analysis). The general principle of model-free linkage analysis is to test whether relatives who have a certain phenotypic resemblance (e.g., affected relatives) share more marker alleles inherited from the same ancestor, i.e., alleles identical by descent (IBD), than expected under random segregation. The most commonly used model-free linkage analysis approach is the sib-pair method. As shown in Fig. 2.2, two sibs can share 0, 1, or 2 parental alleles IBD at any locus, and the respective proportions of this IBD sharing under random segregation are simply 0.25, 0.5, and 0.25 (overall, two sibs are expected to share 50% of their parental alleles).

When the phenotype under study is binary (affected/unaffected), the method tests whether *affected* sibling pairs share more parental alleles at the marker(s) of interest than randomly expected. This excess allele sharing can be tested by a simple χ^2 test, in particular when all parental marker data are known. The rationale for focusing on affected rather than unaffected subjects is that they are more informative and robust since their phenotypes are fixed, whereas we hardly know whether an unaffected subject will develop the disease later. In situations where unaffected status is less ambiguous (especially by using additional information brought by relevant covariates), however, it has been shown that incorporating such individuals into the analysis can increase the power (Alcais and Abel, 2001). Maximum likelihood methods have also been developed to analyse affected sib-pair data, such as a maximum likelihood score (MLS) (Risch, 1990) and a maximum likelihood binomial (MLB) approach (Abel et al., 1998a; Abel and Muller-Myhsok, 1998b), and can lead to more powerful tests. The MLB, which relies on the idea of binomial distributions of parental alleles among offspring, is of particular interest when the sample includes families with more than two affected sibs since it does not need to decompose the sibship into its constitutive sib-pairs. This latter strategy of decomposition can lead to large inflation of type I errors (i.e., false conclusions in favor of linkage) due to the nonindependence of the resulting pairs (Abel et al., 1998a; Holmans, 2001).

When the phenotype under study is quantitative, the general idea of the approach consists in testing whether sibs having close phenotype values share more alleles IBD than sibs having more distant values. Most of the methods differ only on the way that they quantify the phenotypic resemblance. The widely used regression-based approach proposed 30 years ago by Haseman and Elston (1972) regresses the squared difference of the sib-pair phenotypes on the expected proportion of alleles shared IBD by the sib-pair. Numerous developments have been performed on this method to account for multipoint analysis (Olson, 1995) or large pedigrees (Amos and Elston, 1989), and recent works have proposed to use the cross-product of the sib-pair phenotypes in addition to their difference (Elston et al., 2000; Xu et al., 2000). It is noteworthy that this latter approach is nothing more than a reformulation of the weighted pairwise correlation (WPC) method previously developed (Commenges et al., 1994). As a regression-based approach, the Haseman–Elston method relies on the same assumptions on the residuals (i.e., normal distribution, homosedasticity, and independence) of which violation can significantly impact on the statistical properties of the method (both the power and the robustness). Another way of assessing the phenotypic resemblance is through the covariance and this is the core of the variance components methodology that came across increased popularity in the past few years (Goldgar, 1990; Amos, 1994; Almasy and Blangero, 1998). In essence, this method involves first estimating IBD sharing for relative pairs and then estimating covariances between relatives that are a function of the IBD sharing. Estimates and test statistics are obtained under the assumption of a multivariate normal distribution of the trait within families and violation of this assumption (e.g., by ascertainment on the trait values as proposed by Risch and Zhang [1995] to increase the power of analysis) has been shown to inflate the type I error rate (Allison et al., 1999b). By contrast, extension of the MLB approach to a quantitative trait, denoted as MLB-QTL (Alcais and Abel, 1999), is insensitive to nonnormal phenotypic distribution whatever the mechanism underlying this nonnormality (Alcais and Abel, 2000b) and can be used to analyse sibships of any size (Alcais and

Abel, 2000a). Some of these methods are implemented in popular packages such as MAPMAKER/SIBS (Kruglyak and Lander, 1995), GENEHUNTER (Kruglyak et al., 1996), SOLAR (Almasy and Blangero, 1998), or MLBGH (Abel and Muller-Myhsok, 1998b; Alcais and Abel, 1999).

As mentioned previously, model-free methods share some issues with model-based linkage analysis regarding missing parental marker data and testing with multiple markers. In particular, the significance levels of the tests should be adapted to the number of comparisons that are made, and replication studies are required to confirm suggestive linkage. Moreover, it should be noted that the distinction between model-free and model-based approaches may not be so clear. As an example, in the affected sib-pairs method all calculations are carried out on marker alleles and clearly this method does not require specification of a disease inheritance model. However, this method *implies* an inheritance model, as shown by Knapp et al. (1994), since the affected sib-pair approach is equivalent to a model-based linkage analysis carried out assuming a recessive inheritance with complete penetrance, no sporadic cases, and unknown parental phenotypes. This latter point ('assuming unknown parental phenotypes') underlines the important feature that model-free approaches do not consider parental phenotypes, i.e., they assume that both parents are potentially informative for linkage. For this reason, affected sib-pairs methods are less powerful for dominant-like than for recessive-like traits.

2.3 Conclusions

Although it is clear that there is no unique answer to the question of whether to use a model-free or model-based approach, it is possible to propose some practical guidelines (Goldgar, 2001). As already mentioned, when there is some knowledge about the prevalence of the disease under study and the level of familial aggregation, it is possible to generate a limited number of realistic genetic models to use in lod-score analysis (with appropriate correction for the number of models tested). Conversely, when there is a lack of reliable epidemiological information, so that consistent models cannot be generated, model-free approaches should be preferentially used. Also, investigators should be aware of the multiple testing issue when using several methods of analysis unless they carefully account for it when computing the significance criterion.

Finally, another key point when performing a linkage analysis is whether to choose a candidate gene strategy or to go for a whole-genome search. From a statistical point of view, this has striking consequences. The 'candidate gene by hypothesis' approach is motivated by what is known about the trait biologically and can be understood as a classical *hypothesis testing strategy* where the type I (i.e., false positive) and II (i.e., false negative) errors have their standard definitions and interpretations. By contrast, the 'genome scan' strategy is applied without prior knowledge of the biological basis of the disease and therefore aims at *generating hypothesis*. In this context, the central parameter becomes the type II error, whereas the interpretation of type I errors is more difficult and controversial. The debate is still open to decide between (unrealistically) increasing the sample size to allow for both acceptable type I and II error rates and using less stringent nominal significance levels than those initially proposed by Lander and Kruglyak (1995) that were based on complex analytic calculations under very particular assumptions rather than practical considerations (Elston, 1998; Morton, 1998). In the article by Lander and Kruglyak (1995), the P values associated with 'suggestive/significant linkage' were defined as $1.7 \times 10^{-3}/4.9 \times 10^{-5}$ (corresponding to a lod score of 1.9/3.3) and $7.4 \times 10^{-4}/2.2 \times 10^{-5}$(2.2/3.6) for model-based and model-free approaches, respectively. Despite the initial fear that has led to the definition of such rigorous thresholds, it seems that genome scans are rather flooded by false-negative than by false-positive results and several authors now advocate for new criteria (Elston, 1998; Rao, 1998; Sawcer et al., 1997). Although there is no doubt that improvements in statistical methods will increase the power of linkage analysis, it is our feeling, however, that the most efficient way to achieve this goal may be to increase the correlation between the marker locus and the phenotype either theoretically, e.g., by focusing on a mendelian-like phenotype, or experimentally, e.g., by using larger pedigrees and more homogenous samples.

3. ASSOCIATION STUDIES

Association studies are used to investigate the role of polymorphisms (or alleles) of genes that are defined as candidate genes on the basis of their function or location (Fig. 2.1). Classical association studies are population-based case-control studies comparing the frequency of a given allele marker, denoted as M_1, between unrelated affected (cases) and unaffected (controls) subjects (Khoury et al., 1993; Lander and Schork, 1994). A first approach considers the allele frequency per se, and because each individual has two alleles at any autosomal locus, there will be twice as many alleles as people. A second approach considers genotypes, i.e., the differences in disease risk between individuals who do not carry M_1, those who have a single copy and those who are homozygous for M_1. In an appealing article, Sasieni (1997) analytically established that both the odds ratio and the χ^2 statistic computed from the first allelic approach are appropriate provided that the population from which the cases and controls are sampled is in Hardy–Weinberg (HW) equilibrium, i.e., in practice, HW must hold in the *combined* sample. Although some methodological developments have been proposed to overcome this problem (Schaid and Jacobsen, 1999), the interpretation in terms of relative risk of disease remains questionable when using the allelic analysis. Therefore the use of the genotypic approach should be preferred whenever possible and also provides the opportunity of testing some specific allelic effects (dominant/recessive). An understated statistical issue is the number of degrees of freedom of the test based on genotypic distribution. Three cases must be distinguished as follows: (1) assuming a multiplicative model (i.e., the risk of M1M1 versus M2M2 is the square of the risk between M1M2 versus M2M2), inference about the null hypothesis of no association between the disease and the gene is based on Armitage's trend test (Armitage, 1955), which is asymptotically distributed as a χ^2 with 1 *df*; (2) assuming no overdominance effect (i.e., the risk of M1M2 versus M2M2 lies within the range bounded by 1 and the risk of M1M1 versus M2M2) the test statistic is distributed as a mixture of χ^2 distributions with 1 and 2 *df* (see for example Chiano and Clayton, 1998); and (3) when both previous assumptions are regarded as undesirable, then a traditional χ^2 with 2 *df* may be used. Furthermore, checking the Hardy–Weinberg equilibrium in controls must be the first step of any population-based study, as it can also help to detect genotyping inconsistencies (e.g., excess of both homozygous due to the presence of a null allele). Several approaches have been proposed to test whether the HW holds details of which are beyond the scope of this article (e.g., Gomes et al., 1999).

A statistically significant association between a given polymorphism and a given phenotype has several possible explanations as follows: (1) random, i.e., the association has occurred just by chance (type I error); (2) the phenotype causes variation in the marker genotype. However, as noted by Allison et al. (1999a), this point can be ruled out a priori both 'as being logically impossible because genotype precedes phenotype in time and because it is a fundamental axiom of causality that cause must precede effect'; (3) the allelic variation causes variation in the phenotype, i.e., it is the functional variant; (4) the marker allele under study is in linkage disequilibrium with the allele causing variation in the phenotype; and (5) population admixture. In the context of gene identification, we are interested only in associations due to points 3 and 4. For further explanations, we will consider a likely present situation with two SNPs denoted as G and M. G has two alleles, G_1 and G_2, and G_1 is the functional polymorphism increasing the risk of disease (G_1may be understood as the deleterious allele D described previously). M has also two alleles, M_1 and M_2, and corresponds to the SNP, which has been genotyped and will be tested as the marker. The simplest reason that explains association is that allele M_1 is the functional polymorphism G_1 itself (M corresponds to G). A more likely explanation is that M_1 has no direct role on the phenotype but is in linkage disequilibrium with allele G_1. Linkage disequilibrium means two conditions: (1) linkage between M and G (generally tight linkage and in particular M and G can be within the same gene) *and* (2) allele M_1 is preferentially associated with allele G_1, i.e., the M_1-G_1 haplotype is more frequent than expected by the respective frequencies of M_1 and G_1 (e.g., many present cases are due to one ancestral G_1 allele and the ancestor who transmitted this allele was bearing the M_1-G_1 haplotype). It should be noted that linkage alone (only the first condition is fulfilled), even very close, does not lead to association, and therefore, that absence of association does not exclude linkage. Taking into account these two explanations, it seems that association studies should have their greater interest in candidate gene approach by considering markers which are either within or in close linkage with a gene having a known relationship with the phenotype.

Unfortunately, there is one additional, and potentially frequent, situation which can lead to fallacious association between a marker and a phenotype: population admixture. For example, a case-control study conducted in a population which is a mixture of two subpopulations of which one has a higher disease frequency and a higher M_1 frequency than the second will conclude to positive association of allele M_1 with the disease. One way to avoid the problem of population admixture is to condition on parents' genotypes at the marker locus. This is the rationale for the recent development of family-based association methods, such as the transmission disequilibrium test (TDT) (Spielman et al., 1993). The sampling unit in the classical TDT consists in two parents with an affected child, and parental alleles nontransmitted to affected children are used as control alleles. More specifically, the TDT considers affected children born from parents heterozygous for M_1, i.e., M_1M_2 parents, and simply tests whether these children have received M_1 with a probability different from 0.5, the value expected under random segregation. Methods have been subsequently developed to handle families with missing parental data by either using unaffected sibs as controls (Sib-TDT) (Spielman and Ewens, 1998) or reconstructing parental genotypes from children (RC-TDT) (Knapp, 1999), and family data can be efficiently analysed by methods combining these different approaches such as the FBAT software (Horvath et al., 2001; Lake et al., 2000).

Although association studies mostly focused on binary traits (affected/unaffected), numerous developments have been recently performed for the analysis of quantitative phenotypes that rely either on regression (e.g., George et al., 1999; Waldman et al., 1999; Monks and Kaplan, 2000) or variance component (e.g., Fulker et al., 1999; Abecasis et al., 2000; Sham et al., 2000) techniques. These two statistical tools are closely related and therefore display similar limitations. First, they intrinsically assume multivariate normality of the phenotypic distribution, and violation of this hypothesis can lead to large inflation of the type I error rate when compared to the asymptotic expectations. The use of a very large sample could relax this assumption by allowing reliance on the central limit theorem. However, in some cases (e.g., in the context of an extremely selected sample) neither the normality assumption nor the use of a large sample can be achieved. Therefore, the analysis of such data would usually require the use of additional procedures such as the transformation of the data to handle the nonnormality or the computation of empirical *P* values. Furthermore, it is clear that both approaches derive information from phenotypic variability. Consequently, as mentioned by Abecasis et al. (2001) these methods may have low power to detect LD under some extreme selection schemes (e.g., one-tailed selection). However, when their assumptions are valid, these approaches can be very powerful and their flexibility may allow deep insights of genetic mechanisms involved in the trait under study.

Association studies (population-based or family-based) are expected to be very efficient to detect the effect of allele M_1 when M_1 is the functional polymorphism G_1 itself (Risch and Merikangas, 1996). Under this latter hypothesis, Risch and Merikangas (1996) demonstrated that the TDT was more powerful than the sib-pair method even in the context of a genome-wide search involving 500,000 diallelic polymorphisms (5 polymorphisms per gene for an assumed total number of 100,000 genes). However, in the more common situation where M_1 is different from G_1, the power of TDT (or other association studies) is highly dependent on both the respective frequencies of M_1 and G_1 and the strength of the linkage disequilibrium between M_1 and G_1 (Muller-Myhsok and Abel, 1997; Abel and Muller-Myhsok, 1998a). To a large extent, the utility of such genome-wide association testing depends on the existence of marker alleles strongly associated with disease-causing polymorphisms, but as yet the nature and extent of such associations in the human genome are not well understood. In particular, several recent articles have pointed out that the extent of linkage disequilibrium was not uniformly distributed among both chromosomal regions and populations (Lonjou et al., 1999; Reich et al., 2001; Stephens et al., 2001), adding some important elements to account for when trying to define the optimal strategy of analysis. These results clearly indicate that linkage methods are still of interest to identify genes involved in infectious diseases at least until molecular resources become available for full genomic screening of human genes.

4. MYCOBACTERIAL INFECTIONS

In humans, mycobacterial pathogenicity strongly depends on the species. Tuberculosis and leprosy are the most common human mycobacterial diseases and are caused by *M. tuberculosis* and *Mycobacterium leprae*, respectively. Numerous other species present in the environment, denoted as nontuberculous mycobacteria (NTM), are generally less pathogenic, although they can be responsible for a variety of infections under certain conditions. There is now clear evidence that the intrinsic virulence of a mycobacterial species is not the unique pathogenic factor and that the outcome of mycobacterial infection depends to a large extent on the genetic background of the infected subject (Casanova and Abel, 2002). Many experimental studies have demonstrated the role of genetic factors in mycobacterial infections (reviewed in Wakelin and Blackwell, 1988; Blackwell et al., 1994; McLeod et al., 1995). As detailed below, genetic epidemiological studies have shown that human genes have an important role in the expression of leprosy and tuberculosis, although the molecular basis of this genetic control still remains largely unknown. However, major advances have been made in the genetic dissection of disseminated infections because of poorly virulent mycobacteria such as NTM and detailed in Chapter 9 of this book. After presenting the two genome screens recently reported in tuberculosis and leprosy, the following sections review the main candidate genes that have been investigated by linkage and/or association studies in both diseases. The most important findings are summarised in Tables 2.1 and 2.2 for linkage and association studies, respectively.

5. GENOME SCREENS

The two genome screens performed in tuberculosis and leprosy were based on affected sib-pair linkage studies. The first study was conducted on pulmonary tuberculosis (PTB) in families originating from Gambia (85 sib-pairs) and South Africa (88 sib-pairs) (Bellamy et al., 2000). Most affected subjects had smear-positive pulmonary disease and the remaining patients were diagnosed using clinical and X-ray criteria. No region of the genome showed significant evidence for linkage, ruling out the presence of a gene with a strong effect on tuberculosis in these families. In particular, no linkage was found with the regions containing the *NRAMP1* and *VDR* genes for which associations were previously reported in the Gambian population (Bellamy et al., 1998b, 1999), indicating that those genes did not have a major effect on PTB in this population. Two other regions located on chromosome 15q ($P < 0.001$) and *X*q ($P < 0.005$) provided only suggestive evidence for linkage (as

Table 2.1. *Common mycobacterial infections: most consistent results obtained by case-control association studies*

Candidate genes	Phenotype	Population	Sample	Alleles	OR (95%CI)[a]	Reference
DRB1[b]	Pulmonary tuberculosis	North India	20 cases/46 controls	DRB1*1501	4.8 (1–23.3)	Mehra et al. (1995)
		Mexico	50 cases/95 controls	DRB1*1501	7.9 (2.7–23.1)	Teran-Escandon et al. (1999)
	Tuberculoid leprosy	North India	39 cases/46 controls	DRB1*1502	2.7 (1.1–6.7)	Mehra et al. (1995)
		North India	28 cases/47 controls	DRB1*1501 + DRB1*1502	5.7 (2–16)	Rani et al. (1993)
		North India	54 cases/44 controls	DRB1*1501 + DRB1*1502	4.7 (2–11)	Zerva et al. (1996)
NRAMP1	Pulmonary tuberculosis	Gambia	410 cases/417 controls	INT4 *C* + 3′UTR *del*	4.1 (1.9–9.1)	Bellamy et al. (1998b)
		Korea	192 cases/192 controls	3′UTR *del*	1.8 (1.1–3)	Ryu et al. (2000)
		Japan	267 cases/202 controls	5′$(GT)_n$ *203bp* (homozygous)	0.5 (0.3–0.7)	Gao et al. (2000)

[a] Odds ratios with 95% confidence intervals.

[b] Only association studies using molecular HLA typing are presented, whereas numerous other serological studies found that DR2 (DRB1*1501 and DRB1*1502 are DR2 alleles) was associated with pulmonary tuberculosis, tuberculoid leprosy, and lepromatous leprosy, respectively (see text).

Table 2.2. *Common mycobacterial infections: most consistent results obtained by linkage (candidate region or genome-wide) studies*

Design	Chromosomal region	Phenotype	Population	Sample	P value	Reference
Candidate region	*6p21 (MHC region)*	Tuberculoid leprosy	Surinam	16 multicase families	<0.005	de Vries et al. (1976)
			South India	72 multicase families	<0.05	Fine et al. (1979)
			Central India	13 multicase families	<0.05	de Vries et al. (1980)
			Venezuela	28 multicase families	<0.05	van Eden et al. (1985)
			Egypt	15 multicase families	<0.0005	Dessoukey et al. (1996)
		Lepromatous leprosy	Venezuela	28 multicase families	<0.006	Dessoukey et al. (1996)
			China	26 multicase families	<0.05	Xu et al. (1985)
Candidate region	2q35 (NRAMP1 region)	Pulmonary tuberculosis	Canada	1 large pedigree of Aboriginal Canadians	$<2 \times 10^{-5}$	Greenwood et al. (2000)
		Granulomatous reaction to intradermal lepromin	Vietnam	20 leprosy families	<0.002	Alcais et al. (2000)
Genomewide	10p13	Paucibacillary leprosy	South India	224 multicase families	$<2 \times 10^{-5}$	Siddiqui et al. (2001)

discussed in Section 2.3, *P* values around 0.001 are considered as suggestive in genome screens because of the problem of multiple testing with numerous markers).

The second genome screen was performed on leprosy in a total of 245 affected sib-pairs from South India (Siddiqui et al., 2001). Patients were diagnosed using WHO guidelines and all siblings except four were affected by paucibacillary leprosy. Significant evidence for linkage ($P < 2 \times 10^{-5}$) was found with chromosome 10p13 indicating that a gene located within this region can have a substantial influence in paucibacillary leprosy. Surprisingly, no linkage was observed with the HLA region, whereas several previous sib-pair studies, including those in South India, provided some evidence for HLA-linked genes playing a role in paucibacillary/tuberculoid leprosy (see next section).

6. THE MAJOR HISTOCOMPATIBILITY COMPLEX

Since Human Leukocyte Antigens (HLA) class I and class II molecules are highly polymorphic and play an important role in presenting antigenic peptides to cytotoxic CD8 and helper CD4 T cells, respectively, during the cellular immune response, their genes have been extensively studied in tuberculosis and leprosy but the results have been controversial. Most studies were population-based association surveys in adults, which involve the comparison of HLA class I and/or class II alleles between unrelated cases and unrelated controls. As the number of compared alleles can be quite large, a key methodological point that was often overlooked is to correct the observed *P* values for the number of tests that were performed. Overall, in both tuberculosis and leprosy, the most convincing associations were found with HLA class II alleles. The main chains of antigenic peptides are thought to form hydrogen bonds with HLA residues conserved in most class II alleles, whereas the side chains are accommodated in the binding site by polymorphic pockets which appear to determine the peptide specificity of different class II proteins (Stern et al., 1994). Although little is known about the molecular nature of HLA-restricted mycobacterial antigens, it has been suggested that the variability in the class II genes which commonly occur in or near the peptide binding pockets can affect peptide binding and presentation and may explain some observed associations between HLA class II alleles and tuberculosis (Goldfeld et al., 1998) or leprosy (Zerva et al., 1996).

Several case-control studies reported an increased frequency of HLA-DR2 in PTB patients originating from Indonesia (Bothamley et al., 1989) and India (Brahmajothi et al., 1991; Rajalingam et al., 1996) with an associated odds ratio (OR) estimated between 1.8 and 2.7. In the latter study (Rajalingam et al., 1996) the association was stronger (OR = 3.7, $P < 0.0001$) in patients who did not respond to drug treatment. It is worth noting that when the disease prevalence is low (in practice lower than 10% which may not be the case for tuberculosis in some highly endemic areas), the OR is a valid estimate of the relative risk, i.e., the risk of developing the disease with a particular genotype compared to the risk of developing the disease without this particular genotype. Another strong argument in favor of the role of HLA-DR2 in PTB was provided by a family-based association study (Singh et al., 1983), which showed a skewed transmission of DR2 (around 80%) to affected offspring from DR2 heterozygous parents in a sample of 25 multiple-case Indian families. However, some other case-control studies failed to replicate the HLA-DR2 association in Chinese (Hawkins et al., 1988), Mexican (Cox et al., 1988), and Indian (Sanjeevi et al., 1992) populations. In addition to classical explanations such as ethnic heterogeneity, phenotype definition, or lack of power, these discrepant results can be due, at least in part, to the use in most of these studies of HLA serologic techniques which lead to a lower resolution of HLA class II types (Schreuder et al., 1999). Recent studies that used molecular DNA-based typing methods found an increased frequency of DRB1*1501 (which is part of the DR2 serotype) in PTB patients from India (Mehra et al., 1995; Ravikumar et al., 1999) and Mexico (Teran-Escandon et al., 1999) with estimated ORs varying from 2.7 to 8. Two DQB1 alleles (which are both part of the DQ1 serology specificity) were also found to be associated with pulmonary tuberculosis, DQB1*0503 in Cambodia (Goldfeld et al., 1998), and DQB1*0501 in Mexico (Teran-Escandon et al., 1999), this latter association being independent of that with DRB1*1501.

In leprosy, the role of the HLA system was shown in the two main subtypes, tuberculoid and lepromatous leprosy, but not in the leprosy per se phenotype. Some association studies reviewed by Ottenhoff and de Vries (1987) found an inverse relationship between leprosy subtypes and HLA-DR3, which was increased in tuberculoid and decreased in lepromatous patients. However, the most consistent reported results were an increased frequency of HLA-DR2 in both tuberculoid and lepromatous patients (Meyer et al., 1998). As in PTB, two family-based association studies in India (de Vries et al., 1980) and Egypt (Dessoukey et al., 1996) found a skewed distribution of the DR2 allele in tuberculoid siblings. Using molecular HLA typing, a positive association was found between Indian tuberculoid leprosy patients and alleles *DRB1* 1501* and *DRB1* 1502* (both of which are DR2 alleles) (Rani et al., 1993; Zerva et al., 1996). As these alleles were characterised by arginines at positions 13 or 70–71, a stronger association was found when considering all HLA-DRB1 alleles that contain Arg13 or Arg70–Arg71, suggesting that the presence of a positive charge generated by Arg at these positions may influence a critical site within the binding groove of the DR chain that affects peptide binding and/or T-cell interactions (Zerva et al., 1996).

The role of some other genes located within the MHC was also assessed by means of case-control studies. The first was the gene encoding the tumor necrosis factor-α(TNF-α) which plays a key role in the granulomatous response against mycobacteria. All studies investigated a *TNF-α* promoter polymorphism located at position −308 (a G/A transition denoted as *TNF2*) that, after some debate, is thought to increase both transcription of the TNF-α gene and protein production (Allen, 1999). No association was found between PTB and *TNF2* in populations from Cambodia (Goldfeld et al., 1998) and Brazil (Blackwell et al., 1997). In leprosy, an increased frequency of *TNF2* was reported in Indian lepromatous leprosy patients with an OR around 2 (Roy et al., 1997). This result was independent of the HLA-DR2 association also found in this population and is consistent with previous reports of high serum TNF levels observed in lepromatous patients (Barnes et al., 1992). Located between HLA-DQ and HLA-DP are two genes designated Transporter associated with Antigen-Processing (TAP)1 and TAP2. The products of these genes form a TAP complex which transports antigenic peptides to the endoplasmic reticulum for binding to HLA class I molecules for presentation to CD8 cytotoxic T cells (Spies et al., 1990). In a population from North India no association was found between three *TAP1* polymorphisms and PTB or tuberculoid leprosy (Rajalingam et al., 1997). Five TAP2 polymorphisms were identified, and, as compared to controls, an increased frequency of TAP2-A/F (OR = 4.3, $P < 0.002$) and TAP2-B (OR = 3.5, $P < 0.006$) was found in PTB patients and tuberculoid leprosy patients, respectively. These TAP2 polymorphisms were not found in strong linkage disequilibrium with any DRB1, DQA1, or DQB1 alleles indicating that the putative role of these TAP2 variants can be independent of previously described HLA–class II associations.

6.1 The NRAMP1 Gene

The human *NRAMP1* (natural resistance associated macrophage protein 1) gene, is an excellent candidate gene, as it is the human ortholog (Cellier et al., 1994) of the murine *Nramp1* gene, which was the first mouse mycobacterial susceptibility gene identified (Vidal et al., 1993). In mice, a single nonconservative amino acid substitution in *Nramp1* regulates the early phase of infection to several intracellular mycobacteria (Govoni et al., 1996; Vidal et al., 1996), including *M. bovis* (BCG) and *M. lepraemurium*, the rodent-tropic equivalent of *M. leprae*. The *Nramp1* gene does not appear to affect the susceptibility of mice to *M. tuberculosis* (North et al., 1999), but *M. tuberculosis* is not naturally pathogenic in rodents. *Nramp1* codes for an integral membrane protein which is recruited to the membrane of phagosomes soon after the completion of phagocytosis, and a recent work suggests that Nramp1 contributes to defense against infection by extrusion from the phagosomal space of divalent metal cations essential for microbial activity (Jabado et al., 2000).

In a case-control study performed in Gambia (Bellamy et al., 1998b), four *NRAMP1* variants were found to predispose subjects to PTB with an estimated OR around 1.8. However, these polymorphisms were in strong linkage disequilibrium two by two, so that only two variants were independently associated with the disease. The first is a single nucleotide change in intron 4 (INT4) and the other is a 4-basepair deletion in the 3′ untranslated region (3′UTR). Heterozygosity for both variants was associated with the highest risk of tuberculosis (OR = 4.1, $P < 0.001$). A recent small family-based association study in Guinea-Conakry found a skewed transmission for the INT4 variant among PTB offspring but not for the 3′UTR variant (Cervino et al., 2000). Association of PTB with the 3′UTR variant was reported in a case-control study in Korea (Ryu et al., 2000), whereas another pattern of *NRAMP1* allelic association was found in a Japanese population (Gao et al., 2000). This heterogeneity of results together with the absence of a known functional role of these alleles suggest that the reported associations can reflect the effect of another variant not yet identified and in linkage disequilibrium with the alleles tested. Another convincing source of evidence for the role of *NRAMP1* (or a NRAMP1-linked gene) in PTB came from a recent linkage study performed in a large Aboriginal Canadian pedigree after a tuberculosis outbreak (Greenwood et al., 2000). This study found a major locus of susceptibility to clinical tuberculosis which maps to chromosome 2q35, including the *NRAMP1* gene. The most significant results ($<2 \times 10^{-5}$) were obtained with a dominant susceptibility allele having a major effect, as carriers of at least one copy of this allele have a risk of tuberculosis ten times higher than that of wild-type homozygotes. Whether this linkage reflects the role of *NRAMP1* itself or that of a closely linked gene remains to be established. However, these results indicate that, at least in certain contexts (e.g., populations that do not have an extensive history of exposure to *M. tuberculosis* such as Aboriginal Canadians), Mendelian-like subentities can be involved in the genetic control of common mycobacterial diseases.

The possible role of NRAMP1 in leprosy has been mostly investigated so far through linkage studies. A sib-pair study in Vietnam showed significant linkage ($P < 0.005$–0.02) between leprosy *per se* and *NRAMP1* haplotypes corresponding to six intragenic variants of *NRAMP1* and four polymorphic flanking markers (Abel et al., 1998b). Combined with the segregation analysis performed in the same population (Abel et al., 1995), this study also suggested a genetic heterogeneity according to the ethnic origin of the families (Vietnamese or Chinese). Genetic heterogeneity may account, at least in part, for the results of two previous reports that failed to detect linkage between leprosy and the *NRAMP1* region in families from Pakistan and Brazil (Shaw et al., 1993) and French Polynesia (Levee et al., 1994). In the same Vietnamese study, the *NRAMP1* region was also found to be linked ($P < 0.002$) with the *in vivo* Mitsuda reaction measuring the delayed immune response against intradermally injected lepromin (Alcais et al., 2000). This latter result, which may account for the first linkage observed with leprosy per se, is consistent with the view that *NRAMP1* may be involved in the development of immune responses to mycobacterial antigens with a putative role in the regulation of type 1/type 2 cytokine responses (Soo et al., 1998). There is an increasing body of evidence that lepromatous leprosy (generally displaying negative Mitsuda reactions) is associated with a predominantly type 2 cytokine response, whereas a more type 1 response is observed in tuberculoid leprosy (Yamamura et al., 1991; Yamauchi et al., 2000).

6.2 The VDR Gene

The active form of Vitamin D, 1α,25-$(OH)_2D_3$, modulates the differentiation, growth, and function of a broad range of cells, including cells of the immune system. Although vitamin D deficiency was commonly associated with susceptibility to tuberculosis (Chan, 2000), recent data underlined the immunosuppressive effects of this hormone, showing that 1α,25-$(OH)_2D_3$ inhibited both differentiation of dendritic cells into potent antigen-presenting cells (Piemonti et al., 2000) and IL-12 production in already differentiated dendritic cells and monocytes (D'Ambrosio et al., 1998). The effects of 1α, 25-$(OH)_2D_3$ are exerted through the vitamin D receptor (VDR), and experimental studies on VDR-deficient mice confirmed the inhibitory role of the 1α, 25-$(OH)_2D_3$/VDR pathway on maturation of dendritic cells (Griffin et al., 2001). Several allelic variants of the VDR gene have been described in humans, and the influence of a single base polymorphism in codon 352 with two alleles designated as T and t, respectively, has been investigated in several studies.

In the same Gambian population which was analysed to study *NRAMP1*, a lower proportion of tt homozygous at the VDR gene was found among PTB patients than controls (OR = 0.5, $P < 0.02$) (Bellamy et al., 1999). In another case/control study performed on Gujarati Asians living in England

(Wilkinson et al., 2000), tuberculosis patients were found to have lower serum 25-hydroxycalciferol levels than healthy contacts but no convincing association was found with VDR polymorphisms (Stene, 2000). Another study was performed on leprosy patients from India (Roy et al., 1999). Compared to the control group, the homozygous tt genotype frequency was higher among tuberculoid (OR = 3.2, $P < 0.001$), whereas the TT genotype was found at increased frequency in lepromatous patients (OR = 1.7, $P < 0.04$). Overall, these results can be consistent with the recent data on the immunosuppressive effects of the 1α, 25-$(OH)_2D_3$ under the hypothesis that the VDR function is impaired in tt homozygous. This remains to be established especially with respect to previous studies associating the t allele with higher levels of mRNA expression in transient transfection assays (Morrison et al., 1992).

6.3 The MBL Gene

Mannose binding lectin (MBL), also called mannose binding protein, is a serum protein of hepatic origin which plays an important role in innate immune defense. MBL binds to the repeating sugar arrays on many microbial surfaces through multiple lectin domains and, following binding, is able to activate complement in an antibody- and C1q-independent manner using a specific protease (Neth et al., 2000). Three common functional mutations within exon 1 of the MBL gene at codons 52, 54, and 57 lead to reduced or extremely low serum MBL levels in heterozygotes or homozygotes, respectively, and genetically determined low levels of the protein have been associated with predisposition to various infections, especially in children (Summerfield et al., 1997).

The role of these MBL variants in tuberculosis have been investigated in three case/control studies with contrasting results. A first study in the previously mentioned Gambian sample showed no strong association (Bellamy et al., 1998c). In an Indian population, the frequency of MBL mutant homozygotes (including codons 52, 54, and 57) was higher in PTB patients than in controls (OR = 6.5, $P < 0.009$) (Selvaraj et al. 1999). Conversely, in a South African community, the *MBL* mutant allele at codon 54 was found to be protective against pulmonary tuberculosis (OR = 0.4, $P < 0.02$) and especially against tuberculous meningitis (OR = 0.2, $P < 0.002$) (Hoal-Van Helden et al., 1999). Whereas the results of the Indian study are consistent with the common knowledge about the MBL role, the findings in the South African population need to make the hypothesis that, under certain circumstances, MBL genotypes that usually lead to low MBL levels might be protective (Garred et al., 1994).

6.4 The IL-1 Genes

The proinflammatory cytokine IL-1β is strongly induced by *M. tuberculosis* infection and is likely to play an important role in host defense against this mycobacteria as shown by studies in IL-1 type I receptor-deficient mice (Juffermans et al., 2000). This proinflammatory response is downregulated by the IL-1 receptor antagonist (IL-1Ra), a pure antagonist of the IL-1 type 1 receptor. The genes coding for IL-1β, IL-1Ra, as well as that for IL-1α, are located in a cluster on chromosome 2q (Nicklin et al., 1994). Two single nucleotide polymorphisms in the *IL-1β* gene at positions -511 and $+3953$, respectively, and one VNTR marker in the *IL-1Ra* gene have been described. The *IL-1Ra* VNTR allele A2 (*IL-1Ra A2*) was shown to produce higher levels of IL1-Ra (Hurme and Santtila, 1998).

The functional role of these polymorphisms on *in vitro* IL-1β and IL1-Ra secretion in response to *M. tuberculosis,* and their association with tuberculosis were recently investigated in a population of Gujarati Asians living in England (Wilkinson et al., 1999). Healthy subjects carrying the *IL-1Ra A2* allele were found to produce 1.9-fold more IL-1Ra, and to have higher molar ratios of IL-1Ra/IL-1β than the remaining subjects. The two polymorphisms of the IL-1β gene were not clearly associated with the level of *M. tuberculosis*-stimulated IL-1β production. No differences were observed in the frequency of *IL-1β* and *IL1-Ra* polymorphisms between 89 tuberculosis patients (pulmonary and extrapulmonary) and 114 controls. When combining the genotypes at these polymorphisms, subjects carrying allele 1 of the *IL-1β + 3593* variant and not carrying the *IL-1Ra A2* allele were overrepresented in the twelve patients with tuberculous pleurisy ($P < 0.03$). Finally, the *IL-1Ra A2* allele was found to be associated with a reduced Mantoux response to purified protein derivative of *M. tuberculosis.* The role of variants in IL-1 genes was also investigated in the Gambian sample (Bellamy et al., 1998a). No clear association was found between PTB and alleles of *IL-1β − 511* (*IL-1β + 3593* was not tested), the *IL-1Ra* VNTR, and a *IL1-α* microsatellite. However, and although not significant ($P < 0.09$), subjects not carrying the *IL-1Ra A2* allele were overrepresented in PTB patients.

7. DISCUSSION

Reviewing all these population-based association studies, it is important to recall the inherent limitations of these approaches which may lead to both false-positive and false-negative results. A major methodological problem is multiple testing (e.g., of twenty tests, one is expected to be significant at the 0.05 level just by chance). In particular, looking at several candidate genes (and several polymorphisms within a gene) on the same disease and in the same population requires to correct for multiple tests (or at least to discuss this issue) even if the results are published in successive papers. In addition, it should be noted that negative (nonsignificant) results are often not reported making more difficult to assess the actual meaning of a so-called significant association. A second issue relates to linkage disequilibrium. As explained in Chapter 3, in the presence of an actual association between an allele and a disease, it is possible that the associated allele is not the functional variant but is in linkage disequilibrium with it. In this latter case, the absence of replication of the association with this specific allele in different populations can be explained by heterogeneity in linkage disequilibrium between populations, whereas this explanation cannot hold when the assumed functional variant is tested. Conversely, if no association is found, one explanation could be that the tested allele is not in strong linkage disequilibrium with the functional one although the functional allele is within the same gene. The power of association studies highly depends on several factors detailed in Section 3, and it is important to note that the absence of association with some alleles of a given gene does not always rule out the role of the gene.

Taking into account these remarks, the main convincing results so far can be considered to be the association of HLA-DR2 (or some molecular subtypes) with PTB and leprosy subtypes; the role of NRAMP1 (or a closely linked gene), which may have strong effects under certain circumstances (specific population, specific phenotype); and the presence of a locus predisposing to paucibacillary leprosy on chromosome 10p. Most other findings need additional studies to confirm or not the previous results. It is clear that a minority of candidate genes have been tested hitherto, whereas many others are quite relevant, such as those involved in Mendelian predisposition to mycobacterial infections and/or in the numerous pathways influencing the immune response to common mycobacteria (Yamauchi et al., 2000; Flynn and Chan, 2001). In any case, to be convincing, the role of a given allele, especially when it is supposed to have a moderate effect, should be replicated in several studies and validated by functional studies.

8. CONCLUSION

Essential tools for identifying genes that influence human infectious diseases have been developed recently. These tools include a dense human genetic map, a growing number of candidate genes with intragenic SNPs, and genetic epidemiology methods to optimise analysis of these data. At this point, there is strong evidence for the major role of genetic factors in most infectious diseases, but the molecular basis of this genetic susceptibility/resistance remains largely unknown, except in the situation of rare Mendelian disorders such as disseminated mycobacterial infections. It is likely that progress in the genetic dissection of infectious diseases will come from the complementary analysis of different phenotypes accounting for the different steps of the process going from the exposure to a given infectious agent to the eventual development of a clinical disease (Casanova and Abel, 2002). Indeed, a plausible reason for the difficulty of identifying genes in common infectious diseases is that the link between the DNA sequence and the trait under study is mediated by highly complex biochemical pathways. One way to overcome this problem could come from the development of a *functional map* by means of a new and promising tool, the microarrays or DNA chips. These technologies permit the expression monitoring of thousands of genes by measuring the quantity of mRNA given different conditions (human populations, infecting strains, disease expressions, tissue types) and could be used to define sets of coexpressed genes (also known as pathway fractions). This could lead to the construction of a functional map under the likely hypothesis that there is a strong correlation between gene function and gene expression. A 'functional distance' between genes is then defined by means of correlation analysis, cluster analysis, multidimensional scaling analysis, or classification tree (for a detailed review see Quackenbush, 2001). Following the attractive idea developed by Horvath and Baur (2000), once a functional map exists functional mapping could parallel meiotic mapping: observed functional distances could be compared to expected differences (analogous to linkage analysis) and different variants of pathway fractions could be associated with the trait of interest (analogous to association studies). To make functional mapping successful, however, adequate statistical models need to be developed that relate the traits of interest to these functional maps. This will certainly be one of the most challenging areas in genetic epidemiology during the next few years. Finally, we cannot fully appreciate how genetic information will modify our approach to the prevention and treatment of infectious diseases. However, the identification of susceptibility/resistance genes in malaria, schistosomiasis, mycobacterial, or HIV infections have already opened new avenues for understanding of pathogenesis mechanisms, screening genetically predisposed subjects, designing vaccines, and developing novel drugs.

REFERENCES

Abecasis, G. R., Cardon, L. R. and Cookson, W. O. (2000). A general test of association for quantitative traits in nuclear families. *Am. J. Hum. Genet.* **66**, 279–292.

Abecasis, G. R., Cookson, W. O. and Cardon, L. R. (2001). The power to detect linkage disequilibrium with quantitative traits in selected samples. *Am. J. Hum. Genet.* **68**, 1463–1474.

Abel, L. (2002). Human genetic variability and susceptibility to infectious diseases. In: *Chemokines Receptors, Human Genetics, and AIDS: Insights into Pathogenesis and New Therapeutic Options*. O'Brien, T. R. (ed.). New York: Marcel Dekker, Inc., pp. 105–132.

Abel, L., Alcais, A. and Mallet, A. (1998a). Comparison of four sib-pair linkage methods for analyzing sibships with more than two affecteds: interest of the binomial maximum likelihood approach. *Genet. Epidemiol.* **15**, 371–390.

Abel, L. and Casanova, J. L. (2000). Genetic predisposition to clinical tuberculosis: bridging the gap between simple and complex inheritance. *Am. J. Hum. Genet.* **67**, 274–277.

Abel, L. and Dessein, A. J. (1998). Genetic epidemiology of infectious diseases in humans: design of population-based studies. *Emerg. Infect. Dis.* **4**, 593–603.

Abel, L., Lap, V. D., Oberti, J., et al. (1995). Complex segregation analysis of leprosy in southern Vietnam. *Genet. Epidemiol.* **12**, 63–82.

Abel, L. and Muller-Myhsok, B. (1998a). Maximum-likelihood expression of the transmission/disequilibrium test and power considerations. *Am. J. Hum. Genet.* **63**, 664–647.

Abel, L. and Muller-Myhsok, B. (1998b). Robustness and power of the maximum-likelihood-binomial and maximum-likelihood-score methods, in multipoint linkage analysis of affected-sibship data. *Am. J. Hum. Genet.* **63**, 638–647.

Abel, L., Sanchez, F. O., Oberti, J., et al. (1998b). Susceptibility to leprosy is linked to the human NRAMP1 gene. *J. Infect. Dis.* **177**, 133–145.

Alcais, A. and Abel, L. (1999). Maximum-Likelihood-Binomial method for genetic model-free linkage analysis of quantitative traits in sibships. *Genet. Epidemiol.* **17**, 102–117.

Alcais, A. and Abel, L. (2000a). Linkage analysis of quantitative trait loci: sib pairs or sibships? *Hum. Hered.* **50**, 251–256.

Alcais, A. and Abel, L. (2000b). Robustness of the Maximum-Likelihood-Binomial approach for linkage analysis of quantitative trait loci with non-normal phenotypic data. *Genescreen* **1**, 47–50.

Alcais, A. and Abel, L. (2001). Incorporation of covariates in multipoint model-free linkage analysis of binary traits: how important are unaffecteds? *Eur. J. Hum. Genet.* **9**, 613–620.

Alcais, A., Sanchez, F. O., Thuc, N. V., et al. (2000). Granulomatous reaction to intradermal injection of lepromin (Mitsuda reaction) is linked to the human NRAMP1 gene in Vietnamese leprosy sibships. *J. Infect. Dis.* **181**, 302–308.

Allen, R. D. (1999). Polymorphism of the human TNF-alpha promoter – random variation or functional diversity? *Mol. Immunol.* **36**, 1017–1027.

Allison, D. B., Heo, M., Kaplan, N., et al. (1999a). Sibling-based tests of linkage and association for quantitative traits. *Am. J. Hum. Genet.* **64**, 1754–1763.

Allison, D. B., Neale, M. C., Zannolli, R., et al. (1999b). Testing the robustness of the likelihood-ratio test in a variance-component quantitative-trait loci-mapping procedure. *Am. J. Hum. Genet.* **65**, 531–544.

Almasy, L. and Blangero, J. (1998). Multipoint quantitative-trait linkage analysis in general pedigrees. *Am. J. Hum. Genet.* **62**, 1198–1211.

Amos, C. I. (1994). Robust variance-components approach for assessing genetic linkage in pedigrees. *Am. J. Hum. Genet.* **54**, 535–543.

Amos, C. I. and Elston, R. C. (1989). Robust methods for the detection of genetic linkage for quantitative data from pedigrees. *Genet. Epidemiol.* **6**, 349–360.

Armitage, P. (1955). Test for linear trend in proportions and frequencies. *Biometrics* **11**, 375–386.

Barnes, P. F., Chatterjee, D., Brennan, P. J., et al. (1992). Tumor necrosis factor production in patients with leprosy. *Infect. Immun.* **60**, 1441–1446.

Barnwell, J. W., Nichols, M. E. and Rubinstein, P. (1989). In vitro evaluation of the role of the Duffy blood group in erythrocyte invasion by *Plasmodium vivax*. *J. Exp. Med.* **169**, 1795–1802.

Bellamy, R., Beyers, N., McAdam, K. P., et al. (2000). Genetic susceptibility to tuberculosis in Africans: a genome-wide scan. *Proc. Natl. Acad. Sci. USA* **97**, 8005–8009.

Bellamy, R., Ruwende, C., Corrah, T., et al. (1999). Tuberculosis and chronic hepatitis B virus infection in Africans and variation in the vitamin D receptor gene. *J. Infect. Dis.* **179**, 721–724.

Bellamy, R., Ruwende, C., Corrah, T., et al. (1998a). Assessment of the interleukin 1 gene cluster and other candidate gene polymorphisms in host susceptibility to tuberculosis. *Tuber. Lung Dis.* **79**, 83–89.

Bellamy, R., Ruwende, C., Corrah, T., et al. (1998b). Variations in the NRAMP1 gene and susceptibility to tuberculosis in West Africans. *N. Engl. J. Med.* **338**, 640–644.

Bellamy, R., Ruwende, C., McAdam, K. P., et al. (1998c). Mannose binding protein deficiency is not associated with malaria, hepatitis B carriage nor tuberculosis in Africans. *Q. J. Med.* **91**, 13–18.

Blackwell, J. M., Barton, C. H., White, J. K., et al. (1994). Genetic regulation of leishmanial and mycobacterial infections: the Lsh/Ity/Bcg gene story continues. *Immunol. Lett.* **43**, 99–107.

Blackwell, J. M., Black, G. F., Peacock, C. S., et al. (1997). Immunogenetics of leishmanial and mycobacterial infections: The Belem Family Study. *Philos. Trans. R. Soc. Lond. B Biol. Sci.* **352**, 1331–1345.

Bonney, G. E. (1984). On the statistical determination of major gene mechanisms in continuous human traits: regressive models. *Am. J. Med. Genet.* **18**, 731–749.

Bonney, G. E. (1986). Regressive logistic models for familial disease and other binary traits. *Biometrics* **42**, 611–625.

Bothamley, G. H., Beck, J. S., Schreuder, G. M., et al. (1989). Association of tuberculosis and *M. tuberculosis*-specific antibody levels with HLA. *J. Infect. Dis.* **159**, 549–555.

Brahmajothi, V., Pitchappan, R. M., Kakkanaiah, V. N., et al. (1991). Association of pulmonary tuberculosis and HLA in south India. *Tubercle* **72**, 123–132.

Casanova, J. L. and Abel, L. (2002). Genetic dissection of immunity to mycobacteria: The human model. *Annu. Rev. Immunol.* **20**, 581–620.

Cellier, M., Govoni, G., Vidal, S., et al. (1994). Human natural resistance-associated macrophage protein: cDNA cloning, chromosomal mapping, genomic organization, and tissue-specific expression. *J. Exp. Med.* **180**, 1741–1752.

Cervino, A. C., Lakiss, S., Sow, O., et al. (2000). Allelic association between the NRAMP1 gene and susceptibility to tuberculosis in Guinea-Conakry. *Ann. Hum. Genet.* **64**, 507–512.

Chan, T. Y. (2000). Vitamin D deficiency and susceptibility to tuberculosis. *Calcif. Tissue. Int.* **66**, 476–478.

Chiano, M. N. and Clayton, D. G. (1998). Genotypic relative risks under ordered restriction. *Genet. Epidemiol.* **15**, 135–146.

Clerget-Darpoux, F., Bonaiti-Pellie, C. and Hochez, J. (1986). Effects of misspecifying genetic parameters in lod score analysis. *Biometrics* **42**, 393–399.

Commenges, D., Olson, J. and Wijsman, E. (1994). The weighted rank pairwise correlation statistic for linkage analysis: simulation study and application to Alzheimer's disease. *Genet. Epidemiol.* **11**, 201–212.

Cox, R. A., Downs, M., Neimes, R. E., et al. (1988). Immunogenetic analysis of human tuberculosis. *J. Infect. Dis.* **158**, 1302–1308.

D'Ambrosio, D., Cippitelli, M., Cocciolo, M. G., et al. (1998). Inhibition of IL-12 production by 1,25-dihydroxyvitamin D3: involvement of NF-kappaB down-regulation in transcriptional repression of the p40 gene. *J. Clin. Invest.* **101**, 252–262.

de Vries, R. R., Fat, R. F., Nijenhuis, L. E., et al. (1976). HLA-linked genetic control of host response to *Mycobacterium leprae*. *Lancet* **2**, 1328–1330.

de Vries, R. R., Mehra, N. K., Vaidya, M. C., et al. (1980). HLA-linked control of susceptibility to tuberculoid leprosy and association with HLA-DR types. *Tissue Antigens* **16**, 294–304.

Dessoukey, M. W., el-Shiemy, S. and Sallam, T. (1996). HLA and leprosy: segregation and linkage study. *Int. J. Dermatol.* **35**, 257–264.

Dib, C., Faure, S., Fizames, C., et al. (1996). A comprehensive genetic map of the human genome based on 5,264 microsatellites. *Nature* **380**, 152–154.

Elston, R. C. (1998). Methods of linkage analysis – and the assumptions underlying them. *Am. J. Hum. Genet.* **63**, 931–934.

Elston, R. C., Buxbaum, S., Jacobs, K. B., et al. (2000). Haseman and Elston revisited. *Genet. Epidemiol.* **19**, 1–17.

Fine, P. E., Wolf, E., Pritchard, J., et al. (1979). HLA-linked genes and leprosy: a family study in Karigiri, South India. *J. Infect. Dis.* **140**, 152–161.

Flynn, J. L. and Chan, J. (2001). Immunology of tuberculosis. *Annu. Rev. Immunol.* **19**, 93–129.

Fulker, D. W., Cherny, S. S., Sham, P. C., et al. (1999). Combined linkage and association sib-pair analysis for quantitative traits. *Am. J. Hum. Genet.* **64**, 259–267.

Gao, P. S., Fujishima, S., Mao, X. Q., et al. (2000). Genetic variants of NRAMP1 and active tuberculosis in Japanese populations. *Clin. Genet.* **58**, 74–76.

Garred, P., Harboe, M., Oettinger, T., et al. (1994). Dual role of mannan-binding protein in infections: another case of heterosis? *Eur. J. Immunogenet.* **21**, 125–131.

George, V., Tiwari, H. K., Zhu, X., et al. (1999). A test of transmission/disequilibrium for quantitative traits in pedigree data, by multiple regression. *Am. J. Hum. Genet.* **65**, 236–245.

Goldfeld, A. E., Delgado, J. C., Thim, S., et al. (1998). Association of an HLA-DQ allele with clinical tuberculosis. *J. Am. Med. Assoc.* **279**, 226–228.

Goldgar, D. E. (1990). Multipoint analysis of human quantitative genetic variation. *Am. J. Hum. Genet.* **47**, 957–967.

Goldgar, D. E. (2001), Major strengths and weaknesses of model-free methods. *Adv. Genet.* **42**, 241–251.

Gomes, I., Collins, A., Lonjou, C., et al. (1999). Hardy–Weinberg quality control. *Ann. Hum. Genet.* **63**, 535–538.

Govoni, G., Vidal, S., Gauthier, S., et al. (1996). The Bcg/Ity/Lsh locus: genetic transfer of resistance to infections in C57BL/6J mice transgenic for the Nramp1 Gly169 allele. *Infect. Immun.* **64**, 2923–2929.

Greenwood, C. M., Fujiwara, T. M., Boothroyd, L. J., et al. (2000). Linkage of tuberculosis to chromosome 2q35 loci, including NRAMP1, in a large aboriginal Canadian family. *Am. J. Hum. Genet.* **67**, 405–416.

Griffin, M. D., Lutz, W., Phan, V. A., et al. (2001). Dendritic cell modulation by 1alpha,25 dihydroxyvitamin D3 and its analogs: a vitamin D receptor-dependent pathway that promotes a persistent state of immaturity in vitro and in vivo. *Proc. Natl. Acad. Sci. USA* **98**, 6800–6805.

Haseman, J. K. and Elston, R. C. (1972). The investigation of linkage between a quantitative trait and a marker locus. *Behav. Genet.* **2**, 3–19.

Hawkins, B. R., Higgins, D. A., Chan, S. L., et al. (1988). HLA typing in the Hong Kong Chest Service/British Medical Research Council study of factors associated with the breakdown to active tuberculosis of inactive pulmonary lesions. *Am. Rev. Respir. Dis.* **138**, 1616–1621.

Hoal-Van Helden, E. G., Epstein, J., Victor, T. C., et al. (1999). Mannose-binding protein B allele confers protection against tuberculous meningitis. *Pediatr. Res.* **45**, 459–464.

Holmans, P. (2001). Likelihood-ratio affected sib-pair tests applied to multiply affected sibships: issues of power and type I error rate. *Genet. Epidemiol.* **20**, 44–56.

Horvath, S. and Baur, M. P. (2000). Future directions of research in statistical genetics. *Stat. Med.* **19**, 3337–3343.

Horvath, S., Xu, X. and Laird, N. M. (2001). The family based association test method: strategies for studying general genotype–phenotype associations. *Eur. J. Hum. Genet.* **9**, 301–306.

Hurme, M. and Santtila, S. (1998). IL-1 receptor antagonist (IL-1Ra) plasma levels are coordinately regulated by both IL-1Ra and IL-1beta genes. *Eur. J. Immunol.* **28**, 2598–2602.

Iwamoto, S., Omi, T., Kajii, E., et al. (1995). Genomic organization of the glycoprotein D gene: Duffy blood group Fya/Fyb alloantigen system is associated with a polymorphism at the 44-amino acid residue. *Blood* **85**, 622–626.

Jabado, N., Jankowski, A., Dougaparsad, S., et al. (2000). Natural resistance to intracellular infections: natural resistance-associated macrophage protein 1 (Nramp1) functions as a pH-dependent manganese transporter at the phagosomal membrane. *J. Exp. Med.* **192**, 1237–1248.

Jarvik, G. P. (1998). Complex segregation analyses: uses and limitations. *Am. J. Hum. Genet.* **63**, 942–946.

Jouanguy, E., Altare, F., Lamhamedi, S., et al. (1996). Interferon-gamma-receptor deficiency in an infant with fatal bacille Calmette-Guerin infection. *N. Engl. J. Med.* **335**, 1956–1961.

Juffermans, N. P., Florquin, S., Camoglio, L., et al. (2000). Interleukin-1 signaling is essential for host defense during murine pulmonary tuberculosis. *J. Infect. Dis.* **182**, 902–908.

Khoury, M. J., Beaty, T. H., and Cohen, B. H. (1993). *Fundamentals of Genetic Epidemiology Monographs in Epidemiology and Biostatistics* (*Vol. 19*) New York: Oxford University Press, pp. vi, 383.

Knapp, M. (1999). The transmission/disequilibrium test and parental-genotype reconstruction: the reconstruction-combined transmission/disequilibrium test. *Am. J. Hum. Genet.* **64**, 861–870.

Knapp, M., Seuchter, S. A. and Baur, M. P. (1994). Linkage analysis in nuclear families. 2. Relationship between affected sib-pair tests and lod score analysis. *Hum. Hered.* **44**, 44–51.

Kruglyak, L. (1999). Prospects for whole-genome linkage disequilibrium mapping of common disease genes. *Nat. Genet.* **22**, 139–144.

Kruglyak, L., Daly, M. J., Reeve-Daly, M. P., et al. (1996). Parametric and nonparametric linkage analysis: a unified multipoint approach. *Am. J. Hum. Genet.* **58**, 1347–1363.

Kruglyak, L. and Lander, E. S. (1995). Complete multipoint sib-pair analysis of qualitative and quantitative traits. *Am. J. Hum. Genet.* **57**, 439–454.

Lake, S. L., Blacker, D., and Laird, N. M. (2000). Family-based tests of association in the presence of linkage. *Am. J. Hum. Genet.* **67**, 1515–1525.

Lander, E. and Kruglyak, L. (1995). Genetic dissection of complex traits: guidelines for interpreting and reporting linkage results. *Nat. Genet.* **11**, 241–247.

Lander, E. S. and Botstein, D. (1987). Homozygosity mapping: a way to map human recessive traits with the DNA of inbred children. *Science* **236**, 1567–1570.

Lander, E. S. and Schork, N. J. (1994). Genetic dissection of complex traits. *Science* **265**, 2037–2048.

Levee, G., Liu, J., Gicquel, B., et al. (1994). Genetic control of susceptibility to leprosy in French Polynesia: no evidence for linkage with markers on telomeric human chromosome 2. *Int. J. Lepr. Other Mycobact. Dis.* **62**, 499–511.

Lonjou, C., Collins, A. and Morton, N. E. (1999). Allelic association between marker loci. *Proc. Natl. Acad. Sci. USA* **96**, 1621–1626.

MacLean, C. J., Bishop, D. T., Sherman, S. L., et al. (1993). Distribution of lod scores under uncertain mode of inheritance. *Am. J. Hum. Genet.* **52**, 354–361.

McLeod, R., Buschman, E., Arbuckle, L. D., et al. (1995). Immunogenetics in the analysis of resistance to intracellular pathogens. *Curr. Opin. Immunol.* **7**, 539–552.

Mehra, N. K., Rajalingam, R., Mitra, D. K., et al. (1995). Variants of HLA-DR2/DR51 group haplotypes and susceptibility to tuberculoid leprosy and pulmonary tuberculosis in Asian Indians. *Int. J. Lepr. Other Mycobact. Dis.* **63**, 241–248.

Meyer, C. G., May, J. and Stark, K. (1998). Human leukocyte antigens in tuberculosis and leprosy. *Trends Microbiol.* **6**, 148–154.

Miller, L. H., Mason, S. J., Clyde, D. F., et al. (1976). The resistance factor to *Plasmodium vivax* in blacks: the Duffy-blood-group genotype, FyFy. *N. Engl. J. Med.* **295**, 302–304.

Monks, S. A. and Kaplan, N. L. (2000). Removing the sampling restrictions from family-based tests of association for a quantitative-trait locus. *Am. J. Hum. Genet.* **66**, 576–592.

Morrison, N. A., Yeoman, R., Kelly, P. J., et al. (1992). Contribution of trans-acting factor alleles to normal physiological variability: vitamin D receptor gene polymorphism and circulating osteocalcin. *Proc. Natl. Acad. Sci. USA* **89**, 6665–6669.

Morton, N. E. (1955). Sequential tests for the detection of linkage. *Am. J. Hum. Genet.* **7**, 277–318.

Morton, N. E. (1998). Significance levels in complex inheritance. *Am. J. Hum. Genet.* **62**, 690–697.

Muller-Myhsok, B. and Abel, L. (1997). Genetic analysis of complex diseases. *Science* **275**, 1328–1329.

Neth, O., Jack, D. L., Dodds, A. W., et al. (2000). Mannose-binding lectin binds to a range of clinically relevant microorganisms and promotes complement deposition. *Infect. Immun.* **68**, 688–693.

Newport, M. J., Huxley, C. M., Huston, S., et al. (1996). A mutation in the interferon-gamma-receptor gene and susceptibility to mycobacterial infection. *N. Engl. J. Med.* **335**, 1941–1949.

Nicklin, M. J., Weith, A. and Duff, G. W. (1994). A physical map of the region encompassing the human interleukin-1 alpha, interleukin-1 beta, and interleukin-1 receptor antagonist genes. *Genomics* **19**, 382–384.

North, R. J., LaCourse. R., Ryan L., et al. (1999). Consequence of Nramp1 deletion to *Mycobacterium tuberculosis* infection in mice. *Infect. Immun.* **67**, 5811–5814.

Olson, J. M. (1995). Robust multipoint linkage analysis: an extension of the Haseman–Elston method. *Genet. Epidemiol.* **12**, 177–193.

Ott, J. (1999). *Analysis of Human Genetic Linkage.* Baltimore: Johns Hopkins University Press, pp. xxiii, 382

Ottenhoff, T. H. and de Vries, R. R. (1987). HLA class II immune response and suppression genes in leprosy. *Int. J. Lepr. Other Mycobact. Dis.* **55**, 521–534.

Piemonti, L., Monti, P., Sironi, M., et al. (2000). Vitamin D3 affects differentiation, maturation, and function of human monocyte-derived dendritic cells. *J. Immunol.* **164**, 4443–4451.

Quackenbush, J. (2001). Computational analysis of microarray data. *Nat. Rev. Genet.* **2**, 418–427.

Rajalingam, R., Mehra, N. K., Jain, R. C., et al. (1996). Polymerase chain reaction–based sequence-specific oligonucleotide hybridization analysis of HLA class II antigens in pulmonary tuberculosis: relevance to chemotherapy and disease severity. *J. Infect. Dis.* **173**, 669–676.

Rajalingam, R., Singal, D. P. and Mehra, N. K. (1997). Transporter associated with antigen-processing (TAP) genes and susceptibility to tuberculoid leprosy and pulmonary tuberculosis. *Tissue Antigens* **49**, 168–172.

Rani, R., Fernandez-Vina, M. A., Zaheer, S. A., et al. (1993). Study of HLA class II alleles by PCR oligotyping in leprosy patients from north India. *Tissue Antigens* **42**, 133–137.

Rao, D. C. (1998). CAT scans, PET scans, and genomic scans. *Genet. Epidemiol.* **15**, 1–18.

Ravikumar, M., Dheenadhayalan, V., Rajaram, K., et al. (1999). Associations of HLA-DRB1, DQB1 and DPB1 alleles with pulmonary tuberculosis in south India. *Tuber. Lung Dis.* **79**, 309–317.

Reich, D. E., Cargill, M., Bolk, S., et al. (2001). Linkage disequilibrium in the human genome. *Nature* **411**, 199–204.

Risch, N. (1990). Linkage strategies for genetically complex traits. III. The effect of marker polymorphism on analysis of affected relative pairs. *Am. J. Hum. Genet.* **46**, 242–253.

Risch, N. and Merikangas, K. (1996). The future of genetic studies of complex human diseases. *Science* **273**, 1516–1517.

Risch, N. and Zhang, H. (1995). Extreme discordant sib pairs for mapping quantitative trait loci in humans. *Science* **268**, 1584–1589.

Roy, S., Frodsham, A., Saha, B., et al. (1999). Association of vitamin D receptor genotype with leprosy type. *J. Infect. Dis.* **179**, 187–191.

Roy, S., McGuire, W., Mascie-Taylor, C. G., et al. (1997). Tumor necrosis factor promoter polymorphism and susceptibility to lepromatous leprosy. *J. Infect. Dis.* **176**, 530–532.

Ryu, S., Park, Y. K., Bai, G. H., et al. (2000). 3′UTR polymorphisms in the NRAMP1 gene are associated with susceptibility to tuberculosis in Koreans. *Int. J. Tuberc. Lung. Dis.* **4**, 577–580.

Sanjeevi, C. B., Narayanan, P. R., Prabakar, R., et al. (1992). No association or linkage with HLA-DR or -DQ genes in south Indians with pulmonary tuberculosis. *Tuber. Lung Dis.* **73**, 280–284.

Sasieni, P. D. (1997). From genotypes to genes: doubling the sample size. *Biometrics* **53**, 1253–1261.

Sawcer, S., Jones, H. B., Judge, D., et al. (1997). Empirical genomewide significance levels established by whole genome simulations. *Genet. Epidemiol.* **14**, 223–229.

Schaid, D. J. and Jacobsen, S. J. (1999). Biased tests of association: comparisons of allele frequencies when departing from Hardy–Weinberg proportions. *Am. J. Epidemiol.* **149**, 706–711.

Schreuder, G. M., Hurley, C. K., Marsh, S. G., et al. (1999). The HLA dictionary 1999: a summary of HLA-A, -B, -C, -DRB1/3/4/5, -DQB1 alleles and their association with serologically defined HLA-A, -B, -C, -DR and -DQ antigens. *Tissue Antigens* **54**, 409–437.

Selvaraj, P., Narayanan, P. R. and Reetha, A. M. (1999). Association of functional mutant homozygotes of the mannose binding protein gene with susceptibility to pulmonary tuberculosis in India. *Tuber. Lung Dis.* **79**, 221–227.

Sham, P. C., Cherny. S. S., Purcell. S., et al. (2000). Power of linkage versus association analysis of quantitative traits, by use of variance-components models, for sibship data. *Am. J. Hum. Genet.* **66**, 1616–1630.

Shaw, M. A., Atkinson, S., Dockrell, H., et al. (1993). An RFLP map for 2q33-q37 from multicase mycobacterial and leishmanial disease families: no evidence for an Lsh/Ity/Bcg gene homologue influencing susceptibility to leprosy. *Ann. Hum. Genet.* **57**, 251–271.

Siddiqui, M. R., Meisner, S., Tosh, K., et al. (2001). A major susceptibility locus for leprosy in India maps to chromosome 10p13. *Nat. Genet.* **27**, 439–441.

Singh, S. P., Mehra, N. K., Dingley, H. B., et al. (1983). Human leukocyte antigen (HLA)-linked control of susceptibility to pulmonary tuberculosis and association with HLA-DR types. *J. Infect. Dis.* **148**, 676–681.

Soo, S. S., Villarreal-Ramos, B., Anjam Khan, C. M., et al. (1998). Genetic control of immune response to recombinant antigens carried by an attenuated *Salmonella typhimurium* vaccine strain: Nramp1 influences T-helper subset responses and protection against leishmanial challenge. *Infect. Immun.* **66**, 1910–1917.

Spielman, R. S. and Ewens, W. J. (1998). A sibship test for linkage in the presence of association: the sib transmission/disequilibrium test. *Am. J. Hum. Genet.* **62**, 450–458.

Spielman, R. S., McGinnis, R. E. and Ewens, W. J. (1993). Transmission test for linkage disequilibrium: the insulin gene region and insulin-dependent diabetes mellitus (IDDM). *Am. J. Hum. Genet.* **52**, 506–516.

Spies, T., Bresnahan, M., Bahram, S., et al. (1990). A gene in the human major histocompatibility complex class II region controlling the class I antigen presentation pathway. *Nature* **348**, 744–747.

Stene, L. C. (2000). Vitamin D deficiency and tuberculosis. Lancet **356**, 73–74; discussion 74–75.

Stephens, J. C., Schneider, J. A., Tanguay, D. A., et al. (2001). Haplotype variation and linkage disequilibrium in 313 human genes. *Science* **293**, 489–493.

Stern, L. J., Brown, J. H. and Jardetzky, T. S. (1994). Crystal structure of the human class II MHC protein HLA-DR1 complexed with an influenza virus peptide. *Nature* **368**, 215–221.

Summerfield, J. A., Sumiya, M., Levin, M., et al. (1997). Association of mutations in mannose binding protein gene with childhood infection in consecutive hospital series. *Br. Med. J.* **314**, 1229–1232.

Teran-Escandon, D., Teran-Ortiz, L., Camarena-Olvera, A., et al. (1999). Human leukocyte antigen-associated susceptibility to pulmonary tuberculosis: molecular analysis of class II alleles by DNA amplification and oligonucleotide hybridization in Mexican patients. *Chest* **115**, 428–433.

Tournamille, C., Colin, Y., Cartron, J. P., et al. (1995). Disruption of a GATA motif in the Duffy gene promoter abolishes erythroid gene expression in Duffy-negative individuals. *Nat. Genet.* **10**, 224–228.

van Eden, W., Gonzalez, N. M., de Vries, R. R., et al. (1985). HLA-linked control of predisposition to lepromatous leprosy. *J. Infect. Dis.* **151**, 9–14.

Vidal, S. M., Malo, D., Vogan, K., et al. (1993). Natural resistance to infection with intracellular parasites: isolation of a candidate for Bcg. *Cell* **73**, 469–485.

Vidal, S. M., Pinner, E., Lepage, P., et al. (1996). Natural resistance to intracellular infections: Nramp1 encodes a membrane phosphoglycoprotein absent in macrophages from susceptible (Nramp1 D169) mouse strains. *J. Immunol.* **157**, 3559–3568.

Wakelin, D. and Blackwell, J. M. (1988). *Genetics of Resistance to Bacterial and Parasitic Infection.* London: Taylor & Francis. pp. ix, 287.

Waldman, I. D., Robinson, B. F. and Rowe, D. C. (1999). A logistic regression based extension of the TDT for continuous and categorical traits. *Ann. Hum. Genet.* **63**, 329–340.

Wang, D. G., Fan, J. B., Siao, C. J., et al. (1998). Large-scale identification, mapping, and genotyping of single-nucleotide polymorphisms in the human genome. *Science* **280**, 1077–1082.

Wilkinson, R. J., Llewelyn, M., Toossi, Z., et al. (2000). Influence of vitamin D deficiency and vitamin D receptor polymorphisms on tuberculosis among Gujarati Asians in west London: A case-control study. *Lancet* **355**, 618–621.

Wilkinson, R. J., Patel, P., Llewelyn, M., et al. (1999). Influence of polymorphism in the genes for the interleukin (IL)-1 receptor antagonist and IL-1beta on tuberculosis. *J. Exp. Med.* **189**, 1863–1874.

Xu, K. Y., de Vries, R. R., Fei, H. M., et al. (1985). HLA-linked control of predisposition to lepromatous leprosy. *Int. J. Lepr. Other Mycobact. Dis.* **53**, 56–63.

Xu, X., Weiss, S. and Wei, L. J. (2000). A unified Haseman–Elston method for testing linkage with quantitative traits. *Am. J. Hum. Genet.* **67**, 1025–1028.

Yamamura, M., Uyemura, K., Deans, R. J., et al. (1991). Defining protective responses to pathogens: cytokine profiles in leprosy lesions. *Science* **254**, 277–279.

Yamauchi, P. S., Bleharski, J. R., Uyemura, K., et al. (2000). A role for CD40-CD40 ligand interactions in the generation of type 1 cytokine responses in human leprosy. *J. Immunol.* **165**, 1506–1512.

Zerva, L., Cizman, B., Mehra, N. K., et al. (1996). Arginine at positions 13 or 70–71 in pocket 4 of HLA-DRB1 alleles is associated with susceptibility to tuberculoid leprosy. *J. Exp. Med.* **183**, 829–836.

CHAPTER 3

The diverse genetic basis of immunodeficiencies

Mauno Vihinen

Institute of Medical Technology, Finland and Tampere University Hospital

The immune system is constantly alert to recognise and neutralise invading microorganisms and foreign molecules. To cope with a large spectrum of substances and organisms, nature has developed highly sophisticated response systems. Innate immunity can generate a fast but usually nonspecific response. Adaptive immunity facilitates specific recognition. The recognition components – antigens, B- and T-cell receptors, and major histocompatibility complexes – of the adaptive immunity originate from gene rearrangements, which produce countless combinations of recognition sites. Together the immune system is capable of protecting the body from most microorganisms. When components of the machinery are mutated the affected individuals suffer from immunodeficiencies. Primary immunodeficiencies can arise from numerous mutated genes, which leads to a large number of very different immunodeficiencies, which require different therapeutic approaches. The disorders vary greatly in regard to symptoms, infection-causing organisms, genotype, phenotype, and severity of the disease. The genetic background and disease-causing mutations, symptoms, and therapy are discussed for a number of well-characterised primary immunodeficiencies.

1. INTRODUCTION

Adaptive immune mechanisms recognise and neutralise foreign molecules or microorganisms in a specific manner. B and T lymphocytes can respond selectively to thousands of nonself materials. Adaptation is further acquired with memory of previous infections. The other arm of the immune system, native (innate) immunity is able to respond almost immediately to potentially infectious agents. The major components of innate immunity are natural killer cells, phagocytes, and the complement system. Innate

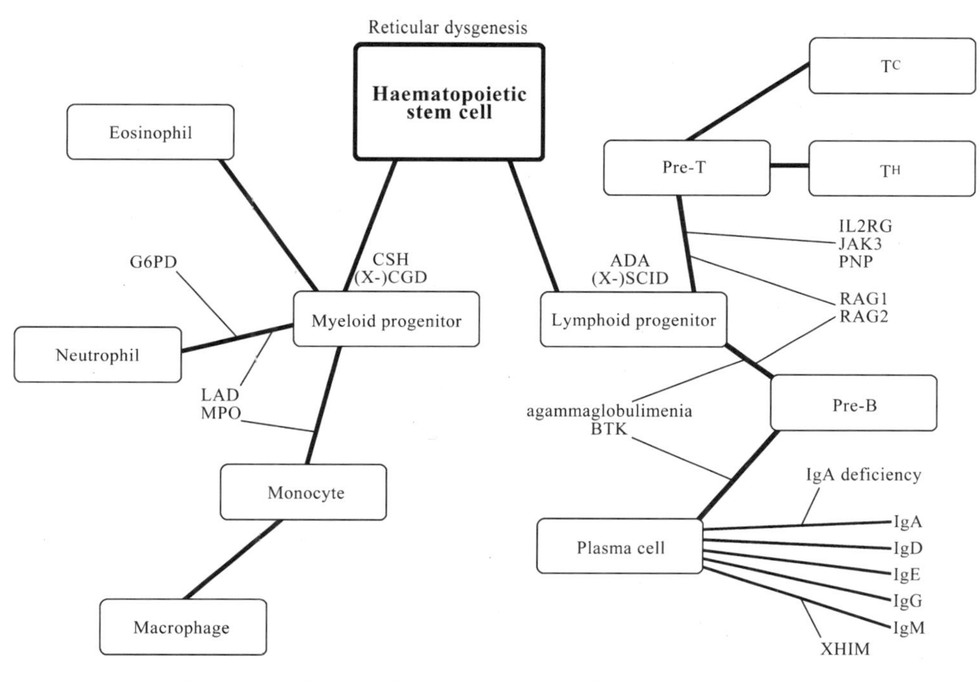

Figure 3.1. Schematic depiction of the differentiation of some cells involved in immunology. The development of the haematopoietic cells from the stem cell are indicated with thick lines. The cellular stages affected by immunodeficiencies are shown above cell types or by thin lines. T^C, cytotoxic T cell; T^H, T helper cell; ADA, adenosine deaminase; BTK; Bruton tyrosine kinase; G6PD, glucose 6-phosphate dehydrogenase; IL2RG, interleukin-2 receptor common γ-chain; JAK3, protein tyrosine kinase JAK3; LAD, leukocyte adhesion deficiency; MPO, myeloperoxidase; PNP, purine nucleoside phosphorylase; RAG1 and −2, recombination activating genes 1 and 2; XHIM, X-linked hyper IgM; X-SCID, X-linked severe combined immunodeficiency.

immunity has only a limited specificity to distinguish one microbe from another. Immunodeficiencies impair the functioning of the immune system (Fig. 3.1). Deficiencies vary widely, e.g., with regard to symptoms, causative microorganisms, genotype, phenotype, and severity, because many types of cells and molecules are required for both natural and adaptive immunity. Increased susceptibility to infections is common to all immunodeficiencies.

Some 100 primary immune deficiencies (PIDs) are known. Most of them are rare disorders. They have been grouped according to the components of the immune system affected (Anonymous, 1997; Ochs et al., 1999) in slightly different ways. We follow the latter classification here (Table 3.1). A constantly updated version is available in the ImmunoDeficiency Resource (IDR) (Väliaho et al., 2000, 2002) at http://bioinf.uta.fi/idr. In this review we describe some common and well-known PIDs with regard to genetics, symptoms, treatment, and disease-causing mutations.

Antibody deficiency disorders are defects in immunoglobulin-producing B cells. T-cell deficiencies affect the capability to kill infected cells or help other immune cells. Both T cells and antibody production are defective in combined immunodeficiencies (CIDs). Life-threatening symptoms can arise in severe combined immunodeficiencies (SCIDs). Other PIDs affect the complement system or phagocytic cells, impairing antimicrobial immunity. Secondary immunodeficiencies, although presenting similar infections as PIDs, arise secondarily to some pathological condition, e.g., age, malnutrition, drugs, infections, or tumours.

The incidence of PIDs varies greatly from about 1:500 births with selective IgA deficiency to only a few known cases for the rarest disorders. In the ESID registry of European PID patients there are close to 9000 patients (Abedi et al., 1995). Antibody deficiencies are by far the most common group of disorders (67%) followed by T-cell and combined deficiencies (18%), phagocytic (7%), and complement disorders (6%). Immunodeficiency mutation databases (IDbases) contain genetic and clinical information for around 2500 individuals (Vihinen et al., 2001).

2. INFECTIONS AND TREATMENT

PID patients have recurrent, serious infections starting early after birth. Immunodeficiencies do not result in prenatal death except when the affected gene is crucial for organs outside the immune system. The immune system normally is not activated prior to birth. Maternal IgG protects children for 6–12 months in immunoglobulin deficiencies. The different immunological

Table 3.1. *Classification of PIDs*

Deficiencies predominantly affecting antibody production
- Agammaglobulinemia
  - X-linked agammaglobulinemia (XLA)
  - X-linked hypogammaglobulinemia with growth hormone deficiency
  - BLNK deficiency
  - Igα-deficiency
  - μ-heavy-chain deficiency
  - λ5-surrogate light-chain deficiency
- κ light-chain deficiency
- Selective deficiency of IgG subclass and IgE and/or IgA class or subclass
  - γ1-isotype deficiency
  - γ2-isotype deficiency
  - Partial γ3-isotype deficiency
  - γ4-isotype deficiency
  - α1-isotype deficiency
  - α2-isotype deficiency
  - ε-isotype deficiency
  - IgG subclass deficiency with or without IgA deficiency
- Common variable immunodeficiency
- IgA deficiency
- Antibody deficiency with normal immunoglobulin levels
- Transient hypogammaglobulinemia of infancy

Combined B- and T-cell immunodeficiencies
- T^-B^-severe combined immunodeficiency (SCID)
  - Reticular dysgenesis
  - RAG1 deficiency
  - RAG2 deficiency
  - Omenn syndrome
  - Artemis deficiency
  - Native American SCID
  - CD45 deficiency
- T^-B^+ SCID
  - X-linked SCID (γc-chain defficiency)
  - JAK3 deficiency
- Deficiencies of purine metabolism
  - Adenosine deaminase (ADA) deficiency
  - Purine nucleoside phosphorylase (PNP) deficiency

Table 3.1. (*cont.*)

- Major histocompatibility complex class II deficiency
  - CIITA, MHCII transactivating protein deficiency
  - RFX-5, MHCII promoter X box regulatory factor 5 deficiency
  - RFXAP, Regulatory factor X-associated protein deficiency
  - RFXANK, Ankyrin repeat containing regulatory factor X-associated protein deficiency
- MHC class I deficiency
- Hyper-IgM syndrome
  - X-linked hyper-IgM syndrome (CD40L deficiency)
  - Non-X-linked hyper-IgM syndrome
  - CD 40 deficiency
  - X-linked hyper-IgM syndrome and hypohydrotic ectodermal dysplasia (NEMO deficiency)
- CD3 deficiency
  - CD3ε-deficiency
  - CD3γ-deficiency
- ZAP-70 deficiency
- IL-2 receptor α-chain deficiency
- CD8α deficiency

Defects in lymphocyte apoptosis
- Autoimmune lymphoproliferative syndrome (ALPS)
  - Apoptosis mediator APO-1/Fas defects
  - APO-1 ligand/Fas ligand defects

Other well-defined immunodeficiency syndromes
- Wiskott–Aldrich syndrome (WAS) and X-linked thrombocytopenia
- Autoimmune polyendocrinopathy with candidiasis and ectodermal dystrophy (APECED)
- X-linked lymphoproliferative syndrome (Duncan's disease)
- DiGeorge-anomaly
- Hyper-IgE recurrent infection syndrome
- Chronic mucocutaneous candidiasis
- Cartilage-hair hypoplasia
- Immunodeficiency, polyendocrinopathy, enteropathy, X-linked (IDEX)
- Immunodeficiency, centromeric instability and facial anomalies (ICF)

(*cont.*)

Table 3.1. (*cont.*)

Defects of phagocyte function
Chronic granulomatous disease
X-linked CGD
$p22^{phox}$ deficiency
$p47^{phox}$ deficiency
$p67^{phox}$deficiency
Leukocyte adhesion defects
LAD1
LAD2
Chediak–Higashi syndrome
Griscelli syndrome
Glucose 6-phosphate dehydrogenase deficiency
Myeloperoxidase deficiency
Glycogen storage disease Ib
Shwachman syndrome
Severe congenital neutropenias, including Kostmann syndrome
Cyclic neutropenia
Interferon-γ (IFN-γ) associated immunodeficiency
IFN-γ1-receptor deficiency
IFN-γ1-receptor deficiency
IFN-γ2-receptor deficiency
Interleukin-12 (IL-12) p40 deficiency
Interleukin-12 receptor β1 deficiency
DNA breakage-associated syndromes
Ataxia-telangiectasia
Nijmegen-breakage syndrome
Bloom syndrome
Defects of the complement cascade proteins
C1q deficiency
C1 α-polypeptide deficiency
C1 β-polypeptide deficiency
C1 γ-polypeptide deficiency
C1r and C1s deficiency
C1r deficiency
C1s deficiency
C2 deficiency

Table 3.1. (*cont.*)

- C3 deficiency
- C4 deficiency
  - C4A deficiency
  - C4B deficiency
- C5 deficiency
- C6 deficiency
- C7 deficiency
- C8 deficiency
  - C8 α-polypeptide deficiency
  - C8 β-polypeptide deficiency
  - C8 γ-polypeptide deficiency
- C9 deficiency
- Factor B deficiency
- Mannose-binding lectin (protein) deficiency

Defects of complement regulatory proteins

- Hereditary angioedema
- C4-binding protein deficiency
  - C4 binding protein-α deficiency
  - C4 binding protein-β deficiency
- Factor D deficiency
- Factor I deficiency
- Properdin factor C deficiency
- Factor H1 deficiency
- Decay-accelerating factor (CD55) deficiency
- CD59 (antigen p18-20) or protectin deficiency

systems recognise and destroy different pathogens. Therefore the symptoms vary depending on the component(s) of the immune system impaired in PIDs.

Patients with antibody deficiencies are especially susceptible to pyogenic infections caused by encapsulated bacteria such as *Haemophilus influenzae*, *Staphylococcus aureus*, and *Streptococcus pneumoniae*. Individuals with T-cell immunodeficiencies and SCIDs have opportunistic infections caused by common environmental microorganisms as well as an increased frequency of viral, parasitic, and fungal infections. In SCIDs, life-threatening symptoms can arise already within the first few days of life. Persons with natural killer (NK) cell disorders are mostly susceptible to viral infections that are treatable with antiviral therapy. NK cells have cytolytic activity towards virus-infected cells, some types of tumour cells and cells infected with the protozoan *Toxoplasma gondii* and microbicidal activity against some bacteria (e.g., *Salmonella* and *Escherichia coli*), and the fungus *Cryptococcus neoformans*. In diseases affecting phagocytes, mainly skin and oral bacterial infections occur. Granulomas are formed when the microorganisms spread to organs. Also fungal infections cause severe complications for affected individuals.

Certain defects lead to an increased susceptibility to a few or even a single pathogenic agent(s). Patients with X-linked lymphoproliferative syndrome (XLP) are selectively prone to Epstein–Barr virus infections, and patients having defects in interleukin 12 (IL-12) or interferon-γ signalling are sensitive to atypical forms of mycobacteria and also to salmonella.

Some PIDs result in primarily autoimmune manifestations and in certain primary immunodeficiencies, patients have an increased incidence of cancer such as Wiskott Aldrich syndrome (WAS) and ataxia telangiectasia (AT).

The management of PIDs requires a number of treatments (Stiehm, 1999). Infections in PID patients require prolonged treatment with high doses of antibiotics. Antibody deficiencies are treated with intravenous immunoglobulin substitution therapy. Leukocytes (B and T cells) are produced in stem cells in bone marrow. In many PIDs, including SCIDs, bone marrow transplantation is the most effective treatment (Fischer et al., 1998). In certain metabolic disorders [adenosine deaminase (ADA) and purine nucleoside phosphorylase (PNP) deficiency] enzyme substitution therapy can be applied.

The human genome project and other advances in genetics have generated a large pool of data on human diseases. Gene therapy is a method for delivering normal copies of genes or fragments of genes for the treatment of patients with inherited diseases. The only successful gene therapy trial with a long-lasting effect has been conducted for a PID, X-linked SCID, caused by mutations in the common γ-chain of several interleukin receptors (Cavazzana-Calvo et al., 2000). This condition is well suited for gene therapy, because γc expression confers a major selective advantage to transduced cells. Initially adenosine deaminase (ADA) deficiency, another immunodeficiency, was extensively studied for gene therapy (Bordignon et al., 1995; Onodera et al., 1999).

The functional γc cDNA was transferred to CD34 cells with a retroviral vector (Cavazzana-Calvo et al., 2000). A similar approach could at least in principle be used for treatment of several other monogenic SCIDs, including

IL-R7α and JAK3 deficiencies as well as RAG1 and RAG2 deficiencies in SCID and Omenn syndrome.

3. IMMUNODEFICIENCY INFORMATION SERVICES

Diagnosis of immunodeficiencies can be very difficult (Chapel and Webster, 1999; Väliaho et al., 2000), because several disorders can have similar symptoms. Further, many PIDs are rare. Early and reliable diagnosis is in many instances crucial for the efficient treatment of these diseases because delayed diagnosis and management can cause severe and irreversible complications, even the death of the patient. The early or possibly prenatal molecular diagnosis would allow enough time for the most suitable management and selection of treatment. The European and Pan-American societies for PIDs have released guidelines for the diagnosis of some common immunodeficiencies (Conley et al., 1999).

IDdiagnostics registry (Väliaho et al., 2000) contains two databases, genetic and clinical. It provides a service for physicians to find quickly the nearest and/or most suitable laboratory conducting PID testing. For many PIDs there are only a few laboratories in the world providing the analysis. The IDdiagnostics service is available at http://bioinf.uta.fi/iddiagnostics.

There is plenty of information available regarding PIDs on the Internet, but it may be difficult to find up-to-date and validated knowledge. The Immunodeficiency Resource (IDR) collects and distributes all the essential information related to immunodeficiencies (Väliaho et al., 2000, 2002). The IDR (http://bioinf.uta.fi/idr) aims at providing comprehensive, integrated knowledge on immunodeficiencies in an easily accessible format offering data for clinical, biochemical, genetic, structural, and computational analysis (Fig. 3.2). IDR provides only validated information. Data for each disease have been evaluated by curators who are established scientists and experts on the particular disease.

4. GENETIC BASIS

The immune system is based on a large number of molecules and processes. A particular PID can orginate from defects in any one of the molecules essential for certain responses (Fig. 3.1), because a defect in any of the sequential steps can impair the complete system (Vihinen et al., 2001). Numerous genes involved in PIDs have been identified (Table 3.1). The majority of

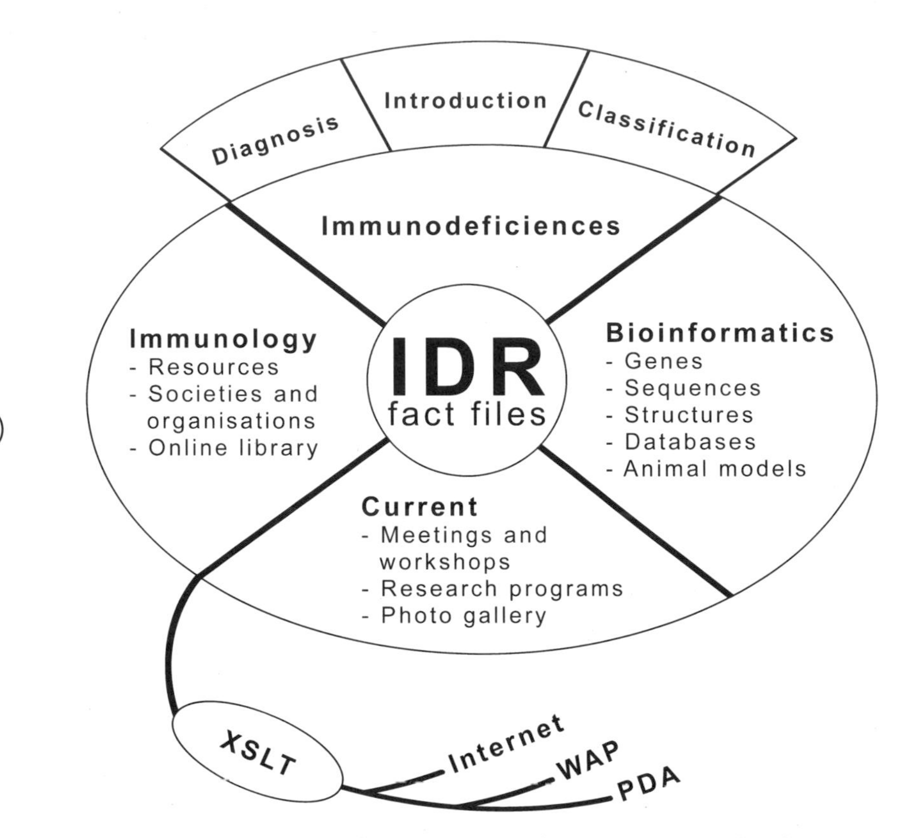

Figure 3.2. The composition of the XML- (eXtensible Markup Language) based ImmunoDeficiency Resource (IDR), which is built around the central fact files. The major categories of distributed information, links, and connections are shown. PDA, personal digital assitant; WAP, wireless application protocol; XSLT, extensible style sheet language transformation.

the PIDs are autosomal recessive (AR), although the best known cases are X-linked forms. In *IFNGR1* dominant mutations are also known. PID causing genes locate in a number of chromosomes. The majority of the genes code for multidomain proteins. Consanguinity is common in families with autosomal recessive forms of immunodeficiencies. In X-linked disorders a single mutated gene is enough to cause the phenotype in males, because X-linked recessive diseases usually have full penetrance.

The functions of the PID related proteins are very diverse in nature indicating the importance of the numerous cellular functions for producing the natural and adaptive immune response (Fig. 3.1). The proteins in PIDs are employed, e.g., in signal transduction, cell surface receptors, nucleotide metabolism, gene diversification, transcription factors, and phagocytosis. The localisation of these proteins ranges from nucleus to cytoplasm and from compartments to cell membrane.

4.1 Immunological Recognition Molecules

Heterogeneous receptors – antibodies, T-cell receptors (TCRs), and the components of the major histocompatibility complexes (MHCs) – form the basis of adaptive immunity by recognising the enormous range of nonself substances. The very highly variable molecules are formed by joining a large number of genetic segments randomly.

Antibodies (immunoglobulins) appear either free in serum or as a part of the B-cell receptor (BCR). The main role of antibodies is to recognise foreign substances and thereby facilitate their destruction. Antibodies are formed of two light and two heavy chains. The gene for each antibody is composed by gene rearrangement from a large number of regions and contains both constant and highly variable regions. From the large number (up to 100) of different segments for each part of the antibody gene, just one is used in each cell. First a diversity region (D) segment is combined with a single joining (J) segment and a single variable region (V). The V(D)J rearrangement is responsible for the diversity of antibodies. Then one of the several constant regions (C) is added to complete the full V(D)JC gene. Antibodies further undergo class switching. Antigen-specific TCRs are produced by related gene rearrangements.

Membrane-bound MHC molecules recognise peptides on the surface of the cell. The peptide-binding cleft is coded by hypervariable gene segments. Class I MHC molecules bind to foreign peptides processed within infected cells and then present them to cytotoxic $CD8^+$ T cells. Class II molecules bind to peptides processed within specialised antigen-presenting cells and present them to helper $CD4^+$ T cells.

Errors in the highly variable antibodies and receptors or defects in their production lead to immunodeficiencies. Immunoglobulin (Ig) gene deletions are usually deletions of the constant heavy chain, although also some κ-light chain gene mutations have been identified. In general, patients with these immunodeficiencies do not have a markedly increased risk of infection. In recombination activating gene 1 (RAG1) and RAG2 deficiencies and the Omenn syndrome as well as Artemis deficiency, both BCR and TCR are deficient, leading into SCID due to defective V(D)J recombination. MHC class II deficiencies are caused by defective promoter-binding proteins affecting the transcription of genes.

5. ANTIBODY DEFICIENCIES

B-cell immunodeficiencies are deficiencies of antibody function (Table 3.1). Either the B lymphocyte development is impaired or B cells fail to respond to T-cell signals. All or only a subset of immunoglobulins may be deficient.

In XLA antibody production is prevented due to a block in the B-cell maturation (Vihinen et al., 2000a; Smith et al., 2001). Serum concentrations of IgG, IgA, and IgM are markedly reduced. Levels of circulating B lymphocytes are significantly decreased and plasma cells are absent, whereas the number of T cells is normal or increased. The clinical outcome varies and even members of the same family can have different symptoms. Patients with XLA have normal response to viral infections and normal V(D)J rearrangement.

Btk belongs to the Tec family of cytoplasmic protein tyrosine kinases (PTKs). They all consist of five distinct structural domains. Btk interacts with several partners (for review see Smith *et al.*, 2001). The PH domain binds to phosphatidyl inositols and can function as a membrane-localising module. The TH domain contains a Zn^{2+}-binding motif (Hyvönen and Saraste, 1997; Vihinen et al., 1997) and a polyproline stretch. The SH2 and SH3 domains bind either to phosphotyrosine (pTyr) residue or polyproline-containing proteins.

Many XLA-causing mutations affect functionally significant, conserved residues (Vihinen et al., 1995, 1999, 2001). The majority of the missense mutations in the PH domain are in the inositol-compound-binding region. In the TH domain the missense mutations affect Zn^{2+} binding. Most of the amino acid substitutions in the SH2 domain impair pTyr binding. In the kinase domain, the mutations are mainly on one face of the molecule, which is in charge of the ATP, Mg^{2+}, and substrate binding.

IgA deficiency can affect selectively only IgA levels. It may also be combined with the lack of other isotypes. It is the most prevalent of PIDs (1:500 Caucasians). The mechanism of the disease is still unknown. Of the affected individuals only about one-third are particularly prone to infections. The patients have a high incidence of autoantibodies although the serum concentrations of the other immunoglobulins are usually normal. Selective deficiencies of IgG subclasses, with or without IgA deficiency, are caused by defects in several genes.

Common variable immunodeficiency (CVI) is a group of disorders of defective antibody formation. CVI patients usually have a normal number of circulating but defective B cells. Their serum levels of IgG and IgA are low. CVI affects females and males equally and it usually has later age of onset than the other antibody deficiencies. Patients frequently have lymphoreticular and gastrointestinal malignancies and also autoimmune disorders are common.

6. SEVERE COMBINED IMMUNODEFICIENCIES

In combined B- and T-cell immunodeficiencies (SCIDs), which are the most severe PIDs, all adaptive immune functions are impaired. SCID is fatal unless the immune system is reconstituted either by transplants of immunocompetent tissue or by enzyme replacement.

6.1 T^-B^+ SCID

In T^-B^+ SCID both T and B cells are lacking. In RAG1 and RAG2 deficiencies and the Omenn syndrome, recombinase-activating proteins are defective (Notarangelo et al., 1999). T^-B^- SCID and Omenn syndrome (OS) are caused by mutations in either of the two recombination activating genes, *RAG1* and *RAG2* (Schwarz et al., 1996a; Villa et al., 1998; 2001). RAGs are crucial because they activate the V(D)J recombination in the antibody and TCR receptor genes required for generation of the diversity of the recognition sites. RAG1 or -2 disruption blocks the initiation of V(D)J recombination and leads further to complete absence of both mature B and T cells. In the Omenn syndrome, recombination is only partially deficient.

The T^-B^- SCID patients exhibit no mature T and B cells and show a complete absence of lymph nodes and tonsils. Infections start in the second or third month after birth. The affected individuals show opportunistic infections, chronic persistent disease of the airways, and local and systemic bacterial infections. The recurrent infections lead to a failure to thrive and virus infections can be lethal.

In addition to SCID, OS patients have a number of other symptoms. The patients have a variable number of circulating T cells that respond poorly to stimulation. B cells are absent or highly reduced in their number. RAG1 core is crucial for DNA binding. This region bears similarity to the homeodomain of the Hin invertase family and it contributes to the heterodimer fomation with RAG2. The three basic regions in the N-terminal half of RAG1 interact with the nuclear transport protein Srp1. Central part of RAG2 has been suggested to contain repeated kelch/mipp motifs. Disease-causing mutations have been found throughout the RAG1 and −2 proteins.

Reticular dysgenesis is a very rare autosomal recessive form of SCID, which generally leads to very early death.

6.2 T^-B^+ SCID

In T^-B^+ SCID, T cells are missing, but there can be B cells even in increased numbers (Buckley *et al.*, 1997). Although B cells are produced, they are defective. X-linked SCID, the most common of SCIDs, is caused by IL-2 receptor γ-chain mutations (Puck, 1999). Patients have very low numbers of T cells and NK cells, whereas B cells are present in high numbers although they are immature and defective (Candotti et al., 1998). Infants with SCID develop severe infections in the first months of life. At the age of 3 to 6 months most boys with XSCID have failure to thrive, chronic diarrhoea, and recurrent and persistent infections, often due to opportunistic pathogens. The γ-chain of the receptor forms part of the receptor also for IL-2, -4, -7, -9, and -15, affecting the differentiation and growth of lymphocytes.

Mutations in *IL2RG* severely compromise the function of γc and result in a severe phenotype. Mutations associated with XSCID are found throughout all the exons of *IL2RG* (Puck et al., 1996). Of the mutations in 220 unrelated families, the most common are missense and nonsense point mutations. The protein contains a signal sequence and a large extracellular domain followed by a transmembrane region and Box1/Box2 domain, the cytoplasmic part of the molecule. WSXWS motif in the extracellular region has been shown to be essential for cytokine binding.

The AR form of T^-B^+ SCID is caused by Janus kinase 3 (JAK3) tyrosine kinase mutations (Candotti et al., 1998). IL-2 stimulates the receptor, induces tyrosine phosphorylation, and activation of JAK3. The γc-JAK3 pathway transmits the signal to the nucleus and effects the expression of genes responding to cytokines. Binding to IL-2 leads to the dimerisation of receptors and into activation of JAK3. Activated JAK family members phosphorylate multiple tyrosine residues in the receptors. Signal transducers and activators of transcription (STATs) recognise with their SH2 domains phosphotyrosines in the receptor. Then activated, dimerised STATs dissociate from the receptor and translocate to the nucleus, where they bind to enhancer regions in DNA and thereby effect transcription of cytokine-responsive genes.

JAKs include altogether seven Jak homology (JH) domains. In the C-terminus there is tyrosine kinase (JH1) domain, which is preceded by an inactive pseudokinase (JH2) domain. The N-terminal domains (JH6 and −7) are necessary and sufficient for γc binding. Mutations have been identified from all the JH domains.

6.3 Deficiencies in Purine Metabolism

Purine nucleoside phosphorylase deficiency is characterised by accumulation of toxic purine metabolites in cells (Osborne and Ochs, 1999). PNP catalyses the phosphorolysis of the purine nucleosides to purine bases and ribose-1-phosphate. dGTP accumulation is particularly toxic to T cells due to the inhibition of ribonucleotide reductase, DNA synthesis, and cell proliferation. The protein consists of a single catalytic domain coded by six exons. The enzyme following PNP in the purine catabolism is adenosine deaminase.

Adenosine deaminase deficiency is the most common of the autosomal recessive forms of SCIDs. ADA deficiency causes even more severe symptoms than PNP deficiency (Hershfield, 1998). ADA degrades toxic adenosine and deoxyadenosine, which accumulate in the cells of patients. Immature lymphoid cells are very sensitive to these nucleotides. In addition to the immunological defect, most ADA deficiency patients have skeletal abnormalities.

About 85–90% of ADA deficient patients have SCID and are diagnosed by 1 year of age. About 15% have milder immunodeficiency diagnosed later. Of all SCID patients, ADA deficiency patients have the most profound lymphopaenia, involving T, B, and NK cells (Buckley et al., 1997). In addition to bone marrow transplantation, ADA deficiency can be treated by enzyme replacement therapy or with a combination of PEG-ADA and gene therapy. The catalytic domain consists of the whole protein. Genotype and both clinical and metabolic phenotype have good correlation.

6.4 Major Histocompatibility Complex Deficiencies

MHC class I deficiency is due to peptide transporter protein 2 (TAP2) mutations. TAP2 transports peptides from the cytoplasm into endoplasmic reticulum, where MHC I molecules can bind to them. Foreign proteins are degraded by proteolysis to peptides. The processed peptides bind to MHC I molecules, which are transported to the cell surface, where cytotoxic T cells recognise the antigen-presenting MHC molecules and kill the infected cells.

MHC class II is a surface molecule on B cells. It presents processed peptide fragments to the TCR of $CD4^+$ T-helper cells triggering the antigen-specific T-cell response. MHC class II deficiencies originate from impaired transcription of MHC II genes (Reith et al., 1999). Four different forms are known.

Patients with MHC class II deficiency generally develop septicaemia and recurrent infections of the gastrointestinal, pulmonary, upper respiratory, and urinary tracts. The patients are prone to bacterial, fungal, viral, and protozoal infections. Infections start within the first year of life. Almost all patients suffer from repeated, severe intestinal infections, diarrhoea, and failure to thrive. The patients usually die before the age of 5 years.

The affected individuals are treated with antibodies and immunoglobulins, but the progressive organ dysfunction and death cannot be prevented. The only curative treatment is bone marrow transplantation, although with a relatively poor success rate.

Regulatory factor (RF) X is a complex that binds to the X-box of MHC II promoters. Class II transcription activator (CIITA) is mutated in complementation group A. CIITA is a positive regulator of MHC class II gene transcription, which does not interact directly with DNA. RFXANK (regulatory factor X, ankyrin repeat-containing) of complementation group B includes three ankyrin repeats and accounts for some 70% of MHC class II mutations. DNA-binding domain containing regulatory factor RFX 5 is mutated in complementation group C deficiency. RFX-associated protein (RFXAP) binding RFX5 is mutated in the complementation group D. $CD4^+$ T-cell level is decreased in all these forms, although patients have normal numbers of circulating lymphocytes. Also immunoglobulin numbers can be decreased. There are only a few identified cases in each complementation group.

6.5 Hyper IgM Syndrome

The majority of hyper IgM (HIM) syndromes are X-linked. XHIM is caused by a defect in the gene for CD40 ligand. The patients have a failure with heavy-chain class switching from IgM to IgG and IgA. Individuals with XHIM show very low or undetectable serum IgG and IgA, with normal to elevated IgM, and have an increased susceptibility to bacterial and opportunistic infections (*Pneumocystis carinii* pneumonia and *Cryptosporidium*-related diarrhoea). They are at high risk for progressive liver disease and liver and intestinal tumours. Neutropenia is also a common manifestation. Treatment is based on regular use of intravenous immunoglobulins and antibodies or bone marrow transplantation.

Interaction between CD40L in T cells and CD40 on B cells is crucial for the formation of germinal centres and the generation of memory B cells. CD40 is also essential for interactions with macrophages and dendritic cells leading to induction of IL-12 secretion and immune response to intracellular microorganisms.

The protein consists of a short intracytoplasmic domain, transmembrane region, and an approximately 200 residues long extracellular domain. Missense mutations comprise the most common mutational event, frameshift mutations are the second most common type of mutations, when added up. Although some of the missense mutations affect the CD40-binding site, most of them disrupt CD40L function by other mechanisms. Some mutations affect residues that participate in the generation of the hydrophobic core and hence interfere with core packing and folding of the monomers, whereas other mutations involve buried residues at the interface between monomers and thus prevent trimer formation. In general, no strict genotype–phenotype correlation has been identified in XHIM. Most patients have a severe phenotype, regardless of the type of mutations.

6.6 Other CIDs

T-cell activation triggers cascades of reactions. Zap-70 (ζ-associated polypeptide of 70 kDa) is a protein tyrosine kinase that binds with its SH2 domains to the phosphorylated immunoreceptor tyrosine-based activation motif (ITAM) sequences of TCR. Signalling through TCR is defective in Zap-70 deficiency, influencing T-cell development.

CD3 is a multicomponent T-cell complex formed of nonidentical subunits that interact with TCR. Interaction with antigen activates cytokine release and cell proliferation. Rare CD3 deficiencies are caused by mutations in the γ- and ϵ-subunits.

7. OTHER WELL-DEFINED PIDs

Wiskott–Aldrich syndrome (WAS), an immunodeficiency of both T and B cells, is characterised by thrombocytopaenia, eczema, and recurrent infections (Ochs, 1998). Patients have progressive lymphopaenia. Without bone marrow transplantation WAS leads to death within the first two decades of life because of viral or bacterial infections. WAS patients with autoimmune manifestations have a high risk of getting malignancies.

WAS protein (WASP) interacts with Cdc42, a GTP-binding protein, which is involved in cytoskeleton reorganisation. WASP includes a number of domains, from the N-terminus WASP homology (WH1) domain, basic domain, GTPase-binding domain, polyproline, cofilin homology, and acidic domain. Defective cell polarisation due to WASP mutations could also affect

B-cell/T-cell interactions. The WASP mutations affecting the coding region are unevenly distributed along the WASP gene (Schwarz et al., 1996b). The WH1 domain accounts for 76% of all mutations, but for only 26% of all amino acids.

DiGeorge syndrome is a congenital PID characterised by lack of embryonic development or underdevelopment. Since the syndrome is associated with other defects, it has also been called CATCH22. Patients usually have a deletion on chromosome 22. The symptoms include cardiac abnormalities, abnormal facies, thymic hypoplasia, cleft palate, and hypocalcaemia. The degree of thymus problems varies.

In X-linked lymphoproliferative disease (XLP), or Duncan's disease, patients are exceptionally susceptible to Epstein–Barr virus (EBV) (Schuster and Kreth, 1999). In XLP, EBV infection causes mononucleosis by vigorous uncontrolled expansion of both T and B cells. XLP is usually associated either with hypogammaglobulinaemia, Burkitt lymphoma, carcinoma, some forms of Hodgkin disease, or several of them. The mortality is complete by the age of 40 years. SH2D1A, also known as DSHP or SAP, is an SH2 domain-containing molecule. SLAM (signalling lymphocyte activation molecule or CDw150) on the surface of T cells is crucial for stimulation. Phosphorylated SLAM can interact with SH2 domain-containing proteins, including protein phosphatase SHP-2. SH2D1A competes for binding to the SLAM. Mutations in SH2D1A affect the interaction between T and B cells and lead into uncontrolled B-cell proliferation in EBV infection. The majority of mutations affect the phosphotyrosine ligand binding region (Lappalainen et al., 2000).

ICF syndrome got its name from immunodeficiency, centromeric instability, facial abnormalities. Patients have mutations in DNA cytosine-5-methyltransferase 3B (DNMT3B).

8. PHAGOCYTE DEFECTS

Adaptive immunity facilitated by B and T cells is complemented by innate, cellular immunity of neutrophils, monocytes, macrophages, and eosinophils.

8.1 Chronic Granulomatous Disease

Phagocytic cells increase remarkably their oxygen consumption when brought into contact with opsonised microorganisms or a number of soluble stimuli. The respiratory burst is released by a NADPH oxidase complex that catalyses the reduction of oxygen to superoxide. Chronic granulomatous disease (CGD) is due to a defective NADPH oxidase. It is manifested usually by a total absence of superoxide production in phagocytes (Roos and Curnutte, 1999). Patients are highly susceptible to life-threatening bacterial and fungal infections, which can be fatal. The most common infections encountered include pneumonia, lymphadenitis, infections of the skin, and hepatic and perirectal abscesses.

In CGD, patients generate granulomas within organs and tissues in an attempt to control and isolate an unclearable infection. Granuloma formation can cause several chronic complications, such as the enlargement of the spleen and liver and obstruction of the urinary or gastrointestinal tracts.

Any one of the genes encoding different phox components can be affected. Mutations in the gene for $p91^{phox}$ cause the X-linked form of the disease, of which nearly two-thirds of CGD patients are suffering. Mutations in genes for $p22^{phox}$, $p47^{phox}$, and $p67^{phox}$ account for autosomal recessive forms of CGD. Aggressive treatment of infections is essential. Bone marrow transplantation has been used successfully to treat the disease.

The phagocyte NADPH oxidase consists of at least five submits. The membrane proteins $p91^{phox}$ and $p22^{phox}$ are integral flavocytochrome b_{558} subunits. $p40^{phox}$, $p47^{phox}$, and $p67^{phox}$ are located in cytosol. The small GTP-binding protein Rac is required for NADPH oxidase activity. When the respiratory burst is stimulated, the cytosolic subunits translocate to the membrane and associate with the flavocytochrome to form the active complex. Superoxide is released either into the phagosome following phagocytosis or into the extracellular space. The produced superoxide is converted to hydrogen peroxide, hydroxyl radicals, and hypohalous acids, all of which are effective antimicrobial agents.

8.2 Leukocyte Adhesion Defects

Free neutrophils and monocytes of blood stream can adhere to the endothelial cell lining of blood vessels to move to sites of infection within the tissues to ingest pathogens.

LAD-I disorder is characterised by recurrent bacterial and fungal infections and the accumulation of very low numbers of neutrophils at sites of infection (Etzioni and Harlan, 1999). Patients with LAD-I disorder have high numbers of circulating neutrophils. The affected gene codes for CD18, that is the β_2-subunit of leukocyte integrins. Integrin subfamilies contain the common β-subunit. In the heterodimeric integrins the α-subunit varies and it is responsible for functional specificity. The patients suffer from increased incidence of bacterial infections and impaired wound healing. Treatment with antibiotics and bone marrow transplantation have excellent results.

LAD-II is a very rare, autosomal recessive disease. Patients suffer from life-threatening recurrent bacterial and fungal infections. The disorder also leads to growth defects, neurological defects, and to abnormal carbohydrate blood group markers on erythrocytes. The levels of CD18 are normal in neutrophils.

The mechanism of LAD-II still remains unknown. It is involved in the biosynthesis of GDP-fucose, which is required for the synthesis of fucose-containing carbohydrates, including sialyl Lewis X and some blood group antigens. The diverse defects of the syndrome originate from the LAD-II related disruption of GDP-fucose and fucosylated surface glycoconjugate biosynthesis.

8.3 Chediak–Higashi Syndrome

The Chediak–Higashi syndrome (CHS) is a rare, autosomal recessive disorder characterised by immunodeficiency, neurologic abnormalities, and hypopigmentation. CHS patients have reduced numbers of neutrophils and other phagocytes and contain large, fused granules. Some granule proteins are missing. Both chemotaxis and phagocytosis are defective. Patients are very prone to bacterial infections.

Many CHS patients die early and the others enter the so-called accelerated phase, which is normally fatal. The phase is characterised by a lymphoma-like syndrome with fever and enlarged liver, spleen, and lymph nodes as well as severely reduced numbers of all blood cells.

CHS is related to the mouse beige phenotype. The human CHS gene encodes a large multidomain protein that contains similarity to motifs associated with vesicle transport and protein interactions.

8.4 Glucose 6-Phosphate Dehydrogenase Deficiency

Glucose-6-phosphate dehydrogenase (G6PD) is the first as well as the rate-limiting enzyme in the hexose monophosphate shunt, which produces 6-phosphogluconate and NADPH. This pathway is important especially for cells which contain only few mitochondria such as neutrophils or from which mitochondria are missing such as erythrocytes. G6PD deficiency is a relatively common X-linked disorder affecting primarily erythrocytes. NADPH is essential for converting the disulphide form of glutathione to the sulphhydryl form (GSH), which together with glutathione peroxidase protects phagocytes from the damaging effects of hydrogen peroxide generated during the respiratory burst.

8.5 Myeloperoxidase Deficiency

Myeloperoxidase catalyses the production of antimicrobial HOCl in neutrophils and monocytes. The majority of affected individuals have normal health without symptoms, but some patients have risk of systemic fungal infections.

9. DNA BREAKAGE-ASSOCIATED SYNDROMES

Ataxia telangiectasia (AT) is a rare, progressive, neurodegenerative childhood disease that affects the nervous and other body systems (Lavin and Shiloh, 1999). AT patients have ataxia, lack of muscle control, and telangiectasia (tiny, red veins). Many patients have an impaired immune system. They are predisposed to leukaemia and lymphoma in addition to being extremely sensitive to radiation exposure. The majority of the patients have IgA deficiency. Some IgG subclasses can also be reduced. There is no cure for AT and the progression cannot be slowed down. The defective protein, ATM, is a protein kinase that reacts to DNA damage and delays the accumulation of a p53 tumour suppressor. Thus, cells can replicate without repair of the damaged DNA lesions and thereby increase the risk of cancer.

ATM is composed of a leucine zipper, a proline-rich region, Rad3 homology, and a PI-3 kinase homology domain, all of which harbour mutations in AT (Vihinen et al., 2001).

10. COMPLEMENT DEFECTS

Complement activities, both classical and alternative pathway, are major recognition systems in innate immunity. Invading microorganisms trigger the complement cascade directly or indirectly (Sullivan and Winkelstein, 1999). Complement activation results in either the opsonisation of the pathogen for phagocytosis or the assembly on its surface for membrane damage and lysis. The classical pathway is antibody-dependent. The alternative pathway is an evolutionarily old defence system, which does not require antibody. Defects in the different complement factors predispose affected individuals for infections by different microorganisms.

Deficiencies of the early components of the classical pathway include C1q, C1r, C1s, C2, and C4 disorders. Patients with any of these deficiencies suffer from systemic lupus erythematosus (SLE), a chronic multisystem autoimmune disease. The risk for SLE is 20- to 30-fold in homozygous individuals and about 3-fold in those being heterozygous for defective alleles.

C1q deficiency presents at very different ages in different individuals. The C1q protein variants are encoded by three closely located genes. The C1r deficiency seems to be associated with at least a partial defect of a closely linked C1s gene. C2 deficiency is the most common complement deficiency, although half of the individuals deficient in C2 are asymptomatic. Because C4 is encoded by two genes, *C4A* and *C4B*, the complete C4 deficiency is very rare.

Of the alternative pathway PIDs, the properdin and factor D deficiencies are associated with recurrent neisserial infections.

The proteolytic cleavage of the C3 to C3a and C3b is common for both pathways of complement activation. C3b is the crucial opsonin that facilitates phagocytosis after binding to the surface. Further, it is also important in erythrocyte-related clearance of immune complexes. C3a mediates local inflammatory responses, including the stimulation of mast cells to release histamine. C3 deficiency is a rare AR PID manifested by frequent pyogenic bacterial infections and some patients also have immune complex disease. Individuals deficient in C5, C6, C7, or C8 have greatly increased susceptibility to infections with *Neisseria* species. The majority of the patients having the homozygous defect have recurrent infections and meningococcal disease caused by *Neisseria meningiditis*. Interestingly, C9-deficient individuals have no increased susceptibility to infections and they are generally in good health.

Both CD59 and decay-accelerating factor (DAF) are GPI-anchored membrane proteins, which protect host cells from complement-mediated lysis by inhibiting the formation of the membrane attack complex. Deficiencies of these proteins usually arise from an acquired mutation in haematopoietic stem cells affecting the synthesis of the GPI linker, but also CD59 and DAF deficiencies are known. Defective cells, particularly erythrocytes, are susceptible to lysis. Other common symptoms include back pain, headache, kidney dysfunction, and iron deficiency.

C1 inhibitor inhibits the protease activity, in addition to that of C1r and C1s, of blood-clotting system proteases and kallikrein. Defects in C1 inhibitor lead to angioneurotic oedema. The episodes of oedema start early in childhood and increase in severity into adolescence but then diminish with age. Oedema of the larynx leading to respiratory obstruction can be fatal. The disease is caused by uninhibited activation of the complement and kallikrein systems.

11. IMMUNODEFICIENCY MUTATION DATABASES (IDBASES)

The number of mutations identified in unrelated families with primary immunodeficiency is close to 2500 (Vihinen et al., 2001) (Table 3.2). Several freely accessible mutation databases have been established to handle and distribute the large amount of available information (Smith and Vihinen, 1996; Vihinen et al., 2001) (Table 3.3).

Information about immunodeficiency patients has been collected into the European Society for Immunodeficiency (ESID) registry (http://213.80.3.170:80/esid/registry.html). The ESID database contains clinical information of almost 9000 patients (Abedi et al., 1995). The registry lists different immunodeficiency disorders, immunoglobulin values, therapy, and family history of patients.

Currently there are more than 20 immunodeficiency mutation databases (Table 3.3). The databases at IMT Bioinformatics distributed by MUTbase system (Riikonen and Vihinen, 1999) are also accessible with mobile devices by using BioWAP service (Riikonen et al., 2001).

All types of mutations have been identified in PIDs (Table 3.3). Point mutations are the most common alterations. As in genetic diseases in general, missense and nonsense mutations comprise some 47% of all the genetic defects. Truncation of the produced protein is the most common effect when summing up nonsense, deletion, insertion, and splice site mutations.

In several forms of PID CpG sites are hot spots for mutations. These dinucleotides are the single most mutated doublet harbouring, e.g., in XLA 30%, in XSCID 22%, and in XCGD 32% of the mutations despite these dinucleotides constituting less than 5% of the genes (Vihinen et al., 2001). Mutations have been found also in single nucleotide repeats, presumably caused by mispairing slippage, e.g., in *BTK* and *IL2RG* genes.

Disease-causing mutations have different effects and consequences. The expression of the proteins can be changed, usually lowered or prevented, and stability, specificity, and activity of the proteins can be altered. Expression level and stability of mRNA are crucial. Despite large numbers of mutations it has not been possible to make genotype–phenotype correlations in many PIDs, at least in some cases because of redundant activities.

The IDbases aim at providing new insights into genotype–phenotype correlations in patients and protein structure–function relationships. The databases have been used to retrieve prospective and retrospective information on the clinical presentation, immunological phenotype, long-term prognosis, and efficacy of therapeutic options. The information may also prove valuable when developing new treatments, including drug design.

Table 3.2. *Immunodeficiency-causing mutations in unrelated families*

Gene name	Missense	Nonsense	Deletion in frame	Deletion frameshift	Insertion in frame	Insertion frameshift undefined	Splice site in frame	Splice site frameshift undefined	Gross deletion	Other	Total
BTK	264	119	14	86	3	49	14	96	17	10	672
CD40LG	35	15	2	14	0	10	3	8	2		89
SH2D1A	21	23		12		1		11	20	1	89
IL2RG	76	44	3	32	1	9	40	4	4	7	220
JAK3	13	7	2	1	0	1	0	5			29
ADA	62	4	1	10			2	8	4	1	92
T^-B^- SCID: RAG1	17	7		6		1					31
T^-B^-SCID: RAG2	9	1	1	1		1				1	14
Omenn: RAG1	19	3		7							29
Omenn: RAG2	6	1									7
MHC2TA	1	1					3				5
RFXANK	1	3		1			1	20			26
RFX5		2						3			5
RFXAP		2		3		1					6

WAS	32	25	1	16	2	13	4	16		6	115
ATM	42	47	27	121	4	41	84	83		6	455
CYBB	103	102	6	45	4	34	46	35	41		416
CYBA	6			1		1	2		1		11
NCF1		3		45							48
NCF2	4	2	2	3		1	1	3	1		17
IFNGR1	3	1	2	5		1	3				15
IFNGR2	1			1							2
IL12RB1		3		1			1				5
IL12B				1							1
ITGB2	14		1	3	1						19
DNMT3B	13	2				1	2	1			19
ZAP70	4			1			3				8
TCIRG1	14	10	4	3		2	1	17			51
Total	760	427	66	419	15	167	210	310	90	32	2496
%	30.5	17.1	2.6	16.8	0.6	6.7	8.4	12.4	3.6	1.3	

Table 3.3. *Immunodeficiency mutation databases*

Database	Internet address	Immunodeficiency	References
ADAbase	http://bioinf.uta.fi/ADAbase	Adenosine deaminase (ADA) deficiency	Vihinen et al. (2001)
AIREbase	http://bioinf.uta.fi/AIREbase	Autoimmune polyendocrinopathy with candidasis and ectodermal dystrophy (APECED)	
ATbase	http://www.cnt.ki.se/ATbase/	Ataxia-telangiectasia	Vihinen et al. (2001)
ATM	http://www.vmreseach.org/atm.htm	Ataxia-telangiectasia	Concannon and Gatti (1997)
BLMbase	http://bioinf.uta.fi/BLMbase	Bloom syndrome	Rong et al. (2000)
BTKbase	http://bioinf.uta.fi/BTKbase	X-linked agammaglobulinemia (XLA)	Vihinen et al. (1999, 2001)
C2base	http://bioinf.uta.fi/C2base	Complement component C2 deficiency	
CD3Ebase	http://bioinf.uta.fi/CD3Ebase	CD3ε deficiency	
CD3Gbase	http://bioinf.uta.fi/CD3Gbase	CD3γ deficiency	
CD40Lbase	http://bioinf.uta.fi/CD40Lbase	X-linked hyper-IgM syndrome (XHIM)	Notarangelo et al. (1996)
CYBAbase	http://bioinf.uta.fi/CYBAbase	Autosomal recessive CGD $p22^{phox}$ deficiency	Vihinen et al. (2001)
CYBBbase	http://bioinf.uta.fi/CYBBbase	X-linked chronic granulomatous disease (XCGD)	Roos et al. (1996) Heyworth et al. (2001)

DNMT3Bbase	http://bioinf.uta.fi/DNMT3Bbase	Immunodeficiency, centromeric instability and facial anomalies (ICF)	
FAA and FAC	http://www.rockefeller.edu/fanconi/mutate/	Fanconi anemia	
IL2RGbase	http://www.nhgri.nih.Gov/DIR/LGT/SCID/	X-linked severe combined immunodeficiency (XSCID)	Puck et al. (1996)
JAK3base	http://bioinf.uta.fi/JAK3base	Jak3 deficiency	Vihinen et al. (2000) Notarangelo et al. (2001)
NCF1base	http://bioinf.uta.fi/NCF1base	Autosomal recessive CGD $p47^{phox}$ deficiency	Vihinen et al. (2001)
NCF2base	http://bioinf.uta.fi/NCF2base	Autosomal recessive CGD $p67^{phox}$ deficiency	Vihinen et al. (2001)
RAG1base	http://bioinf.uta.fi/RAG1base	RAG1 deficiency Omenn syndrome	Villa et al. (2001)
RAG2base	http://bioinf.uta.fi/RAG2base	RAG2 deficiency Omenn syndrome	Villa et al. (2001)
SH2D1Abase	http://bioinf.uta.fi/SH2D1Abase	X-linked lymphoproliferative Syndrome (XLP)	Lappalainen et al. (2000)
ZAP70base	http://bioinf.uta.fi/ZAP70base	ZAP70 deficiency	

REFERENCES

Abedi, M. R., Morgan, G., Paganelli, R., et al. (1995). In: *Progress in Immune Deficiency V*. Caragol, I., Español, T., Fontan, G., and Matomoros, N. (eds.). Springer-Verlag, pp. 113–115.

Anonymous. (1997). Primary Immunodeficiency Diseases. *Clin. Exp. Immunol.* **109**, 1–28.

Bordignon, C., Notarangelo, L. D., Nobili, N., et al. (1995). Gene therapy in peripheral blood lymphocytes and bone marrow for ADA-immunodeficient patients. *Science* **270**, 470–475.

Buckley, R. H., Schiff, R. I., Schiff, S. E., et al. (1997). Human severe combined immunodeficiency: genetic, phenotypic and functional diversity in one hundred eight infants. *J. Pediatr.* **130**, 378–387.

Candotti, F., O'Shea, J. J., and Villa, A. (1998). Severe combined immune deficiencies due to defects of the common γ chain-JAK3 signaling pathway. *Springer Semin. Immunopathol.* **19**, 401–415.

Cavazzana-Calvo, M., Hacein-Bey, S., de Saint Basile, G., et al. (2000). Gene therapy of human severe combined immunodeficiency (SCID)-X1 disease. *Science* **288**, 669–672.

Chapel, H. M. and Webster, A. D. B. (1999). In: *Primary immunodeficiency diseases*. Ochs, H. D., Smith, C. I. E., and Puck, J. M. (eds.). New York: Oxford University Press, p. 419.

Conley, M. E., Notarangelo, L. D., and Etzioni, A. (1993). Diagnostic criteria for immunodeficiencies. *Clin. Immunol.* **93**, 190–197.

Concannon, P. and Gatti, R. A. (1997). Diversity of ATM gene mutations detected in patients with ataxia-telangiectasia. *Hum. Mutat.* **10**, 100–107.

Etzioni, A. and Harlan, J. M. (1999). In: *Primary Immunodeficiency Diseases. A Molecular and Genetic Approach*. Ochs, H. D., Smith, C. I. E., and Puck, J. M. (eds.). New York: Oxford University Press, pp. 375–388.

Fischer, A., Haddad, E., Jabado, N., et al. (1998). Stem cell transplantation for immunodeficiency. *Springer Semin. Immunopathol.* **19**, 479–492.

Hershfield, M. S. (1998). Adenosine deaminase deficiency: Clinical expression, molecular basis, and therapy. *Semin. Hematol.* **35**, 291–298.

Heyworth, P. G., Curnutte, J. T., Rae, J., et al. (2001). Hematologically important mutations: X-linked chronic granulomatous disease (second update). Blood Cells Mol. Dis., **27**(1), 16–26. Available: http://bioinf.uta.fi/CYBBbase/cybbpubs.html

Hyvönen, M. and Saraste, M. (1997). Structure of the PH domain and Btk motif from Bruton's tyrosine kinase: Molecular explanations for X-linked agammaglobulinaemia. *EMBO J.* **16**, 3396–3404.

Lappalainen, I., Giliani, S., Franceschini, R., et al. (2000). Structural basis for SH2D1A mutations in X-linked lymphoproliferative disease. *Biochem. Biophys. Res. Commun.* **269**, 124–130.

Lavin, M. F. and Shiloh, Y. (1999). In: *Primary Immunodeficiency Diseases.* Ochs, H. D., Smith, C. I. E., and Puck, J. M. (eds.). New York: Oxford University Press, pp. 306–323.

Notarangelo, L. D., Peitsch, M. C., Abrahamsen, T. G., et al. (1996). CD40Lbase: A database of CD40L gene mutations causing X-linked hyper-IgM syndrome. *Immunol. Today* **17**, 511–516.

Notarangelo, L. D., Villa, A., and Schwarz, K. (1999). RAG and RAG defects. *Curr. Opin. Immunol.* **11**, 435–442.

Notarangelo, L. D. and Vihinen, M. (2001). JAK3base: Mutation registry for autosomal recessive severe combined JAK3 deficiency. Available: http://bioinf.uta.fi/JAK3base.

Ochs, H. D. (1998). The Wiskott–Aldrich syndrome. *Springer Semin. Immunopathol.* **19**, 435–458.

Ochs, H. D., Smith, C. I. E., and Puck, J. M. (1999). *Primary Immunodeficiency Diseases: A Molecular and Genetic Approach.* Oxford University Press, New York.

Onodera, M., Nelson, D. M., Sakiyama, Y., et al. (1999). Gene therapy for severe combined immunodeficiency caused by adenosine deaminase deficiency: Improved retroviral vectors for clinical trials. *Acta Haematol.* **101**, 89–96.

Osborne, W. R. A. and Ochs, H. D. (1999). In: *Primary Immunodeficiency Diseases: A Molecular and Genetic Approach.* Ochs, H. D., Smith, C. I. E., and Puck, J. M. (eds.). New York: Oxford University Press, pp. 140–145.

Puck, J. M. (1999). In: *Primary Immunodeficiency Diseases: A Molecular and Genetic Approach.* Ochs, H. D., Smith, C. I. E., and Puck, J. M. (eds.). New York: Oxford University Press, pp. 99–110.

Puck, J. M., de Saint Basile, G., Schwarz, K., et al. (1996). IL2RGbase: A database of γc-chain defects causing human X-SCID. *Immunol. Today* **17**, 507–511.

Reith, W., Steimle, V., Lisowska-Grospierre, B. A. F., et al. (1999). In: *Primary Immunodeficiency Diseases: A Molecular and Genetic Approach.* Ochs, H. D., Smith, C. I. E., and Puck, J. M. (eds.). New York: Oxford University Press, pp. 167–180.

Riikonen, P. and Vihinen, M. (1999). MUTbase: Maintenance and analysis of distributed mutation databases. *Bioinformatics* **15**, 852–859.

Riikonen, P., Boberg, J., Salakoski, T., et al. (2001). BioWAP, mobile bioinformatics. *Bioinformatics* **17**, 855–856.

Rong, S. B., Väliaho, J., and Vihinen, M. (2000). Structural basis of Bloom syndrome (BS) causing mutations in the BLM helicase domain. *Mol. Med.* **6**, 155–164.

Roos, D. and Curnutte, J. C. (1999). In: *Primary Immunodeficiency Diseases: A Molecular and Genetic Approach.* Ochs, H. D., Smith, C. I. E., and Puck, J. M. (eds.). New York: Oxford University Press, pp. 353–374.

Roos, D., Curnutte, J. C., Hossle, J. P., et al. (1996). X-CGDbase: A database of X-CGD-causing mutations. *Immunol. Today* **17**, 517–521.

Schuster, V. and Kreth, H. W. (1999). In: *Primary Immunodeficiency Diseases: A Molecular and Genetic Approach.* Ochs, H. D., Smith, C. I. E., and Puck, J. M. (eds.). New York: Oxford University Press, pp. 222–232.

Schwarz, K., Gauss, G. H., Ludwig, L., et al. (1996a). RAG mutations in human B cell-negative SCID. *Science* **274**, 97–99.

Schwarz, K., Nonoyama, S., Peitsch, M. C., et al. (1996b). WASPbase: A database of WAS- and XLT-causing mutations. *Immunol. Today* **17**, 496–502.

Smith, C. I. E., and Vihinen, M. (1996). Immunodeficiency mutation databases – A new research tool. *Immunol. Today* **17**, 495–496.

Smith, C. I. E., Islam, T. C., Mattsson, P. T., et al. (2001). The Tec family of cytoplasmic tyrosine kinases: Mammalian Btk, Bmx, Itk, Tec, Txk and homologs in other species. *BioEssays* **23**, 436–446.

Stiehm, E. R. (1999). In: *Primary Immunodeficiency Diseases: A Molecular and Genetic Approach.* Ochs, H. D., Smith, C. I. E., and Puck, J. M. (eds.). New York: Oxford University Press, pp. 448–458.

Sullivan, K. E. and Winkelstein, J. A. (1999). In: *Primary Immunodeficiency Diseases: A Molecular and Genetic Approach.* Ochs, H. D., Smith, C. I. E., and Puck, J. M. (eds.). New York: Oxford University Press, pp. 397–416.

Väliaho, J., Riikonen, P., and Vihinen, M. (2000). Novel immunodeficiency data servers. *Immunol. Rev.* **178**, 177–185.

Väliaho, J., Pusa, M., Ylinen, T., et al. (2002). IDR: The immunodeficiency resource. *Nucleic Acids Res.* **30**, 232–234.

Vihinen, M., Cooper, M. D., de Saint Basile, G., et al. (1995). BTKbase: A database of XLA-causing mutations. *Immunol. Today* **16**, 460–465.

Vihinen, M., Arredondo-Vega, F. X., Casanova, J. L., et al. (2001). Primary immunodeficiency mutation databases. *Adv. Genet.* **43**, 103–118.

Vihinen, M., Mattsson, P., and Smith, C. I. E. (2000a). Bruton tyrosine kinase (Btk) in X-linked agammaglobulinemia (XLA). *Front. Biosci.* **5**, 917–928.

Vihinen, M., Villa, A., Mella, P., et al. (2000b). Molecular modeling of the Jak3 kinase domains and structural basis for severe combined immunodeficiency. *Clin. Immunol.* **96**, 108–118.

Vihinen, M., Kwan, S. P., Lester, T., et al. (1999). Mutations of the human BTK gene coding for Bruton tyrosine kinase in X-linked agammaglobulinemia. *Hum. Mutat.* **13**, 280–285.

Vihinen, M., Nore, B. F., Mattsson, P. T., et al. (1997). Missense mutations affecting a conserved cysteine pair in the TH domain of Btk. *FEBS Lett.* **413**, 205–210.

Villa, A., Santagata, S., Bozzi, F., et al. (1998). Partial V(D)J recombination activity leads to Omenn syndrome. *Cell* **93**, 885.

Villa, A., Sobacchi, C., Notarangelo, L. D., et al. (2001). V(D)J recombination defects in lymphocytes due to RAG mutations: Severe immunodeficiency with a spectrum of clinical presentations. *Blood* **97**, 81–88.

CHAPTER 4

Genetic diversity in the major histocompatibility complex and the immune response to infectious diseases

Leland J. Yee

Imperial College, Faculty of Medicine at St. Mary's Hospital, London, United Kingdom; London School of Hygiene and Tropical Medicine, London, United Kingdom

Mark R. Thursz

Imperial College, Faculty of Medicine at St. Mary's Hospital, London, United Kingdom

1. INTRODUCTION: THE NATURAL HISTORY OF INFECTIOUS DISEASES

The natural history of many infectious diseases is characterised by a broad range of clinical outcomes. For example, infection with bacterial agents such as *Mycobacterium tuberculosis* may be asymptomatic in some individuals; in others, clinical manifestations of disseminated tuberculosis may develop. Infection with viral agents such as the hepatitis B virus (HBV) may result in spontaneous clearance in some individuals, whereas in others, chronic infection may develop. Among those with chronic HBV infection, disease progression proceeds at varied rates, with some individuals developing liver cirrhosis and others only mild amounts of liver fibrosis despite similar durations of infection. Protozoan organisms, such as *Plasmodium*, are similarly characterised by a broad spectrum of outcomes. Some individuals may suffer from a severe course of malaria with cerebral manifestations or malarial anaemia, whereas others may naturally avoid such severe sequelae.

What factors influence this broad clinical spectrum in so many different diseases? Infectious disease natural history is affected by four main factors. First, pathogen virulence may determine outcome. For example, some strains of influenza virus cause only mild flu symptoms, whereas others have resulted in a global pandemic, such as the strain that killed millions of individuals across the world in 1918. Second, external modulators, or environmental factors, may affect outcome. These factors may be environmental, such as the increased risk of developing liver cancer among chronic carriers of HBV who are exposed to aflatoxins, or they may be behavioural, such as the increased risk of accelerating HBV or hepatitis C virus (HCV) progression by consumption of large amounts of alcohol (Poynard, 1997; Schiff, 1997; Ostapowicz, 1998; Wiley, 1998; Thomas et al., 2000; Harris et al., 2001). Third, host genetic variation between individuals may affect clinical outcome. In particular, genetic diversity in genes encoding key immunomodulatory molecules may affect the ability of individuals to effectively respond to infections. Fourth is the possible interaction between these three aforementioned components.

This chapter will focus on the contribution of host genetics to the natural history of infectious diseases. Specifically, we will address the role of genetic diversity in the major histocompatibility complex (MHC) in mediating this immune response and highlight methods for conducting disease association studies, as well as examine examples of such disease association studies.

2. GENETIC CONTRIBUTIONS TO IMMUNE RESPONSES AGAINST INFECTIOUS AGENTS

Studies of twins suggest that genetic factors may play a significant role in modulating the immune response to infectious agents. Twin heritability studies have found a high genetic contribution for the immune response to several infectious diseases (Kallman and Reisner, 1942; Comstock, 1978; Sorensen et al., 1988; Lin et al., 1989; Malaty et al., 1994). This is in contrast to many autoimmune diseases, where concordance among monozygotic twins is almost always less than 50%, suggesting a more significant role for environmental factors in autoimmune disease pathogenesis (Cooper et al., 1998, 1999). For this reason, much interest has been focused in recent years on the possible contributions of host genetic factors to the often diverse clinical manifestations seen in many infectious diseases.

3. THE HUMAN MAJOR HISTOCOMPATIBILITY COMPLEX

The MHC was originally identified by Gorer and Snell through their work on transplant compatibility in mice. The region conferring transplant compatibility/incompatibility was mapped to mouse chromosome 17 and subsequently the orthologous region in humans was mapped to the short arm of chromosome 6.

The complete DNA sequence of the MHC region was completed ahead of the draft human genome sequence (MHC Sequencing Consortium, 1999). This revealed that the 3.6-Mb region contains 224 genetic loci, of which 128 are predicted to be expressed. Approximately 40% of the expressed genes are thought to participate in immune functions, which is particularly

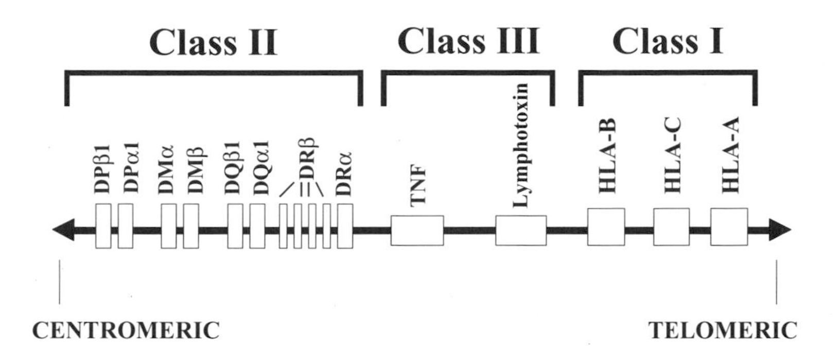

Figure 4.1. Simplified schematic map of the human major histocompatability complex (MHC) on the short arm of chromosome 6. This figure highlights just some of the key polymorphic loci related to immune function in the complex, including the class I and class II regions, which encode the human leukocyte antigens (HLA). (The MHC Sequencing Consortium *Nature* 1999)

extraordinary as the MHC region is thought to predate the emergence of the adaptive immune system (Trachtulec et al., 1997). Clustering of these immune-related genes is unlikely to have occurred by chance. However, duplication, particularly in the class I and II regions, has occurred frequently. The loci in the MHC exhibit unprecedented allelic diversity with variation levels of 5–17%. Highly polymorphic loci, such as the HLA-DRB1 locus or HLA-B, have more than 200 alleles (Marsh et al., 2001).

This polymorphic cluster of genes is divided into three major regions (Fig. 4.1). The class I and class II MHC genes encode the human leukocyte antigen (HLA) molecules that are important in the antigen-processing and antigen presentation pathways. Although the class I molecules interact with cytotoxic T lymphocytes (CTL), the class II molecules interact with T-helper (T_H) cells. The class III region encodes a variety of genes, including cytokines, complement components, and heat shock proteins. Within the class I region there are also a number of loci encoding atypical HLA class I molecules.

The principle of T-cell restriction was identified by Doherty and Zinkernagel (1975) more than 25 years ago. T cells will only respond to specific peptides presented by antigen-presenting cells bearing specific MHC class I or class II alleles. The molecular basis for antigen-MHC restriction was revealed by the high-resolution crystal structure of an MHC class I molecule and associated antigen (Bjorkman et al., 1987; Ghosh et al., 1995; Jardetzky et al., 1996). HLA molecules bind antigenic peptide fragments in a groove on the exterior surface of the molecule. The HLA–peptide complex is recognised by T-cell receptors (TCR) expressed on the surface of CD4+ T_H cells or CD8+ CTLs. Interaction among HLA molecule, peptide, and TCR, in the context of

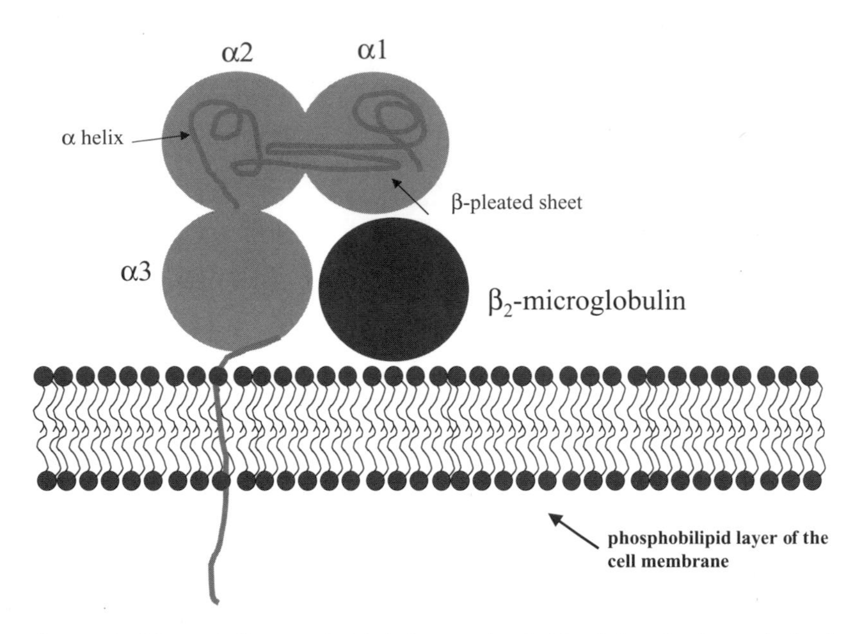

Figure 4.2. Schematic diagram of a HLA class I molecule. The alpha subunits are depicted in complex with β_2-microglobulin. (Reviewed in Williams A. et al. *Tissue Antigens* 2002)

appropriate costimulatory signals, may initiate T-cell activation, proliferation, and cytokine release. Class I molecules are present, or can be induced, on virtually every nucleated cell in the body, whereas class II molecules are usually only expressed by professional antigen-presenting cells. Class I molecules form heterodimers with β_2-microglobulin on the cell membrane and assist in the presentation of 'endogenous antigens' such as those from virally infected cells (Klein and Sato, 2000a, 2000b). Figure 4.2 presents a schematic diagram of HLA class I molecule structure. In Fig. 4.1, one can see the various subunits of the molecule, along with β_2-microglobulin. Endogenous proteins are digested by the proteosome complex and transported to the endoplasmic reticulum (ER) by the transporter associated with antigen-processing (TAP) proteins (Cresswell et al., 1999). In the ER, peptide fragments are loaded into the HLA class I groove. HLA class II molecules (Fig. 4.3), as heterodimers, are encoded by more than one genetic locus and assist in the presentation of "exogenous antigens" such as those from foreign pathogens (Shackelford et al., 1982; Schafer et al., 1995). Figure 4.2 depicts the various subunits of the class II molecule on the cell surface. One locus encodes an α-chain, whereas the other encodes a β-chain. These α/β heterodimers are stabilised by an invariant chain polypeptide and are translocated to the lysosomal compartment

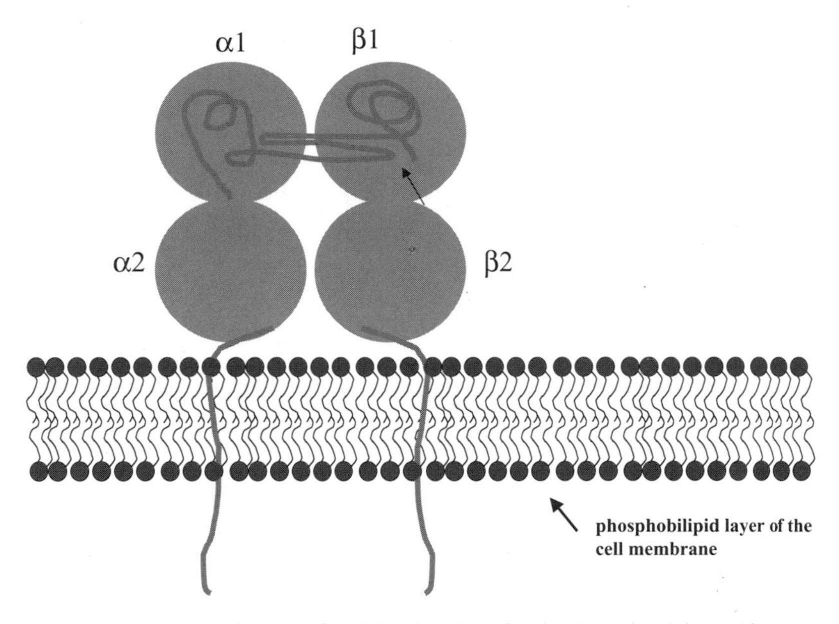

Figure 4.3. Schematic diagram of an HLA class II molecule. Here, the alpha and beta subunits are depicted on the Surface of the cell membrane. (Reviewed in Shackelford 1982 and Schafer 1995)

where they encounter exogenously derived peptides which have been ingested by the antigen-presenting cell. Antigenic peptides competitively bind to the HLA class II molecule, and the trimeric complex is then transported to the cell membrane. Figure 4.4 shows an HLA class I molecule with a peptide in the binding domain (Parker et al., 1995). Other molecules, such as the DM molecules, are also involved with class II antigen presentation.

The genes encoding the major components of the HLA molecules fall into the class I and class II regions (Fig. 4.1). The region between, often referred to as the class III region, encodes a number of immunomodulatory cytokines as well as other genes. The α-class I heavy chains are encoded within this class I region, whereas the β-light chain, β_2-microglobulin, is actually encoded by a gene located on chromosome 15. Although there are over 20 class I and class I-like genes (many of them polymorphic), HLA-A, -B, and -C are the key molecules involved in differential antigen presentation and natural killer (NK) cell activities (Thorsby, 1999; Williams et al., 2002).

In contrast, both the α- and β-portions of the class II genes are encoded within the HLA region of chromosome 6. Nomenclature for molecular genotyping of class II consists of three letters. The first letter, D, designates class II. The second letter designates the family: M, O, P, Q, or R. The third letter

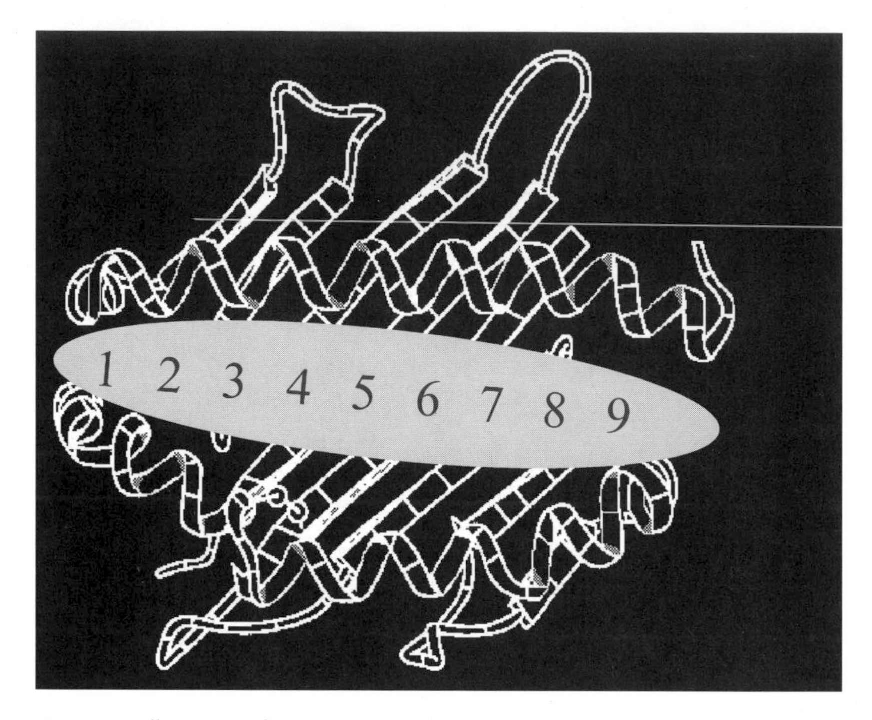

Figure 4.4. Illustration of a nonamer peptide (amino acids are numbered 1–9) bound within the binding cleft of an HLA class I molecule. (Jardetzky 1996 and Ghosh 1995)

designates whether it is the α- or the β-chain of the HLA molecule. This three-letter complex is then followed by an arabic number designating the family member. For example, DRB4 is the fourth member of the Rβ-family. This four-character designation is then followed by an arabic number designating the allele and separated from the prefix by an asterisk. For example DRB1*1501 stands for allele 1501 of gene 1 of the β-chain of the R-family gene in the class II region.

Early classification of HLA alleles was based on serological methods employing the microcytotoxicity assay. Serological methods have relied on HLA-specific antibodies binding to antigens exposed on cell surfaces. As a result, polymorphisms in the internal regions of the HLA molecule (such as those in the binding cleft) have escaped easy classification using this method. Over the past decade, a revolution in molecular genotyping techniques has allowed for DNA sequence-based definition of HLA alleles and these methods have largely replaced the serologically based typing methods for most areas of research. As a result, HLA classification now increasingly relies on molecular methods and many of the older serogroups have been reclassified according

Table 4.1. *Summary of serologically defined DR groups and molecularly defined allele(s) within each sero group*

Serologically defined group		Molecularly defined (DRB1*) allele(s)						
DR1		0101	0102					
DR103		0103						
DR2	DR15	1501	1502	1503				
	DR16	1601	1602					
DR3	DR17	0301						
	DR18	0302	0303					
DR4		0401	0402	0403	0404	0405	0406	0407
		0409	0410	0411	0412			
DR5	DR11	1101	1102	1103	1104	1105		
	DR12	1201	1202					
DR6	DR13	1301	1302	1303	1304	1305	1306	
	DR14	1401	1402	1405	1406	1407	1408	1409
DR1403		1403						
DR1404		1404						
DR7		0701	0702					
DR8		0801	0802	0803	0804	0805		
DR9		0901						
DR10		1001						
DR51		0101	0102	0201	0202			
DR52		0101	0201	0202	0301			
DR53		0101						

Note: N > 200 in total.

to molecular definitions, resulting in additional alleles. Table 4.1 briefly highlights the molecular subdivisions of the serological groups at the DR locus. Similar molecular subdivisions exist for other loci in the HLA as well.

Molecular HLA typing methods include several commonly employed methods. Polymerase chain reaction (PCR) with sequence-specific primers (PCR-SSP) that specifically target amplification of particular alleles is one common method. Here, a set of primers for specific alleles is used, and if a particular allele is present, an amplicon will be identified when the products are run on a gel. A combination of PCR amplification and subsequent sequence-specific oligonucleotide hybridisation (PCR-SSO) is also commonly used. Also, direct sequencing of the gene(s) is often employed. Reference strand-mediated conformational analysis (RSCA) is another new technique. Microarray-based typing has been developed for HLA-B typing as well.

4. LINKAGE DISEQUILIBRIUM AND HLA HAPLOTYPES

Although the HLA region is the most polymorphic region in the human genome, different alleles at different loci often do not sort independently of one another, resulting in greater frequencies of specific combinations of alleles than would be expected by chance. This nonrandom sorting is often referred to as *linkage disequilibrium*. The specific combinations of different alleles at two or more loci are termed *haplotypes*. It is important to note that haplotypes are generally inferred in immunogenetic studies, thus necessitating the need for large study sizes. Linkage disequilibrium may be measured using several techniques and is typically expressed as a delta value (Δ), which can range between 0 and 1. A delta value of 1 indicates complete association. Theoretically, linkage disequilibrium should be a function of the genetic distance between two loci due to the effect of recombination, but in reality the relationship between linkage disequilibrium and genetic distance is complicated by hot spots and cold spots for recombination. Furthermore, differences in the evolutionary age and geography of specific polymorphisms make it impossible to predict the patterns of linkage disequilibrium. Blocks of linkage disequilibrium within the MHC appear to extend over fairly long genetic distances and empirical investigations revealed that particular alleles tend to segregate with others, leading to conserved combinations, or common haplotypes. For example, the class I allele A1 often occurs with the B8 allele. These alleles also commonly occur with the class II alleles DR3 or DR4, leading to the common extended haplotypes: A1-B8-DR3 or A1-B8-DR4. The strength of linkage varies from allele to allele. Although there is a tendency for DR3 to be seen with A1-B8, it also occurs in the absence of A1 and B8.

It is important to consider linkage disequilibrium in immunogenetic studies, as it is difficult to implicate a particular allele in disease association studies without properly considering its linkage disequilibrium with other variants. Using the example above, let us say that the allele B8 was observed more frequently in individuals exposed to pathogen X who developed the disease when compared to those who were exposed to pathogen X but did not develop the disease. Because of linkage disequilibrium, it is difficult to solely implicate the B8 allele as the determinant because B8-carrying chromosomes often have A1, Cw7, and several class II alleles. Alternatively, there may be yet another gene located on the haplotype with B8 that is truly responsible and B8 alone or in combination with A1 is merely reflecting linkage with this other factor. Finally, it may not be B8 alone that is responsible for the observed association. It may be the extended haplotype rather than the individual alleles that truly alters immune function and ultimately plays a role in disease pathogenesis. In fact, this may be the case with the A1-B8-DR3 haplotype as *in vivo* and *in vitro* studies suggest that individuals with this genetic haplotype may have altered immune responses (Caruso et al., 1990, 1993, 1997; Modica et al., 1990; Lio et al., 1995; 1997).

Another important factor to consider in immunogenetic studies is the great variation in linkage disequilibrium patterns. Not only do allele frequencies vary among different racial groups, but linkage disequilibrium patterns as well. For example, Hill et al. (1991) observed the DRB1*1302 allele to be associated with the DQB1*0501 allele in Gambia, whereas DQB1*0604 is the association commonly seen in Caucasians.

Although linkage disequilibrium patterns may make disease associations more difficult to make, they may be simultaneously exploited to further refine observed associations. For example, if you conducted a study and observed an association with the A1-B8 class I haplotype, but are unsure whether most of the association is due to A1-B8 or the class II allele DR3 (with which A1-B8 is often associated), or even the extended A1-B8-DR3 haplotype, it will be possible to conduct subgroup analysis, providing the study is sufficiently large. By comparing individuals who have the A1-B8-DR3 haplotype to those who do not have DR3 (for example, those with the A1-B8-DR4 haplotype) it is possible to determine whether it is the class II allele or the haplotype by comparing any changes in the observed strengths of association between the different groups. Similarly, the different linkage disequilibrium patterns observed in different populations may be exploited to determine the relative effects of a particular allele on disease outcome. In summary, although linkage disequilibrium may make the implication of specific alleles more difficult in disease association studies and complicate the statistical analyses, it may simultaneously be exploited as a mapping aid for the further refinement of observed associations (Ardlie et al., 2002).

5. THE DESIGN OF STUDIES OF HLA AND DISEASE OUTCOME

The majority of studies investigating the relationship between disease outcome and host genetic background use a disease association/case-control design, where a group of individuals with a particular outcome of interest are compared to a group of appropriate and carefully selected controls and differential distributions in the frequencies of alleles, genotypes, and haplotypes are examined. Genes are often chosen for study based on biological plausibility or knowledge of biological function. Because of the relative ease of this type of study design, it is the most commonly employed method for disease association studies. Because of its widespread use, we will focus on this type of study design in this chapter. Associations using this type of design have been implicated in many diseases of varying aetiology, including those of bacterial, viral, and protozoan origin.

There are a few inherent difficulties with respect to disease association studies on MHC that arise from the great allelic diversity of this region and from linkage disequilibrium. Study design should take into account the rigour of phenotypic selection. Controls, in theory, should be similar to cases in every manner except for the outcome under study. If matching on potential confounding factors between cases and controls cannot be made, appropriate measurement and statistical adjustment for these factors should be conducted in the analysis portion of the study (see statistical issues following). Comparison of cases who have a particular disease outcome against healthy controls may not always be appropriate. For example, in HCV infection, a number of studies compare MHC class II allele frequencies between groups of patients with persistent HCV infection and healthy control volunteers. However, if these volunteers were exposed to the virus, 80% would become persistently infected. A more suitable comparison is between phenotypic extremes; so in the example of HCV, comparison is often made between groups with self-limiting infection versus groups with persistent infection. Such comparisons lead to more consistent results (Thursz, 2001).

6. STATISTICAL ISSUES IN MHC ASSOCIATIONS

Because of the highly polymorphic nature of the MHC, it is important to highlight a number of statistical issues inherent in these types of studies. First is the issue of adequate power. This must be addressed in the planning stage of the study. Studies must be of sufficient size to detect meaningful differences between cases and controls for the gene under study. This is particularly important with respect to studies of the MHC due to (1) the highly polymorphic nature of the MHC and (2) the relatively low frequency of many of the alleles in the MHC. Molecular typing has further subdivided the serological allele groups, splitting the categories into smaller numbers (Table 4.1). The low frequency of many of the HLA alleles is further compounded by linkage disequilibrium, an important aspect to consider in the analysis of immunogenetic studies, and in particular those of the MHC. By combining alleles at different loci into haplotypes, the numbers available for study are further decreased. As a result, proper power calculations accounting for molecular classification of alleles and extended haplotypes should be conducted prior to the inception of a study.

The appropriate measurement of and the proper control of potential confounders also forms a necessary part of the analyses of studies of the MHC. Control for such confounders may be planned into the study design and include matching cases and controls on the confounding factor(s), or addressed in the analysis phase of the study, where statistical techniques such as multivariable modelling may be employed. Adjustment must be made for not only other potential genetic confounders, but for environmental and demographic factors as well as differences in the virulence of pathogens. In a study of *Mycobacterium avium* complex infection, an opportunistic infection seen in HIV disease, LeBlanc and colleagues (1999) matched on factors such as initial diagnosis, CD4+ cell counts, and the slope of CD4+ cell decline to account for, among other things, variability in the progression of HIV. Similarly, the study of genetic mediators of the response to anti-HCV therapy must adjust for the fact that genotype 1 viruses are more resilient to treatment with interferon (McHutchison, 1998; Poynard, 1998; Manns et al., 2001). Studies of histologic progression of HCV must account for environmental confounders such as alcohol intake or even co-infection with other hepatitis viruses (Schiff, 1997; Ostapowicz, 1998; Wiley, 1998; Thomas et al., 2000; Harris et al., 2001).

Because the MHC is so highly polymorphic, some researchers advocate making a statistical adjustment for multiple comparisons (Bonferroni correction) to account for the possibility that an observed association is due to chance simply as a result of the numbers of comparisons made. This approach is problematic in some respects. First, the strength of many of the observed genetic associations are relatively modest. Odds ratios (OR) of 2 or 3 are common. The conservative nature of the Bonferroni correction will often reduce the level of statistical significance for such nominal associations. Second, the frequencies of many of the alleles in the MHC are relatively low. A conservative statistical adjustment such as the Bonferroni correction might result in the erroneous rejection of an association, even if it is true, solely based on the *P* value. Many in this camp believe that observed statistical associations should stand alone, irrespective of the numbers of comparisons made (Rothman, 1990). Given the highly polymorphic nature of the MHC complex, such a statistical adjustment will likely be too conservative and unfairly mitigate the significance of the observed effects. Moreover, the Bonferroni correction places undue emphasis on the significance of *P* values, which are highly influenced by both the study size and the observed strength of association (Lang et al., 1998). As a result, some researchers have advocated the use of independent replication in lieu of potentially overly conservative statistical adjustments. In this method, putative associations seen in one cohort are confirmed in a second unrelated cohort (Editor Nature Genetics, 1999). In other words, one cohort is used to generate hypotheses regarding candidate genes and another is used to verify any observed associations. Of course, this method is not without its inherent difficulties as well. Often large, well-characterised cohorts are not readily available; needing two such cohorts is even more difficult.

7. THE MHC, IMMUNE RESPONSE, AND THE NATURAL HISTORY OF INFECTIOUS DISEASES

There are several points in the natural history of infectious diseases in which the varied immune response among different individuals arising from genetic diversity in the MHC can influence disease outcome. First is susceptibility. Not everyone infected with a pathogen will develop persistent infection with that pathogen. For example, only about 10% of individuals infected with the HBV virus develop persistent infections. The rest clear it. The second point is in the progression of the disease. Some individuals with persistent HCV infection, for example, will progress to liver cirrhosis, whereas others infected for similar amounts of time may be asymptomatic. The third stage involves therapy. Diversity in the genes that encode the antigen-processing pathways may influence the outcome(s) of immunomodulatory therapies.

8. DISEASE ASSOCIATION STUDIES

Although studies of twins suggest that most disease susceptibility genes map outside the HLA (Jepson et al., 1997), it is nevertheless important to recognise that the HLA system plays an important role in mediating the immune response to all infectious agents to some degree. Thus, although HLA may not play the major role in mediating the immune response to every infectious agent, it does play a role in the immune response in some capacity. For this reason, many association studies have been conducted to correlate specific alleles within the HLA to various disease outcomes.

In this section we will give examples of disease association studies for all of the major categories of pathogens: bacterial, viral, and protozoan. The summaries of the studies included here are not systematic reviews and are not intended to be exhaustive; rather, they are included merely as illustrative examples of studies that have addressed the role of the HLA in infectious disease outcome. We have deliberately chosen a select and somewhat biased group of studies to illustrate our points.

Table 4.2 summarises some of the studies concerning mycobacterial infections. The most consistent finding has been the association with DR2 and DR3 with risk for developing mycobacterial diseases. This association can be seen in both *M. tuberculosis* and *Mycobacterium leprae* infections. It is worth noting that HLA typing in earlier studies was done by serological methods. A recent study by Rani and colleagues (1993) in India relied on molecular genotyping and confirmed previous associations with DR2. Interestingly, LeBlanc et al. (1999) looked at disseminated *M. avium* complex (DMAC) in HIV-1 disease, as DMAC is an important opportunistic infection associated with HIV in the developed world. Molecular genotyping revealed similar associations with DRB1*1501, part of the older DR2 serogroup. It is important to bear in mind that although *M. tuberculosis, M. leprae,* and DMAC are different diseases, the similar associations with DR2 might reflect common pathways in the class II immune response to mycobacterial antigens. However, such a statement is made *a priori* and remains as yet unproven in the absence of mechanistic data. The mechanism behind these observed associations is completely unknown. It is quite possible that linkage with another gene or set of genes might ultimately explain these observations.

Table 4.3 highlights some studies of HLA and HIV disease. Early studies revealed an association between HLA-B35 or Cw4 (or both) with rapidly progressing HIV disease (Keet, 1996; Carrington et al., 1999). Similar associations were observed for the A1-B8-DR3 haplotype and rapid progression (Kaslow et al., 1990; McNeil et al., 1996). HLA-B57 and B27 were associated with slow HIV progression (McNeil et al., 1996). Using a more comprehensive strategy examining not only individual alleles but also combinations of alleles in multiple class I and class II loci, along with variants in the Tap (transporter associated with antigen processing) gene, Kaslow and colleagues (1996) identified alleles most strongly associated with rapid or slow progression and combined them into an algorithm that summarised the HLA profile for each patient. Observations made in one cohort were validated in another. This HLA profile appears to correlate with serum HIV-RNA levels early in the course of HIV-1 infection as well (Saah et al., 1998). Although the individual effects of each allele are relatively small, this approach has helped highlight the cumulative effects of combinations of multiple HLA alleles on disease outcome. It is worth noting that a lot of the studies listed in Table 4.3 have examined the class I genes as the primary candidates. This is likely to reflect the role of class I HLA molecules in mediating the CTL response – an important component in the immune response to viral replication.

Table 4.2. *Summary of selected association studies of HLA and mycobacterial diseases*

Study	Country	Study size	Finding	Strength of association	Significance level
			Pulmonary Tuberculosis		
Singh et al. (1983b)	India	25 families	DR2	$\chi^2 = 4.8$	$P < 0.05$
Singh et al. (1983a)	India	233	DR2	RR = 1.6	NS
			DRw6	RR = 0.4	$P < 0.05$
Bothamley et al. (1989)	Indonesia	165	DR2	RR = 2.8	$P < 0.01$
			DRw14	RR = 2.7	$P < 0.05$
			DQw1	RR = 2.2	$P < 0.05$
			DQw3	RR = 0.33	$P < 0.01$
Rajalingam et al. (1997)	India	442	DR2, (DRB1*1501 and *1502)	RR = 1.8	$P_c = 0.029$
Pospelov et al. (1996)	Tuvinia	329	DR2	RR = 3.3	$P < 0.01$; $P_c > 0.05$
			DRw53	RR = 11.9	$P < 0.001$; $P_c \leq 0.05$
Khomenko et al. (1990)	Former USSR	1627 of varied ethnicities	DR2	RR = 2.2–6.41	$P < 0.05$
			DR3	RR = 0.33–0.52	$P < 0.05$
			Tuberculoid Leprosy		
Dessoukey et al. (1996)	Egypt	15 families	DR2	$\chi^2 = 13.29$	$P < 0.001$
Izumi et al. (1982)	Japanese	479	DR2	RR = 5.9–8.7	$P_c = 0.000–0.008$
			DRw9	RR = 0.2	$P_c = 0.000$
			B7	RR = 3.8	$P_c = 0.044$
			Bw54	RR = 0.4	$P_c = 0.016$
van Eden et al. (1982)	Suriname	165	DR3	$\chi^2 = 13.14–4.2$	$P = 0.0003–0.04$
Gorodezky et al. (1987)	Mexico	176	DR3	RR = 4.93	$P = 0.0007$; $P_c = 0.04$

Todd et al. (1990)	Pooled analyses	572	DR2	RR = 2.65	$P < 1 \times 10^{-8}$
			DQw1	RR = 2.73	$P = 3.6 \times 10^{-8}$
		Lepromatous Leprosy			
Rani et al. (1993)	India	140	DRB1*1501, DRB1*1502	RR = 5.7–16.3	$P < 0.05$
			DQA1*0103, DQA1*0102	RR = 2.8–2.9	$P < 0.05$
			DQB1*0601	RR = 7.4	$P < 0.00001$
Rani et al. (1992)	India	355	Bw60, DR2, DRw8, DQw1, DQw7	—	$P = 0.0003–0.031$
		Lepromatous and Tuberculoid Leprosy Combined			
Kim et al. (1987)	Korea	319	A2	RR = 0.6	$P = 0.03$
			A11	RR = 2.0	$P = 0.03$
			Aw33	RR = 2.3	$P = 0.003$
			Cw5	RR = 0.2	$P = 0.001$
			DR1	RR = 2.5	$P = 0.02$
			DR2	RR = 2.6	$P < 0.0001$
			DR4	RR = 0.5	$P < 0.0001$
			DRw9	RR = 2.6	$P = 0.02$
			DRw53	RR = 0.4	$P < 0.0001$
			DQw1	RR = 2.6	$P < 0.0001$
			DQw3	RR = 0.4	$P < 0.0001$
		Disseminated Mycobacteriam Avium Complex in HIV-1 Infection			
LeBlanc et al. (2000)	USA	352	DRB1*1501	1.46	$P = 0.04$

OR, odds ratio; RR, relative risk; P_c, P value with correction for multiple comparisons; NS, reported as being not significant; χ^2, chi-square.

Table 4.3. *Summary of selected association studies of HLA and HIV infection*

Study	Country	Study size	Finding	Strength of association	Significance level
Kaslow et al. (1996)	United States	139, 102*	B27, B57 or B51 +/− Tap 2.3	RH = 0.32	$P = 0.002$
			A25, A26, A32 or B18 +/− Tap 2.3	RH = 0.39	$P = 0.003$
			DRB1*0401-DQA1*0300-DQB1*0301 or DRB1*1200-DQA*0501-DQB1*0301 or DRB1*1300-DQA1*0102-DQB1*0604 or DRB1*1400-DQA1*0101-DQB1*0503 +/− Tap 1.2	RH = 2.04	$P = 0.004$
Keet et al. (1999)	United States and The Netherlands	375	A29-33 + Tap 2 Ala 665	RH = 0.46	$P = 0.006$
			B27	RH = 0.40	$P = 0.003$
			B57	RH = 0.54	$P = 0.02$
			DRB1*1300-DQB1*0603	RH = 0.67	$P = 0.07$
			A25/26 (A10) + Tap 2 Ala 665	RH = 0.31	$P = 0.02$
			A24	RH = 1.57	$P = 0.004$
			B8 + Tap 2 Ile 379	RH = 1.88	$P = 0.02$
			Cw4 without Tap 2 Ala 665	RH = 1.79	$P = 0.001$
			DRB1*1200-DQB1*0301	RH = 1.83	$P = 0.04$
			A23 without Tap 2 Ala 665	RH = 2.04	$P = 0.02$
			A28(68) + Tap 2 Ala 665	RH = 1.88	0.08
			B40/60 + Tap 2 Ile 379	RH = 2.24	$P = 0.005$

Hendel et al. (1999)	France	275	A29	OR = 2.91	$P = 0.0076$
			B22	OR = 13.07	$P = 0.0001$
			B14	OR = 0.16	$P = 0.001$
			C8	OR = 0.19	$P = 0.004$
Costello et al. (1999)	Rwanda	202	B*5703	OR = 0.37	$P = 0.02$
Migueles et al. (2000)	United States	213	B*5701	OR = 52	$P < 0.001$ (95%CI: 10–501)
Carrington et al. (1999)	United States	474	B*35	RH = 2.34	$P_c = 1 \times 10^{-4}$
			Cw*04	RH = 2.41	$P_c = 1 \times 10^{-5}$
Steel et al. (1988)	UK	18	A1-B8-DR3	RR = 28	$P < 0.05$
Kaslow et al. (1990)	United States	108	A1-Cw7-B8	OR = 10.3	$P = 0.0005$
			A1-Cw7-B8-DR3	OR = 7.6	$P = 0.01$
			A24	OR = 4.3	$P = 0.006$
Flores-Villanueva et al. (2001)	United States	39	Homozygosity for Bw4	RH = 0.29–0.39	$P_{adj} = 0.007–0.02$
McNeil et al. (1996)[41]	UK	313	A1-B8-DR3	RR = 3.7	95%CI: 1.9–7.2
			B27	RR = 0.3	95%CI: 0.1–0.9

Abbreviations: OR = odds ratio; P, P value; P_c, P value with adjustment for multiple comparisons; RH, Relative Hazard; RR, Relative Risk; 95% CI, 95% confidence interval; P_{adj}, adjusted P value.

* This study involved two cohorts.

** Adjusted values accounting for effects of CCR5.

Disease association studies have yielded some consistent results with respect to viral hepatitis (HCV and HBV) (Table 4.4). The class II allele DQB1*0301 has been associated with natural clearance of viraemia in several studies conducted in largely Caucasian cohorts (Alric, 1997; Cramp, 1998; Minton et al., 1998; Mangia, 1999; Thursz et al., 1999). Associations have not been very consistent with respect to the severity of HCV-related liver disease (Table 4.4). In hepatitis B virus infection, alleles of the DRB1*13 group have been associated with persistent hepatitis B infections, defined as hepatitis B surface antigen (HBsAg) persistence, in several studies conducted in different ethnic populations (Thursz, 1995; Hohler et al., 1997; Ahn et al., 2000). A number of studies conducted prior to the availability of molecular genotyping techniques reported inconsistent results which may be accounted for either by the lack of specificity of the serotyping and/or ethnic stratification between cases and controls.

Compared to the mycobacterial diseases, with regard to HIV and viral hepatitis, which we discussed above, not as much has been done with respect to disease associations and protozoan diseases (Table 4.5). We have highlighted studies of three different diseases: *Plasmodium falciparum*, which causes malaria; *Trypanosoma cruzi*, which causes Chagas' disease; and *Leishmania*, which causes leishmaniasis. The most interesting of these studies is the large study of malaria in Gambia showing an association between the class I antigen Bw53 and the haplotype DRB1*1302-DQB1*0501 and protection from severe malaria. Interestingly, these alleles and haplotypes are common in this West African population, but not in others (Hill et al., 1991). This observation further supports the notion that HLA diversity evolved from selection by infectious pathogens (Hill et al., 1991).

Although the above highlights associations between particular MHC alleles and infectious disease outcomes, it is important to bear in mind that the MHC is not the only genetic locus influencing disease outcome. Other loci that work in conjunction with the MHC may also exert pressure on an infectious pathogen. In many cases these genes may have an even greater impact than the MHC, for example, the *NRAMP* gene in affecting tuberculosis outcome and the effects of the 32-basepair deletion in the *CCR5* gene on the outcome of HIV disease (both are discussed elsewhere in this book).

9. MECHANISM OF MHC ASSOCIATIONS

As highlighted above, consistent and reproducible associations between MHC loci polymorphisms and the outcome of infectious diseases have been established over the past 5 years but little progress has been made in

Table 4.4. *Summary of selected association studies of HLA and viral hepatitis*

Study	Country	Study size	Finding	Strength of association	Significance level
			Natural (Spontaneous) Clearance of Hepatitis C Virus (HCV) Infection		
Congia et al. (1996)	Italy	752	DRB1*1601-DQB1*0502	2.3	$P = 0.009$
Alric et al. (1997)	France	128	DQB1*0301	2.7	$P < 0.01$
			DRB1*1101	4.1	$P < 0.02$
Cramp et al. (1998)	United Kingdom	238	DQA1*0301	4.7	$P = 0.001$
			DQB1*0301	5.1	$P = 0.008$
			DRB1*04-	4.5	$P = 0.002$
Minton et al. (1998)	United Kingdom	170	DRB1*11-	5.2	$P = 0.001$
			DQB1*0301	2.75	$P = 0.002$
Mangia et al. (1999)	Italy	384	DRB1*1104	4.5	$P = 0.054$
			DQB1*0301	4.5	$P = 0.004$
Lechmann et al. (1999)	Germany	70	DRB1*1501	6.5	$P = 0.02$
Barrett et al. (1999)	Ireland	157	DRB1*01	4.9	$P = 0.01$
Thursz et al. (1999)	European multicentre	897, 209*	DQB1*0301	2.32	$P = 0.00075$
			DRB1*1101	2.48	$P = 0.0015$
Fanning et al. (2000)	Ireland	156	DRB1*01	3.97	$P < 0.05$
			DRB1*0702 without DQB1*0501	—	$P < 0.05$
			Severity of HCV-Related Liver Disease		
Alric et al. (1997)	France	128	DQB1*0301	—	$P = 0.045$
Aikawa et al. (1996)	Japan	1400	DRB1*0405	2.8	$P < 0.05$
			DRB1*0401	2.1	$P < 0.05$
			DRB1*0901	0.3	$P < 0.05$
			DQB1*0303	0.2	$P < 0.01$

(cont.)

Table 4.4. *(cont.)*

Study	Country	Study size	Finding	Strength of association	Significance level
Mangia et al. (1999)	Italy	384	DRB1*0502	8.2	$P_c = 0.09$
Czaja et al. (1996)	United States	144	No associations found	—	—
Kuzushita et al. (1996)	Japan	130	DR13	0.092	
Asti et al. (1999)	Italy	319	DRB1*1104	4.8	$P = 0.058$
			DRB3*03	16.47	$P = 0.042$
Natural Variance of HCV Viral Load					
Fanning et al. (2001)	Ireland	57	DRB1*15	—	$P = 0.036$
			DRB1*0701	—	$P = 0.036$
			DQB1*0602	—	$P = 0.026$
			DQB1*0201	—	$P = 0.026$
Hepatitis B Surface Antigen (HBsAg) Persistance					
Jeannet and Farquet (1974)	Switzerland	98	None	—	—
Patterson et al. (1977)	United States	199	None	—	—
van Hattum et al. (1987)	Netherlands	396	DRw6	0.4	$P = 0.02$
			DQw1	3.5	$P = 0.001$
Almarri and Batchelor (1994)	Qatar	152	DR2	0.1	$P = 0.013$
			DR7	3.73	$P = 0.05$
Thursz et al. (1995)	Gambia	1604	DRB1*1302	0.53	$P = 0.01$
Höhler et al. (1997)	Germany	176	DRB1*1301/2	0.12	$P = 0.0004$
Ahn et al. (2000)	Korea	1272	DR13	0.14	$P = 0.002$
Thio et al. (1999)	United States	91	DQA1*0501	2.6	$P = 0.05$
			DQB1*0301	3	$P = 0.005$

Abbreviations: *P*, *P* value.

* This study involved two cohorts.

Table 4.5. *Summary of selected association studies of HLA and parasitic diseases*

Study	Country	Study size	Finding	Strength of association	Significance level
			Plasmodium Falciparum		
Hill et al. (1991)	Gambia	2000	Bw53	$\chi^2 = 4.3$	$P = 0.04$
			DRB1*1302-DQB1*0501	$\chi^2 = 14.0$	$P = 0.00018$
			Trypanosoma Cruzi		
Deghaide et al. (1998)	Brazil	624	DQ1	RR = 2.2	$P_c = 0.026$
			DQ7	RR = 0.21	$P_c < 0.001$
			A30	RR = 5.16	$P_c < 0.001$
			Leishmania Spp.		
Lara et al. (1991)	Venezuela	24 families	A28	—	$P = 0.0018$
			Bw22	—	$P = 0.0122$
			DQw8	—	$P = 0.0364$
			B15	—	$P = 0.0076$

OR, odds ratio; RR, relative risk; *P*, *P* value; P_c, *P* value with correction for multiple comparisons; χ^2, chi-square.

identification of the mechanisms underlying these associations. Although this is disappointing, it should be recalled that the association of HLA-B27 with ankylosing spondylitis has been established for more than 25 years and a complete explanation of the association remains elusive.

At a hypothetical level it is possible to provide mechanistic explanations for the observed associations in a number of ways.

1. Associated MHC alleles induce quantitatively superior/inferior T-cell responses than nonassociated alleles.
2. Associated MHC alleles induce qualitatively superior/inferior T-cell responses.
3. The peptide repertoire presented by associated MHC alleles is broader/narrower than the repertoires of nonassociated alleles.
4. Immunodominant peptide epitopes are only/never presented by the associated MHC allele.
5. The T-cell lines bearing the T-cell receptor required to respond to the immunodominant peptide epitope presented by the associated MHC allele is deleted during thymic maturation.
6. The MHC allele is in linkage disequilibrium with the true disease susceptibility allele.

In support of the first hypothesis Diepolder et al. (1998) have demonstrated that the CD4+ T-cell response to hepatitis B core antigen (HBcAg) is greater in individuals with HLA-DRB1*1301/2 compared to those who do not have these alleles. These data suggest that the HLA DRB1*13 alleles were more proficient than other alleles in presenting the HbcAg-derived epitopes. Surprisingly, the majority of T-cell lines generated from HLA-DRB1*13 patients with self-limiting HBV infection recognise a single epitope (Cao, T. unpublished observations). These data support the fourth hypothesis implicating the presentation of an immunodominant epitope.

Analysis of the peptide binding repertoire of MHC alleles may be revealing. Comparing the binding affinities of peptides eluted from MHC class II molecules, Davenport and Hill (1996) demonstrated that HLA-DRB1*1302 bound a broader range of peptides with higher affinity than seen with other MHC class II alleles. If this property is extrapolated to antigenic peptides in the context of an acute infection it is unlikely that sequence variation in pathogen derived antigens would allow the pathogen to evade T-cell-mediated immune responses. This property of HLA-DRB1*1302 may therefore underlie the ubiquitous benefit of the allele which is associated with self-limiting HBV infection, protection from HPV-associated cervical carcinoma, long-term survival with HIV infection, reduced severity of HCV associated hepatitis, and protection from severe malaria (Hill et al., 1991; Apple et al., 1994; Thursz et al., 1995; Kuzushita et al., 1996; Höhler et al., 1997; Ahn et al., 2000).

There is evidence that qualitative differences in T-cell responses may occur with different MHC alleles. MHC class II antigen presentation to CD4+ T helper cells may induce T-cell activation with two different phenotypes: a Th1 response characterised by interferon-γ secretion or a Th2 response characterised by interleukin 4 and interleukin 5 secretion from the T helper cells. In a mouse model of immunological responses to HBV-derived antigens Milich and colleagues have demonstrated that MHC haplotypes are one of the determinants of the phenotype of the CD4+ T helper cell response (Milich et al., 1995). However, similar *in vivo* human examples of this phenomenon do not exist and it is not a particularly plausible hypothesis for the mechanism underlying MHC associations.

T-cell receptors responding to HLA–self peptide complexes in the thymus are induced into apoptosis to avoid autoimmunity. It is possible that a narrow range of T cells which respond to self peptides presented by disease-associated MHC alleles are deleted in this system, leaving a hole in the T-cell receptor repertoire for immunodominant-pathogen-derived antigens. There is no data to support this theoretical mechanism presumably because there is no suitable model system to investigate this possibility. Nevertheless, it remains an attractive hypothesis to explain the association of MHC alleles with negative outcomes of infection such as viral persistence or inability to control pathogen replication.

Linkage disequilibrium is a universal criticism levelled against disease association studies for disease susceptibility or resistance. Clearly MHC disease associations could be explained by linkage disequilibrium to other loci within or adjacent to the MHC region as discussed above. This has previously been demonstrated in the case of the haemochromatosis gene, where there had been a strong association with the HLA-A3 allele. The association of an antigen presentation gene with an iron storage disorder clearly lacked plausibility, thus fuelling the search for non-MHC loci. In contrast the associations of antigen presentation genes with the outcome of infectious or autoimmune diseases are entirely credible even if the precise mechanism of association cannot be provided. Autoimmune chronic active hepatitis may provide a salutary warning. This is one of many autoimmune disorders associated with the HLA-A1-B8-DR3 haplotype. More recently, however, investigations suggest that the complement component C4-null allele might be the more important causative variant on this haplotype rather than the MHC class I or II loci (Scully et al., 1993; Doherty et al., 1994).

Patterns of linkage disequilibrium in the MHC region extend further than in many other regions of the human genome and may be complex. In the HBV study conducted in Gambia, association was excluded at loci immediately centromeric and telomeric to the HLA-DRB1 locus but these data do not entirely exclude a causative association at a greater distance (Thursz, 1995).

10. HOST–PARASITE INTERACTIONS

The loci within the MHC region display a unique degree of allelic diversity which is thought to relate to the function of the gene products in immune response to infection. Allelic diversity is thought to arise through pathogen-driven selection pressure. Pathogen-driven selection could, theoretically, operate through one or more of three proposed mechanisms: overdominant selection, (heterozygote advantage), frequency-dependent selection, and fluctuating selection (Snell, 1968; Doherty and Zinkernagel, 1975; Gillespie, 1985). Based on their discovery of MHC restriction of T-cell responses, Doherty and Zinkernagel (1975) proposed a mechanism whereby heterozygosity for MHC antigens might be beneficial in resistance to infectious pathogens. They envisaged that this might operate at either or both of two levels: against a single infectious pathogen or against a range of different infectious pathogens. Heterozygosity at MHC class I or class II loci has been shown to confer advantage in several different infections (Table 4.6). It is logical to assume that the repertoire of pathogen-derived epitopes presented to the T-cell population is broader in heterozygotes than in homozygotes, circumventing the possibility that the pathogen could escape from T-cell-mediated immune responses.

When T-cell responses are inadequate to successfully eliminate viral replication, host–pathogen interaction may be observed as sequence variation in antigen sequences driven by the weak host response. Individuals with chronic HBV infection may spontaneously, or with the assistance of interferon therapy, mount an immune response to the virus which partially controls high levels of viral replication. This is accompanied by seroconversion from HBeAg positivity to anti-HBe positivity (Thursz and Thomas, 1999). Sequence variation in the HBcAg of the dominant viral strain may be observed with the frequency of nucleotide and amino acid substitutions being disproportionately high in MHC class II epitope regions (Carman et al., 1997). This imprint of the host MHC on the virus may also be observed in HBV when transmitted within families. Viral sequences from pairs of MHC identical family members display closer homology than between pairs who have inherited different MHC haplotypes (Zampino et al., 2002).

Sequence variation in pathogen genomes is clearly driven by the need to evade host T-cell responses. This may be achieved in two ways; substitution of amino acid residues critical for epitope binding to HLA molecules and T-cell receptors or through epitope antagonism. In the second mechanism, sequence variation in an antigen may generate a peptide which binds to the HLA molecule but is not recognised by the host T-cell receptor repertoire. Loading of peptides into the cleft of HLA molecules is governed by the parameters of the Henderson–Hasselbach equation. Therefore an unrecognised peptide with high binding affinity may displace an immunodominant peptide from the HLA cleft preventing efficient recognition by T cells. Examples of epitope antagonism have been cited in both HBV and HIV infection but the relevance of these to viral persistence is uncertain.

11. EVOLUTIONARY ORIGINS OF THE HLA COMPLEX

So why is there such diversity in the MHC region? One hypothesis is that greater diversity conferred some evolutionary advantage. According to this theory, the high degree of polymorphism, and accordingly the allele frequencies, in these loci are a reflection of selective pressures from common and familiar pathogens (Doherty and Zinkernagel, 1975; Jeffery and Bangham, 2000). Hill and colleagues (1991) observed an association of the common West African allele Bw53 with protection from severe malaria in a malaria-endemic region. Because this allele is not common in other racial groups, it has been hypothesised that selection for this allele in this population has been largely due to evolutionary pressure from malaria (Hill et al., 1991). Thus, "beneficial" alleles are selected by pressure from infectious pathogens over many centuries. For newly emerging pathogens such as HIV infection, the selection process has just begun and it is therefore more likely that the influence of HLA in response to these newly "emerging pathogens" will be more apparent (Kaslow et al., 1996).

Another theory suggests that there is an element of innate self-selection for genetic diversity in the mate selection process (Penn and Potts, 1999). A number of studies conducted in both mice and rats suggest that these animals can distinguish individuals that are virtually genetically identical except in the MHC region and at a single locus (Yamazaki et al., 1979; 1983; Brown et al., 1987; Penn and Potts, 1998a). These studies suggest a role for the MHC in determining individual body odour (Yamazaki et al., 2000; Schaefer et al., 2001). Although the mechanisms for this phenomena are unclear, one possibility is that the MHC genes influence microbial flora and volatile acids (Singh et al., 1987; Singer et al., 1997). Studies in both animals and humans suggest that there is indeed an ability for individuals to detect individual body odours and that preference for these odours favours greater HLA mismatch (Ober et al., 1997; Penn and Potts, 1998b, 1999; Yamazaki et al., 2000; Jacob et al., 2002). In short, there is an innate predisposition for individuals to select others with greater HLA differences. Presumably, this encourages outbreeding and greater genetic diversity leading to more immunocompetent offspring.

12. CLINICAL USE OF MHC DISEASE ASSOCIATION STUDIES

The primary aim of disease association studies has been to define key variables which determine the outcome of infectious diseases. Knowledge of such variables should theoretically provide prognostic guidance, prediction of treatment outcome, or novel therapeutic targets which could be used in the clinic.

12.1 Prognostic Information

There is an inherent disadvantage to the exceptional allelic diversity of the MHC loci: the predictive value of many of the associations is relatively weak. The strength of MHC disease associations, estimated using the crude odds ratio, usually fall in the range of 2–7. Calculation of the population attributable risk takes into account the allele frequency which is inevitably low due to the allelic diversity. Population attributable risks and predictive values are, in general, too low for clinical utility in contrast to the use of HLA-B27 positivity in the diagnosis of ankylosing spondylitis.

It is likely that a number of genetic loci exert influence over the outcome of infectious diseases and therefore accurate prognostic information will require a comprehensive knowledge of as many of these as possible.

In HBV infection, DRB1*13 alleles and, in HCV infection, DRB1*1101 and DQB1*0301 (Table 4.4) are consistently associated with spontaneous resolution of infection. However, there is little clinical utility to this information as patients usually present after the division between these outcomes has occurred. However, understanding host factors in the response to therapeutics may aid in more individually tailored therapies in the future.

In chronic HCV infection, progression of hepatic fibrosis has been associated with TAP polymorphisms and with the HLA-DRB1*13 alleles (Kuzushita, 1996, 1999). Progression of hepatic fibrosis is difficult to monitor in the clinic and liver biopsy, an invasive and hazardous technique, can only reveal the current stage of fibrosis development. Therapy for chronic

Table 4.6. *Studies suggesting an advantage for heterozygosity of HLA alleles*

Study	Disease	Finding
	Heterozygote Advantage for Class I	
Carrington et al. (1999)	HIV	Disease progression
Tang et al. (1999)	HIV	Disease progression
MacDonald et al. (1998)	HIV	Vertical transmission*
Jeffrey et al. (2000)	HTLV-1	Proviral viral load
	Heterozygote Advantage for Class II	
Thursz et al. (1997)	Hepatitis B	Class II

* This study examined class I concordance between mother and child.

HCV infection is currently based on interferon therapy, which is expensive, unpleasant for the patient, and relatively ineffective. Only 20–30% of patients with chronic HCV infection will progress to a degree of fibrosis associated with increased morbidity or mortality and the infection risk to others can be readily controlled by changes to lifestyle. Prediction of the individuals with highest risk of fibrosis progression would be a rational method of targeting therapy. However, MHC alleles are not consistently associated with disease progression (Table 4.4).

13. RESPONSES TO THERAPY AND VACCINATION

Factors which predict the outcome of therapy are valuable when treatment is expensive, associated with high frequency of severe side effects, and effective in only a small proportion of patients. Response to interferon-α therapy has been associated with specific MHC alleles in HCV infection but there is little consistency in the studies and the strength of the associations is insufficient to be clinically useful at present.

HBV envelope proteins are encoded by three overlapping reading frames in the viral genome generating a large, middle, and small envelope protein. In HBV vaccination with the small envelope protein (HBsAg) failure to mount a satisfactory antibody response has been consistently associated with HLA-DR3 (Table 4.7). Although this is not particularly useful to the individual patient there have been advances in the design of HBV vaccine which have improved response rates. New generation HBV vaccines incorporate additional epitopes from the middle and large envelope protein which overcome vaccine nonresponse in HLA-DR3 individuals (McDermott et al., 1999).

Table 4.7. *Summary of selected studies concerning HLA and the response to therapeutics*

Country	Finding
*Response to Hepatitis B Surface Antigen (HBsAg) Vaccination**	
Japan	DRw6-DQ1 and response
	DR4-DRw53-DQ3/4 and nonresponse
	Bw54-DR4-DR53 and nonresponse
China	DQ3 and response
	DR14 and nonresponse
	DR14-DRw52 and nonresponse
United States	DR3 and nonresponse
	DR4 and nonresponse
	DR7 and nonresponse
	B8-DR3 and nonresponse
	B44-DR7 and nonresponse
Europe	A3 and response
	A11 and response
	DQ1 and response
	DR1 and response
	DR3 and nonresponse
	DR4 and nonresponse
	DR7 and nonresponse
	A1-B8-DR3 and nonresponse
Response to Interferon-α Therapy for Hepatitis C Virus (HCV) Infection	
France (Alric et al., 1999)	DRB1*07 and nonresponse
	DQB1*06 and sustained response
Egypt (Almarri et al., 1998)	DR2 and ALT normalisation
Poland (Wawrzynowicz-Syczewska, 2000)	DRB1*0701-DQA1*0201-DQB1*02 and response
Response to Canarypox AIDS Vaccines or Drug Side Effects	
United States (Kaslow et al., 2001)	B*27 and response to Gag protein
	B*57 and response to Gag protein
	B*57 and response to Env protein
Drug trials (Hetherington et al., 2002; Mallal et al., 2002)	B*5701, DR7 and DQ3 and hypersensitivity to the reverse transcriptase inhibitor, abacavir

* Reviewed in Thursz (2001).

Response to vaccine is also a problem in HIV treatment, where novel vaccines based on the canarypox virus bearing HIV-env or HIV-gag proteins are in development. Response to these vaccines has been associated with HLA-B27 and HLA-B57 (Table 4.7). Similarly, HLA variants have also been associated with the side effects of reverse transcriptase inhibitors (Table 4.7). Such observations lead to the hope that one day genetic information may be included in the clinical management of patients (Telenti et al., 2002).

14. VACCINE DEVELOPMENT

We have argued (above) that one of the mechanisms underlying the association of MHC alleles with better outcomes of infection is the optimal presentation of immunodominant epitopes from pathogenic antigens. If these epitopes and antigens could be identified then, in theory, vaccination could be used to boost the cognate T-cell responses in individuals who do not have the ideal MHC allele. *Reverse immunogenetics* is a term used to describe the process of identifying T-cell epitopes based on MHC disease association data (Davenport and Hill, 1996). The process involves the elution and sequencing of peptides from the cleft in the HLA molecule. Inevitably the majority of eluted peptides will be derived from host proteins but the sequences of these peptides may be used to identify a binding motif for MHC class I molecules (Lanzavecchia, 1985; Lanzavecchia et al., 1992). The "rules" of peptide binding are not as well defined for MHC class II molecules but sequence data from eluted peptides can be used to create a scoring system based on the frequency of each amino acid at each position in the eluted peptides. These scoring systems or binding motifs can be incorporated in computer-based algorithms to predict epitopes from antigen sequence data. Epitopes predicted *in silico* can be tested *in vitro* in both HLA–peptide-binding assays and in T-cell assays.

This approach was successfully used to identify a T-cell epitope in gliadin which is involved in the pathogenesis of coeliac disease (Anderson et al., 2000) and has recently been used to identify key HCV epitopes which are presented by HLA-DRB1*1101 (Godkin et al., 2001). It remains to be determined whether these epitopes can be used as the basis of therapeutic or prophylactic vaccines.

15. CONCLUSIONS

Understanding of the contribution of genetic diversity to the outcome of infectious diseases consists of several steps: first is the identification of new polymorphisms; second, understanding the biological function of these polymorphisms; and third, the correlation of these polymorphisms with disease outcomes. In turn, this can lead to the study of genetic and biologic function *in vivo*.

The MHC is the richest region for human immune response genes and is inevitably the first source of candidates in the investigation of genetic loci contributing to susceptibility or resistance to infectious diseases. Well-designed case-control studies have generated consistent associations in a number of infections and the challenge is now to make use of this work to investigate immunopathogenesis in greater depth and to provide clinical utility.

ACKNOWLEDGMENTS

We wish to thank Dr. Sara Marshall and J. Eric Tongren for their critical review of this chapter and offering their helpful comments.

REFERENCES

Ahn, S. H., Han, K. H., Park, J. Y., et al. (2000). Association between hepatitis B virus infection and HLA-DR type in Korea. *Hepatology* **31**, 1371–1373.

Aikawa, T., Kojima, M., Onishi, H., et al. (1996). HLA DRB1 and DQB1 alleles and haplotypes influencing the progression of hepatitis C. *J. Med. Virol.* **49**, 274–278.

Almarri, A. and Batchelor, J. R. (1994). HLA and hepatitis B infection. *Lancet* **344**, 1194–1195.

Almarri, A., El Dwick, N., Al Kabi, S., et al. (1998). Interferon-alpha therapy in HCV hepatitis: HLA phenotype and cirrhosis are independent predictors of clinical outcome. *Hum. Immunol.* **59**, 239–242.

Alric, L., Fort, M., Izopet, J., et al. (1997). Genes of the major histocompatibility complex class II influence the outcome of hepatitis C virus infection. *Gastroenterology* **113**, 1675–1681.

Alric, L., Izopet, J., Fort, M., et al. (1999). Study of the association between major histocompatibility complex class II genes and the response to interferon alpha in patients with chronic hepatitis C infection. *Hum. Immunol.* **60**, 516–523.

Anderson, R. P., Degano, P., Godkin, A. J., et al. (2000). *In vivo* antigen challenge in celiac disease identifies a single transglutaminase-modified peptide as the dominant A-gliadin T-cell epitope. *Nat. Med.* **6**, 337–342.

Apple, R. J., Erlich, H. A., Klitz, W., et al. (1994). HLA DR-DQ associations with cervical carcinoma show papillomavirus-type specificity. *Nat. Genet.* **6**, 157–162.

Ardlie, K. G., Kruglyak, L. and Seielstad, M. (2002). Patterns of linkage disequilibrium in the human genome. *Nat. Rev. Genet.* **3**, 299–309.

Asti, M., Martinetti, M., Zavaglia, C., et al. (1999). Human Leukocyte Antigen Class II and III alleles and severity of hepatitis C virus-related chronic liver disease. *Hepatology* **29**, 1272–1279.

Barrett, S., Ryan, E. and Crowe, J. (1999). Association of the HLA-DRB1*01 allele with spontaneous viral clearance in an Irish cohort infected with hepatitis C virus via contaminated anti-D immunoglobulin. *J. Hepatol.* **30**, 979–983.

Bjorkman, P. J. Saper, M. A., Samraoui, B., et al. (1987). Structure of the human class I histocompatibility antigen, HLA-A2. *Nature* **329**, 506–512.

Bothamley, G. H., Beck, J. S., Schreuder, G. M., et al. (1989). Association of tuberculosis and M. tuberculosis-specific antibody levels with HLA. *J. Infect. Dis.* **159**, 549–555.

Brown, R. E., Singh, P. B. and Roser, B. (1987). The major histocompatibility complex and the chemosensory recognition of individuality in rats. *Physiol. Behav.* **40**, 65–73.

Candore, G., Cigna, D., Gervasi, F., et al. (1994). *In vitro* cytokine production by HLA-B8,DR3 positive subjects. *Autoimmunity* **18**, 121–132.

Candore, G., Cigna, D., Todaro, M., et al. (1995). T-cell activation in HLA-B8, DR3-positive individuals: early antigen expression defect in vitro. *Hum. Immunol.* **42**, 289–294.

Candore, G., Romano, G. C., D'Anna, C., et al. (1998). Biological basis of the HLA-B8,DR3-associated progression of acquired immune deficiency syndrome. *Pathobiology* **66**, 33–37.

Carman, W. F., Boner, W., Fattovich, G., et al. (1997). Hepatitis B virus core protein mutations are concentrated in B cell epitopes in progressive disease and in T helper cell epitopes during clinical remission. *J. Infect. Dis.* **175**, 1093–1100.

Carrington, M., Nelson, G. W., Martin, M. P., et al. (1999). HLA and HIV-1: heterozygote advantage and B*35-Cw*04 disadvantage. *Science* **283**, 1748–1752.

Caruso, C., Candore, G. and Modica M. A. (1990). HLA-DR3 and immunoresponsiveness. *Lancet* **336**, 506–507.

Caruso, C., Candore, G., Colucci, A. T., et al. (1993). Natural killer and lymphokine-activated killer activity in HLA-B8,DR3-positive subjects. *Hum. Immunol.* **38**, 226–230.

Caruso, C., Bongiardina, C., Candore, G., et al. (1997). HLA-B8,DR3 haplotype affects lymphocyte blood levels. *Immunol. Invest.* **26**; 333–340.

Comstock, G. W. (1978). Tuberculosis in twins: a re-analysis of the Prophit survey. *Am. Rev. Respir. Dis.* **117**, 621–624.

Congia, A., Clemente, M. G., Dessi, C., et al. (1996). HLA class II genes in chronic hepatitis C virus-infection and associated immunological disorders. *Hepatology* **24**, 1338–1341.

Cooper, G., Dooley, M., Treadwell, E., et al. (1998). Hormonal, environmental and infectious risk factors for developing systemic lupus erythematosus. *Arthritis Rheum.* **41**, 1714–1724.

Cooper, G. S., Miller, F. W. and Pandey, J. P. (1999). The role of genetic factors in autoimmune disease: implications for environmental research. *Environ. Health Perspect.* **107**(Suppl. 5), 693–700.

Costello, C., Tang, J., Rivers, C., et al. (1999). HLA-B*5703 independently associated with slower HIV-1 disease progression in Rwandan women. *AIDS* **13**, 1990–1991.

Cramp, M. E., Carucci, P., Underhill, J., et al. (1998). Association between class II genotype and spontaneous clearance of hepatitis C viraemia. *J. Hepatol.* **29**, 207–213.

Cresswell, P., Bangia, N., Dick, T., et al. (1999). The nature of the MHC class I peptide loading complex. *Immunol. Rev.* **172**: 21–28.

Czaja, A. J., Carpenter, H., Santrach, P. J., et al. (1996). DR human leukocyte antigens and disease severity in chronic hepatitis C. *J. Hepatol.* **24**, 666–673.

Davenport, M. P. and Hill, A. V. (1996). Reverse immunogenetics: from HLA-disease associations to vaccine candidates. *Mol. Med. Today* **2**, 38–45.

Deghaide, N. H., Dantas, R. O. and Donadi, E. A. (1998). HLA class I and II profiles of patients presenting with Chagas' disease. *Dig. Dis. Sci.* **43**, 246–252.

Dessoukey, M. W., El-Shiemy, S. and Sallam, T. (1996). HLA and leprosy: segregation and linkage study. *Int. J. Dermatol.* **35**, 257–264.

Diepolder, H. M., Jung, M. C., Keller, E., et al. (1998). A vigorous virus-specific CD4+ T cell response may contribute to the association of HLA-DR13 with viral clearance in hepatitis B. *Clin. Exp. Immunol.* **113**, 244–251.

Doherty, D. G., Underhill, J. A., Donaldson, P. T., et al. (1994). Polymorphism in the human complement C4 genes and genetic susceptibility to autoimmune hepatitis. *Autoimmunity* **18**, 243–249.

Doherty, P. C. and Zinkernagel, R. M. (1975). A biological role for the major histocompatibility antigens. *Lancet* **1**, 1406–1409.

Editor. (1999). Freely associating. *Nat. Genet.* **22**, 1–2.

Fanning, L. J., Levis, J., Kenny-Walsh, E., et al. (2000). Viral clearance in hepatitis C (1b) infection: relationship with human leukocyte antigen class II in a homogeneous population. *Hepatology* **31**, 1334–1337.

Fanning, L. J., Levis, J., Kenny-Walsh, E., et al. (2001). HLA class II genes determine the natural variance of hepatitis C viral load. *Hepatology* **33**, 224–230.

Flores-Villanueva, P. O., Yunis, E. J., Delgado, J. C., et al. (2001). Control of HIV-1 viremia and protection from AIDS are associated with HLA-Bw4 homozygosity. *Proc. Natl. Acad. Sci. USA* **98**, 5140–5145.

Ghosh, P., Amaya, M., Mellins, E., et al. (1995). The structure of an intermediate in class II MHC maturation: CLIP bound to HLA-DR3. *Nature* **378**, 457–462.

Gillespie, J. H. (1985). The interaction of genetic drift and mutation with selection in a fluctuating environment. *Theor. Popul. Biol.* **27**, 222–237.

Godkin, A., Jeanguet, N., Thursz, M., et al. (2001). Characterization of novel HLA-DR11-restricted HCV epitopes reveals both qualitative and quantitative differences in HCV-specific CD4+ T cell responses in chronically infected and non-viremic patients. *Eur. J. Immunol.* **31**, 1438–1446.

Gorodezky, C., Flores, J., Arevalo, N., et al. (1987). Tuberculoid leprosy in Mexicans is associated with HLA-DR3. *Lepr. Rev.* **58**, 401–406.

Harris, D. R., Gonin, R., Alter, H. J., et al. (2001). The relationship of acute transfusion-associated hepatitis to the development of cirrhosis in the presence of alcohol abuse. *Ann. Intern. Med.* **134**, 120–124.

Hendel, H., Caillat-Zucman, S., Lebuanec, H., et al. (1999). New class I and II HLA alleles strongly associated with opposite patterns of progression to AIDS. *J. Immunol.* **162**, 6942–6946.

Hetherington, S., Hughes, A. R., Mosteller, M., et al. (2002). Genetic variations in HLA-B region and hypersensitivity reactions to abacavir. *Lancet* **359**, 1121–1122.

Hill, A. V., Allsopp, C. E., Kwiatkowski, D., et al. (1991). Common west African HLA antigens are associated with protection from severe malaria. *Nature* **352**, 595–600.

Höhler, T., Gerken, G., Notghi, A., et al. (1997). HLA-DRB1*1301 and *1302 protect against chronic hepatitis B. *J. Hepatol.* **26**, 503–507.

Höhler, T., Gerken, G., Notghi, A., et al. (1997). MHC class II genes influence the susceptibility to chronic active hepatitis C. *J. Hepatol.* **27**, 259–264.

Izumi, S., Sugiyama, K., Matsumoto, Y., Ohkawa, S. (1982). Analysis of the immunogenetic background of Japanese leprosy patients by the HLA system. *Vox. Sang.* **42**, 243–247.

Jacob, S., McClintock, M. K., Zelano, B., et al. (2002). Paternally inherited HLA alleles are associated with women's choice of male odor. *Nat. Genet.* **30**, 175–179.

Jardetzky, T. S., Brown, J. H., Gorga, J. C., et al. (1996). Crystallographic analysis of endogenous peptides associated with HLA-DR1 suggests a common, polyproline II-like conformation for bound peptides. *Proc. Natl. Acad. Sci. USA* **93**, 734–738.

Jeannet, M. and Farquet, J. J. (1974). Letter: HL-A antigens in asymptomatic chronic HBAg carriers. *Lancet* **2**, 1383–1384.

Jeffery, K. J. and Bangham, C. R. (2000). Do infectious diseases drive MHC diversity? *Microbes Infect.* **2**, 1335–1341.

Jeffery, K. J., Siddiqui, A. A., Bunce, M., et al. (2000). The influence of HLA class I alleles and heterozygosity on the outcome of human T-cell lymphotropic virus type I infection. *J. Immunol.* **165**, 7278–7284.

Jepson, A., Banya, W., Sisay-Joof, F., et al. (1997). Quantification of the relative contribution of major histocompatibility complex (MHC) and non-MHC genes to human immune responses to foreign antigens. *Infect. Immun.* **65**, 872–876.

Kallman, F., and Reisner, D. (1942). Twin studies on the significance of genetic factors in tuberculosis. *Am. Rev. Tuberc.* **47**, 549–574.

Kaslow, R. A., Duquesnoy, R., Van Raden, M., et al. (1990). A1, Cw7, B8, DR3 HLA antigen combination associated with rapid decline of T-helper lymphocytes in HIV-1 infection. A report from the Multicenter AIDS Cohort Study. *Lancet* **335**, 927–930.

Kaslow, R. A., Carrington, M., Apple, R., et al. (1996). Influence of combinations of human major histocompatibility complex genes on the course of HIV-1 infection. *Nat. Med.* **2**, 405–411.

Kaslow, R. A., Rivers, C., Tang, J., et al. (2001). Polymorphisms in HLA class I genes associated with both favorable prognosis of human immunodeficiency virus (HIV) type 1 infection and positive cytotoxic T-lymphocyte responses to ALVAC-HIV recombinant canarypox vaccines. *J. Virol.* **75**, 8681–8689.

Keet, I. P. M., Klein, M., Just, J. J., et al. (1996). The role of host genetics in the natural history of HIV-1 infection: the needles in the haystack. *AIDS* **10**(Suppl. A), 1–9.

Keet, I. P., Tang, J., Klein, M. R., et al. (1999). Consistent associations of HLA class I and II and transporter gene products with progression of human immunodeficiency virus type 1 infection in homosexual men. *J. Infect. Dis.* **180**, 299–309.

Khomenko, A. G., Litvinov, V. I., Chukanova, V. P., et al. (1990). Tuberculosis in patients with various HLA phenotypes. *Tubercle* **71**, 187–192.

Kim, S. J., Choi, I. H., Dahlberg, S., et al. (1987). HLA and leprosy in Koreans. *Tissue Antigens* **29**, 146–153.

Klein, J. and Sato, A. (2000a). The HLA system: First of two parts. *N. Engl. J. Med.* **343**, 702–709.

Klein, J. and Sato, A. (2000b). The HLA system: Second of two parts. *N. Engl. J. Med.* **343**, 782–786.

Kuzushita, N., Hayashi, N., Katayama, K., et al. (1996). Increased frequency of HLA DR13 in hepatitis C virus carriers with persistently normal ALT levels. *J. Med. Virol.* **48**, 1–7.

Kuzushita, N., Hayashi, N., Kanto, T., et al. (1999). Involvement of transporter associated with antigen processing 2 (TAP2) gene polymorphisms in hepatitis C virus infection. *Gastroenterology* **116**, 1149–1154.

Lang, J. M., Rothman, K. J. and Cann, C. I. (1998). That confounded *P*-value. *Epidemiology* **9**, 7–8.

Lanzavecchia, A. (1985). Antigen-specific interaction between T and B cells. *Nature* **314**, 537–539.

Lanzavecchia, A., Reid, P. A. and Watts, C. (1992). Irreversible association of peptides with class II MHC molecules in living cells. *Nature* **357**, 249–252.

Lara, M. L., Layrisse, Z., Scorza, J. V., et al. (1991). Immunogenetics of human American cutaneous leishmaniasis. Study of HLA haplotypes in 24 families from Venezuela. *Hum. Immunol.* **30**, 129–135.

LeBlanc, S., Naik, E. G., Jacobson, L., et al. (1999). Association of DRB1*1501 with disseminated *Mycobacterium avium* complex infection in North American AIDS patients. *Tissue Antigens* **55**, 17–23.

Lechmann, M., Schneider, E. M., Giers, G., et al. (1999). Increased frequency of the HLA-DR15 (B1*15011 allele in German patients with self-limited hepatitis C virus infection. *Eur. J. Clin. Invest.* **29**, 337–343.

Lin, T. M., Chen, C. J., Wu, M. M., et al. (1989). Hepatitis B virus markers in Chinese twins. *Anticancer Res.* **9**, 737–741.

Lio, D., D'Anna, C., Gervasi, F., et al. (1995). In vitro impairment of interleukin-5 production in HLA-B8, DR3-positive individuals implications for immunoglobulin A synthesis dysfunction. *Hum. Immunol.* **44**, 170–174.

Lio, D., Candore, G., Romano, G. C., et al. (1997). Modification of cytokine patterns in subjects bearing the HLA-B8,DR3 phenotype: implications for autoimmunity. *Cytokines Cell Mol. Ther.* **3**, 217–224.

MacDonald, K. S., Embree, J., Njenga, S., et al. (1998). Mother–child class I HLA concordance increases perinatal human immunodeficiency virus type 1 transmission. *J. Infect. Dis.* **177**, 551–556.

Malaty, H. M., Engstrand, L., Pedersen N. L., et al. (1994). Helicobacter pylori infection: genetic and environmental influences. A study of twins. *Ann. Intern. Med.* **120**, 982–986.

Mallal, S., Nolan, D., Witt, C., et al. (2002). Association between presence of HLA-B*5701, HLA-DR7, and HLA-DQ3 and hypersensitivity to HIV-1 reverse-transcriptase inhibitor abacavir. *Lancet* **359**, 727–732.

Mangia, A., Gentile, R., Cascavilla, I., et al. (1999). HLA class II favors clearance of HCV infection and progression of the chronic liver damage. *J. Hepatol.* **30**, 984–989.

Manns, M. P., McHutchison, J. G., Gordon, S. C., et al. (2001). Peginterferon alfa-2b plus ribavirin compared with interferon alfa-2b plus ribavirin for initial treatment of chronic hepatitis C: a randomised trial. *Lancet* **358**, 958–965.

Marsh, S. G., Bodmer, J. G., Albert, E. D., et al. (2001). Nomenclature for factors of the HLA system, 2000. *Eur. J. Immunogenet.* **28**, 377–424.

McDermott, A., Cohen, S., Zuckerman, J., et al. (1999). Human leukocyte antigens influence the immune response to a pre-S/S hepatitis B vaccine. *Vaccine* **17**, 330–339.

McHutchison, J. G., Gordon, S. C., Schiff, E., et al. (1998). Interferon alfa-2b alone or in combination with ribavirin as initial treatment for chronic hepatitis C. *N. Engl. J. Med.* **339**, 1485–1492.

McNeil, A. J., Yap, P. L., Gore, S. M., et al. (1996). Association of HLA types A1-B8-DR3 and B27 with rapid and slow progression of HIV disease. *Q. J. Med.* **89**, 177–185.

MHC Sequencing Consortium. (1999). Complete sequence and gene map of a human major histocompatibility complex. The MHC sequencing consortium. *Nature* **401**, 921–923.

Migueles, S. A., Sabbaghian, M. S., Shupert, W. L., et al. (2000). HLA B*5701 is highly associated with restriction of virus replication in a subgroup of HIV-infected long term nonprogressors. *Proc. Natl. Acad. Sci. USA* **97**, 2709–2714.

Milich, D. R., Peterson, D. L., Schodel, F., et al. (1995). Preferential recognition of hepatitis B nucleocapsid antigens by Th1 or Th2 cells is epitope and major histocompatibility complex dependent. *J. Virol.* **69**, 2776–2785.

Minton, E. J., Smillie, D., Neal, K. R., et al. (1998). Association between MHC class II alleles and clearance of circulating hepatitis C virus. Members of the Trent Hepatitis C Virus Study Group. *J. Infect. Dis.* **178**, 39–44.

Modica, M. A., Zambito, A. M., Candore, G., et al. (1990). Markers of T lymphocyte activation in HLA-B8, DR3 positive individuals. *Immunobiology* **181**, 257–266.

Ober, C., Weitkamp, L. R., Cox, N., et al. (1997). HLA and mate choice in humans. *Am. J. Hum. Genet.* **61**, 497–504.

Ostapowicz, G., Watson, K. J. R., Locarnini, S. A., et al. (1998). Role of alcohol in the progression of liver disease caused by hepatitis C virus infection. *Hepatology* **27**, 1730–1735.

Parker, K. C., Shields, M., DiBrino, M., et al. (1995). Peptide binding to MHC class I molecules: implications for antigenic peptide prediction. *Immunol. Res.* **14**, 34–57.

Patterson, M. J., Hourani, M. R. and Mayor, G. H. (1977). HLA antigens and hepatitis B virus. *N. Engl. J. Med.* **297**, 1124.

Penn, D. and Potts, W. (1998). How do major histocompatibility complex genes influence odor and mating preferences? *Adv. Immunol.* **69**, 411–436.

Penn, D. and Potts, W. K. (1998). Untrained mice discriminate MHC-determined odors. *Physiol. Behav.* **64**, 235–243.

Penn, D. and Potts, W. (1999). The evolution of mating preferences and major histocompatibility genes. *Am. Nat.* **153**, 145–164.

Pospelov, L. E., Matrakshin, A. G., Chernousova, L. N., et al. (1996). Association of various genetic markers with tuberculosis and other lung diseases in Tuvinian children. *Tuber. Lung Dis.* **77**, 77–80.

Poynard, T., Bedossa, P., Opolon, P., et al. (1997). Natural history of liver fibrosis progression in patients with chronic hepatitis C. *Lancet* **349**, 825–832.

Poynard, T., Marcellin, P., Lee, S. S., et al. (1998). Randomized trial of interferon α 2b plus ribavirin for 48 weeks or for 24 weeks versus interferon α2b plus placebo for 48 weeks for treatment of chronic infection with hepatitis C virus. *Lancet* **352**, 1426–1432.

Rajalingam, R., Mehra, N. K., Mehra, R. D., et al. (1997). HLA class I profile in Asian Indian patients with pulmonary tuberculosis. *Ind. J. Exp. Biol.* **35**, 1055–1059.

Rani, R., Zaheer, S. A. and Mukherjee, R. (1992). Do human leukocyte antigens have a role to play in differential manifestation of multibacillary leprosy: a study on multibacillary leprosy patients from north India. *Tissue Antigens* **40**, 124–127.

Rani, R., Fernandez-Vina, M. A., Zaheer, S. A., et al. (1993). Study of HLA class II alleles by PCR oligotyping in leprosy patients from north India. *Tissue Antigens* **42**, 133–137.

Rothman, K. J. (1990). No adjustments are needed for multiple comparisons. *Epidemiology* **1**, 43–6.

Saah, A. J., Hoover, D. R., Weng, S., et al. (1998). Association of HLA profiles with early plasma viral load, CD4+ cell count and rate of progression to AIDS following acute HIV-1 infection: Multicenter AIDS Cohort Study. *AIDS* **12**, 2107–2113.

Schaefer, M. L., Young, D. A. and Restrepo, D. (2001). Olfactory fingerprints for major histocompatibility complex-determined body odors. *J. Neurosci.* **21**, 2481–2487.

Schafer, P. H., Pierce, S. K. and Jardetzky, T. S. (1995). The structure of MHC class II: a role for dimer of dimers. *Semin. Immunol.* **7**, 389–398.

Schiff, E. R. (1997). Hepatitis C and alcohol. *Hepatology* **26**(Suppl. 1), 39S–42S.

Scully, L. J., Toze, C., Sengar, D. P., et al. (1993). Early-onset autoimmune hepatitis is associated with a C4A gene deletion. *Gastroenterology* **104**, 1478–1484.

Shackelford, D. A., Kaufman, J. F., Korman, A. J., et al. (1982). HLA-DR antigens: structure, separation of subpopulations, gene cloning and function. *Immunol. Rev.* **66**, 133–187.

Singer, A. G., Beauchamp, G. K. and Yamazaki, K. (1997). Volatile signals of the major histocompatibility complex in male mouse urine. *Proc. Natl. Acad. Sci. USA* **94**, 2210–2214.

Singh, S. P., Mehra, N. K., Dingley, H. B., et al. (1983a). HLA-A, -B, -C and -DR antigen profile in pulmonary tuberculosis in North India. *Tissue Antigens* **21**, 380–384.

Singh, S. P., Mehra, N. K., Dingley, H. B., et al. (1983b). Human leukocyte antigen (HLA)-linked control of susceptibility to pulmonary tuberculosis and association with HLA-DR types. *J. Infect. Dis.* **148**, 676–681.

Singh, P. B., Brown, R. E. and Roser, B. (1987). MHC antigens in urine as olfactory recognition cues. *Nature* **327**, 161–164.

Snell, G. D. (1968). The H-2 locus of the mouse: observations and speculations concerning its comparative genetics and its polymorphism. *Folia Biol. (Praha)* **14**, 335–358.

Sorensen, T. I., Nielsen, G. G., Andersen, P. K., et al. (1988). Genetic and environmental influences on premature death in adult adoptees. *N. Engl. J. Med.* **318**, 727–732.

Steel, C. M., Ludlam, C. A., Beatson, D., et al. (1988). HLA haplotype A1 B8 DR3 as a risk factor for HIV-related disease. *Lancet* **i**, 1185–1188.

Tang, J., Costello, C., Keet, I. P. M., et al. (1999). HLA class I homozygosity accelerates disease progression in human immunodeficiency virus type 1 infection. *AIDS Res. Hum. Retroviruses* **15**, 317–324.

Telenti, A., Aubert, V. and Spertini, F. (2002). Individualising HIV treatment – Pharmacogenetics and immunogenetics. *Lancet* **359**, 722–723.

Thio, C. L., Carrington, M., Marti, D., et al. (1999). Class II HLA alleles and hepatitis B virus persistence in African Americans. *J. Infect. Dis.* **179**, 1004–1006.

Thomas, D. L., Astemborski, J., Rai, R. M., et al. (2000). The natural history of hepatitis C virus infection: host, viral and environmental factors. *J. Am. Med. Assoc.* **284**, 450–456.

Thorsby, E. (1999). MHC structure and function. *Transplant Proc.* **31**, 713–716.

Thursz, M. R., Kwiatkowski, D., Allsopp, C. E. M., et al. (1995). Association between an MHC class II allele and clearance of hepatitis B virus in the Gambia. *N. Engl. J. Med.* **332**, 1065–1069.

Thursz, M. R., Thomas, H. C., Greenwood, B. M., et al. (1997). Heterozygote advantage for HLA class-II type in hepatitis B virus infection. *Nat. Genet.* **17**, 11–12.

Thursz, M., Yallop, R., Goldin, R., et al. (1999). Influence of MHC class II genotype on outcome of infection with hepatitis C virus. *Lancet* **354**, 2119–2124.

Thursz, M. and Thomas, H. (1999). Pathogenesis of HBV infection. In: *Viral Hepatitis, 2nd edition*. Thomas, H. and Zuckerman, A. (eds.). Oxford: Churchill Livingstone.

Thursz, M. (2001). MHC and the viral hepatitides. *Q. J. Med.* **94**, 287–291.

Todd, J. R., West, B. C. and McDonald, J. C. (1990). Human leukocyte antigen and leprosy: study in northern Louisiana and review. *Rev. Infect. Dis.* **12**, 63–74.

Trachtulec, Z., Hamvas, R. M., Forejt, J., et al. (1997). Linkage of TATA-binding protein and proteasome subunit C5 genes in mice and humans reveals synteny conserved between mammals and invertebrates. *Genomics* **44**, 1–7.

van Eden, W., de Vries, R. R., D'Amaro, J., et al. (1982). HLA-DR-associated genetic control of the type of leprosy in a population from surinam. *Hum. Immunol.* **4**, 343–350.

van Hattum, J., Schreuder, G. M. and Schalm, S. W. (1987). HLA antigens in patients with various courses after hepatitis B virus infection. *Hepatology* **7**, 11–14.

Wawrzynowicz-Syczewska, M., Underhill, J. A., Clare, M. A., et al. (2000). HLA class II genotypes associated with chronic hepatitis C virus infection and response to alpha-interferon treatment in Poland. *Liver* **20**, 234–239.

Wiley, T. E., McCarthy, M., Breidi, L., et al. (1998). Impact of alcohol on the histological and clinical progression of hepatitis C infection. *Hepatology* **28**, 805–809.

Williams, A., Peh, C. A. and Elliott, T. (2002). The cell biology of MHC class I antigen presentation. *Tissue Antigens* **59**, 3–17.

Yamazaki, K., Yamaguchi, M., Baranoski, L., et al. (1979). Recognition among mice. Evidence from the use of a Y-maze differentially scented by congenic mice of different major histocompatibility types. *J. Exp. Med.* **150**, 755–760.

Yamazaki, K., Beauchamp, G. K., Wysocki, C. J., et al. (1983). Recognition of H-2 types in relation to the blocking of pregnancy in mice. *Science* **221**, 186–188.

Yamazaki, K., Beauchamp, G. K., Curran, M., et al. (2000). Parent–progeny recognition as a function of MHC odortype identity. *Proc. Natl. Acad. Sci. USA* **97**, 10500–10502.

Zampino, R., Lobello, S., Chiaramonte, M., et al. (2002). Intra-familial transmission of hepatitis B virus in Italy: phylogenetic sequence analysis and amino-acid variation of the core gene. *J. Hepatol.* **36**, 248–253.

CHAPTER 5

The cystic fibrosis transmembrane conductance regulator

Alan W. Cuthbert

Department of Medicine, University of Cambridge, Addenbrooke's Hospital, Cambridge, United Kingdom

1. INTRODUCTION

Cystic fibrosis (CF) patients suffer from persistent airway infections. A hallmark of CF lung disease is chronic infection, initially with *Staphylococcus aureus* or *Haemophilus influenzae* and subsequently with *Pseudomonas aeruginosa* and *Burkholderia cepacia*, in which host defence mechanisms become overwhelmed. Infection is followed by a neutrophil-dominated inflammatory response due to exotoxins released by the bacteria together with enzymes, such as elastase, cathepsin G, and proteinase-3, and release of inflammatory mediators, such as IL8. The latter acts as a powerful neutrophil attractant. Increased mucin formation by the epithelial cells also occurs but does not seem to be a consequence of bacterial infection, as it is found in the airways of CF foetuses (Ornoy et al., 1987). The continual tissue damage and fibrotic changes following infection lead to a gradual loss of lung function, the major cause of morbidity and mortality in this genetic disease.

Curiously, as we shall see later, CF can protect individuals from other types of infectious disease. Also, heterozygotes, carrying a faulty CF gene on one allele, are considered to be entirely normal and the persistence of the gene in the population has led to suggestions that there may be an heterozygote advantage in carriers. However, recent evidence suggests that heterozygotes are more susceptible to sinusitis and chronic rhinitis than noncarriers (Wang et al., 2000). Evidence will also be presented that suggests that heterozygotes are less susceptible to a number of bacterial diseases affecting the gut.

In this chapter the mechanisms involved in increased or reduced susceptibility to infectious diseases in CF will be critically reviewed. This is clearly important as reducing susceptibility to infectious disease in cystic fibrosis may delay the onset of its lethal consequences. This point can be clearly underlined by quoting statistical information given in a recent document issued by the Cystic Fibrosis Trust (2001) with suggestions for prevention and infection control against *P. aeruginosa*. Some facts are as follows: (1) 42% of CF patients with mucoid *P. aeruginosa* and 11% of those with nonmucoid *P. aeruginosa* died, compared to 8% with no *P. aeruginosa* infection, monitored over an 8-year period (Henry et al., 1992); (2) Infection with *P. aeruginosa* plus *S. aureus* significantly increased mortality in the 10 years after diagnosis (Hudson et al., 1993); and (3) Survival of CF patients with chronic *P. aeruginosa* infection was 28 years, with *B. cepacia* 16 years, and in patients with neither infection 39 years (FitzSimmons, 1996).

It is noteworthy that *P. aeruginosa* is a common organism, found in plants, soils, and stagnant surface water. CF patients are more susceptible to infection with hypermutable strains of *P. aeruginosa* than non-CF individuals, with the consequent problems of antibiotic resistance. Therefore there can be little doubt that host genetics has a major effect on the susceptibility to infection with, at least, some of the organisms listed above.

2. WHAT IS CF?

The CF gene codes for a protein called the cystic fibrosis transmembrane conductance regulator (CFTR) (Fig. 5.1). When this protein is dysfunctional it leads to the phenotype characteristic of the disease. In this chapter the focus will be on how the absence of CFTR alters susceptibility to infectious disease.

Cystic fibrosis is an autosomal, recessive disease that affects some 1 in 2,500 live births in Caucasian populations. It is not unknown in other ethnic groups where presumably interracial marriages have spread the genetic mutation. Amongst Caucasians 1 in 25 persons are carriers who show no traits of the disease and who may, as will be seen later, have some advantages over non-carriers in relation to infectious disease. CF is a single gene disease in which mutations of the gene, found on the long arm of chromosome 7, leads to the disease phenotype. The mutation may prevent production of CFTR (through nonsense, frameshift, or splice variant mutations), or may cause problems in processing and delivery to the membrane (as with ΔF508CFTR), or may allow delivery to the membrane but interfere with the regulation or function of CFTR.

The gene was cloned in 1989 (Kerem et al., 1989; Riordan et al., 1989; Rommens et al., 1989) and soon after it was confirmed that the protein behaved as a chloride ion channel when transfected into suitable cells and probed electrophysiologically (Anderson et al., 1991). It could be activated

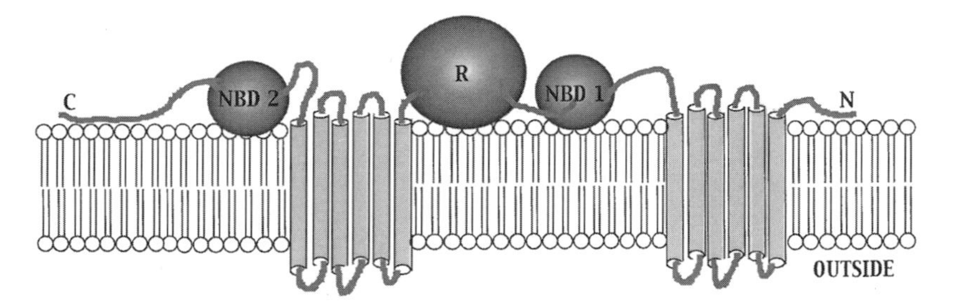

Figure 5.1. Diagram showing the structure of the cystic fibrosis transmembrane conductance regulator (CFTR). The molecule is symmetrical, the two halves each having six membrane spanning domains and a nucleotide binding fold and joined at the R domain. The nucleotide binding domains bind and hydrolyse ATP and regulate channel gating. The R domain has many sites capable of phosphorylation by either protein kinase A or C plus ATP. Note that very little of the protein is exposed to the extracellular environment. The molecules may aggregate as dimers in the cell membrane.

through cAMP, acting through protein kinase A together with ATP. This result was not unexpected as much previous research, notably that by Quinton (1983), had already shown that the chloride conductance of epithelial cells from sweat glands of CF patients was abnormally low when compared to non-CF cells. CFTR is called the transmembrane conductance regulator, rather than the transmembrane chloride channel, because CFTR controls a number of other conductances in epithelial cell membranes as well as chloride conductance. These include those for sodium ions (Mall et al., 1996), potassium ions (Loussouarn et al., 1996) and other types of chloride channel (Wei et al., 1999), together with ion exchangers and transporters, such as the chloride/bicarbonate exchanger (Lee et al., 1999a) and the NaK2Cl cotransporter (Shumaker and Soleimani, 1999) and water channels, the aquaporins (Schreiber et al., 1997).

CFTR is expressed in the membranes of many types of epithelial cells lining hollow organs, but also in the heart and in low amounts in erythrocytes, but it is the specific location in epithelial cells which is responsible for the manifestations of the disease. Around a thousand different mutations of the CF gene have been described which lead to a CF phenotype, but by far the most common is the ΔF508 mutation, due to the deletion of the triplet codon for phenylalanine in position 508. Seventy percent of those with CF have at least one allele with the ΔF508 mutation. In order to understand how the lack of CFTR alters the susceptibility to infectious diseases it is important to understand where CFTR is located, its functions, and the consequences of this lack of function.

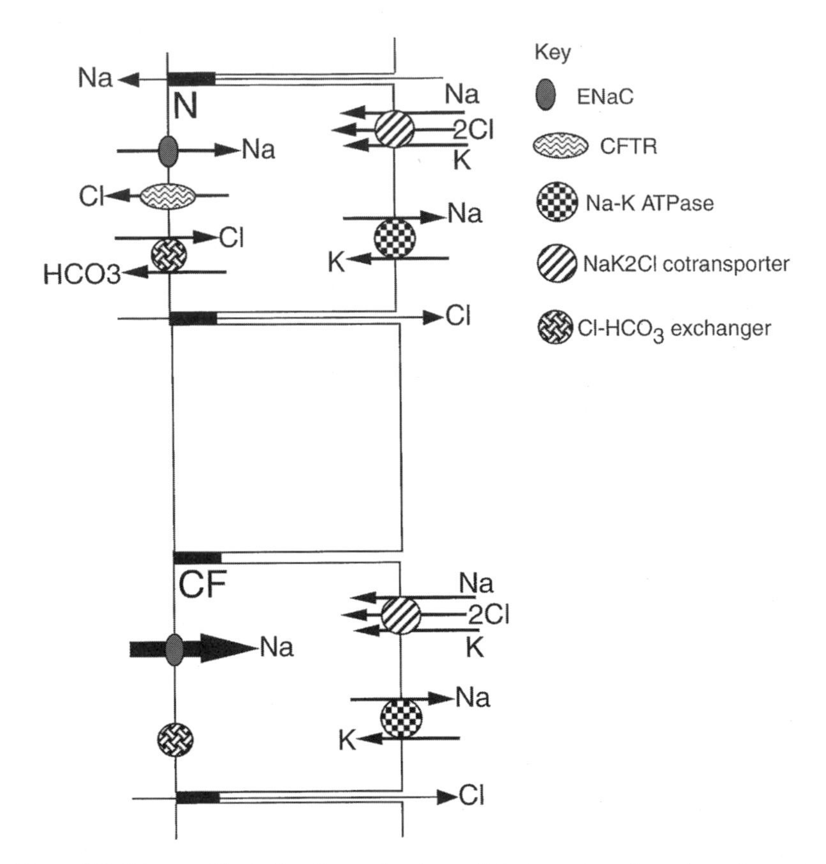

Figure 5.2. Diagram illustrating the mechanisms of electrogenic sodium absorption and electrogenic chloride secretion (top, N) and how these are modified in CF (bottom, CF). In CF chloride secretion (and bicarbonate secretion) fails and sodium transport is enhanced.

3. APICAL MEMBRANE LOCATION OF CFTR

Epithelial cells are asymmetric; the properties of the membrane on the side contacted by tissue fluid are different from those of the opposite face, that is, those facing the lumen of hollow organs. These two faces are known as the basolateral and apical faces, respectively. Tight junctions functionally separate the apical and basolateral membrane domains of epithelial cells (Fig. 5.2). This effectively prevents translational movement of membrane components from one membrane to the other. It is this asymmetric distribution of membrane components which allows the vectorial transport of substances inwardly or outwardly across epithelial surfaces, as clearly the processes at one face must differ from those at the other or no net movement would occur. CFTR is located in the apical membranes of the epithelia lining the airways;

the alimentary canal and associated organs, such as the pancreatic duct and the hepatobiliary system; organs of the genitourinary tract; and ducts of the salivary and sweat glands. The locations of CFTR mirrors the pattern of the disease, characterised by symptoms affecting the systems listed above. The major cause of morbidity and mortality is by disease affecting the airways, particularly the small airways that become filled with thick viscid mucus and are often chronically infected, as described earlier. Similarly, mucus accumulation in the pancreatic duct prevents the secretion of pancreatic enzymes, with consequent problems for digestion. Autodigestion may occur in the pancreas, leading to complete loss of exocrine and sometimes endocrine function. Mucus accumulation in the hepatobiliary tract can give rise to cirrhosis. Mucus plugs in the intestine may lead to compaction and so-called meconium ileus in the neonate, both of which may require surgical intervention. Male CF patients are usually sterile because of blockage or loss of the vas deferens, whereas in females implantation in the uterus may be affected. The loss of these functions cannot be completely tied to mucus accumulation since sweat too is abnormal in CF. The salt concentration of CF sweat is abnormally high, yet the sweat gland produces no mucus.

4. THE ROLE OF CFTR IN EPITHELIAL TRANSPORT PROCESSES

Two ion transporting processes of the airways need to be considered in this context; they are electrogenic sodium ion absorption and electrogenic chloride ion secretion (see Fig. 5.2). In the former sodium ions enter the apical face of epithelial cells moving down an electrochemical gradient. The ions pass through special channels (ENaCs) which are specifically blocked by the drug amiloride. The ions are then moved uphill across the basolateral membrane using the sodium pump, i.e., the Na-K ATPase. This transfer of charge creates the electrical gradient down which an accompanying anion can pass passively, the usual anion being chloride ions passing between the cells. Thus, effectively NaCl has been absorbed across the epithelium with the expenditure of metabolic energy, i.e., the hydrolysis of ATP by the pump.

Electrogenic secretion of chloride ions is similarly powered by the sodium pump, but this process is anion-led, with the accompanying cation passing passively down the electrical gradient created by the active transport of chloride. The chloride ions enter the cells across the basolateral membrane, against a gradient using the NaK2Cl cotransporter. The potential energy stored in the sodium gradient (i.e., there is little Na^+ inside cells, whereas the extracellular fluid is Na^+ rich) provides the driving force for the movement. The sodium ions entering in this way are removed from the cell by the sodium pump and the chloride ions are free to pass outwards across the apical membrane, if the electrochemical gradient is favourable and provided suitable chloride ion channels are present. It is here that CFTR has a crucial function, behaving as a chloride ion channel. Thus together with Na ions that move passively the process results in NaCl secretion. Not shown in Fig. 5.2 are potassium channels both in the apical or basolateral membrane. They are important too, allowing the reequilibration of K^+ that enters on the cotransporter and to compensate for the sodium pump, which has a stoichiometry of 3:2 for Na:K. Also some K ions may leave the cell across the apical membrane along with chloride, moving passively by cation drag.

Briefly, two electrogenic ion transporting processes have been described, one for NaCl absorption and one for NaCl secretion. ENaCs are essential for the first and CFTR chloride channels for the second. One further process must be mentioned, that is the secretion of bicarbonate ions. There are two ways in which bicarbonate secretion may occur. HCO_3 ions may pass out through the apical membrane through CFTR or alternatively movement is out of the cell using a Cl^-/HCO_3^- exchanger, the chloride having exited via CFTR. How does HCO_3 get into the cell from the basolateral side? The most common way is entrance using a $NaHCO_3$ transporter, but some bicarbonate will be formed in the cell by the hydration of CO_2.

In CF two radical changes are present. First there is no CFTR and secondly the activity of ENaC is upregulated, probably by an increase in the open state probability of the channels (Stutts et al., 1997). This is indicated diagrammatically in the lower half of Fig. 5.2. The consequences are that chloride secretion is abolished and sodium absorption is enhanced, the driving force for sodium absorption being enhanced both by increased ENaC activity and the loss of CFTR conductance (Knowles et al., 1981). The epithelium lining the airways has a high hydraulic conductivity or, in other words, the water permeability is high. Water moves across tissues by osmosis and consequently will follow the movement of salt. The consequence of altered ion transporting activities is likely to affect the volume and/or composition of ASL. First, however, it is important to consider the location of CFTR and of ENaC and this is dealt with in the next section.

5. IS AIRWAY INFECTION IN CF DUE TO LACK OF CFTR?

5.1 Effects on Airway Surface Liquid (ASL)

The answer to the question posed in this section is almost certainly yes, but the mechanisms involved are less clear, there being a variety of explanations, each having its own champions. The importance of airway surface

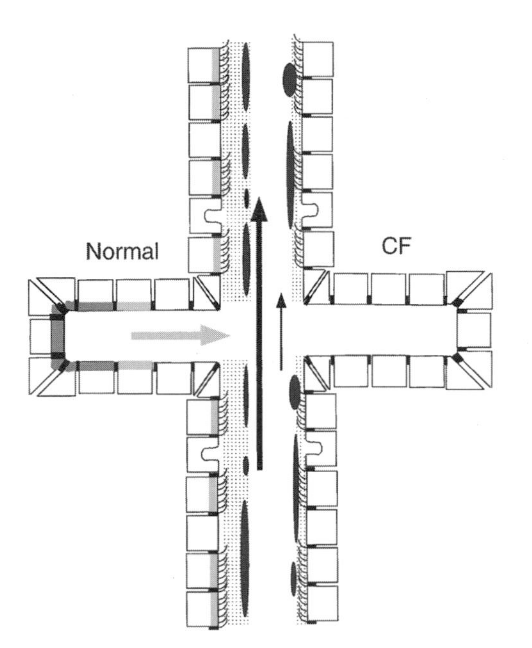

Figure 5.3. Diagram illustrating the mechanisms involved in mucociliary clearance. The normal situation is shown on the left, whereas the right indicates the changes in CF. Airway surface liquid (ASL) consisting of sol and gel phases is moved axially by ciliary action in the normal condition. Secretion from the submucous glands enters the flow. CFTR is shown in grey and is found in highest amounts at the base of the glands. Secretory and absorptive processes throughout the respiratory tree maintain efficient mucociliary clearance. In CF gland secretions are reduced and excessive fluid absorption reduces the depth of the ASL and impairs mucociliary clearance.

liquid (ASL) and how this is affected by the compromised transporting capability of airway epithelia forms the focus of the discussion.

Figure 5.3 is a diagrammatic representation of airway epithelium and indicates the amount and disposition of CFTR in both the surface epithelium and the submucosal glands. One half of the diagram represents the normal situation, whereas the other shows that existing in CF. RNA and immunological methods have revealed a low level of CFTR in the ciliated cells of surface epithelium in the proximal airways (Trapnell et al., 1991). The serous cells at the base of the submucosal glands contain higher levels of CFTR (Engelhardt et al., 1992), whereas in the distal airways a small number of nonciliated surface cells express a high level of CFTR.

A great deal of experimental evidence has been obtained from the study of the surface cells of proximal airways, either cultured as monolayers or alternatively using small areas of resected tissues. It appears that the predominant ion transporting activity in these epithelia is electrogenic sodium absorption (Willumsen and Boucher, 1991). Using microelectrodes to determine the potentials across the apical and basolateral membranes and the electrochemical activities of ions inside the cells has revealed that intracellular chloride is distributed close to equilibrium (Willumsen et al., 1989). Limited *in vivo* measurements confirm these findings. A detailed account of the electrophysiological findings is not appropriate here and readers requiring further information should consult the excellent review by Pilewski and Frizzell (1999). However, the results are of great significance for our discussion. They are, first, that electrogenic sodium absorption is the major transporting activity of the surface ciliated epithelium in the proximal airways. Second, as chloride is distributed close to its equilibrium position, there is little or no driving force for chloride secretion, although there may be some bicarbonate secretion in this site. Finally, in the absence of CFTR, sodium absorption is increased because of lack of regulation of ENaC. As argued elsewhere the major function of CFTR in the surface epithelium of the airways is likely the regulation of sodium absorption (Cuthbert, 1999).

The transporting activity of the submucosal glands is different from that found in the surface epithelium. Technically this activity is more difficult to study and most of the information has been derived from investigations of monolayers grown from a lung adenocarcinoma line, called Calu-3. These cells express many markers present in the serous cells of the submucosal glands. The apical membranes of these cells show a CFTR-dependent increase in chloride conductance. It seems clear that these cells are capable of electrogenic anion transport, but the major anion contained in the secretion is HCO_3^- (Devor et al., 1999). As yet there is no clear evidence to show whether CFTR acts as a bicarbonate ion conductance, whether CFTR in parallel with a Cl^-/HCO_3^- exchanger is responsible, or if CFTR can act as an exchanger under some conditions. It is proposed that the secretion of the serous cells from the base of the glands washes the glycoconjugates secreted from the mucus-producing cells onto the surface. Furthermore, from the localisation of CFTR and ENaCs in the glands (Burch et al., 1995) it is possible that the primary secretion is modified in its passage to the surface by sodium absorption to produce a hypotonic secretion.

I have attempted to depict the proposed mechanism on the left-hand side of Fig. 5.3. Thus secretions from the glands contribute to both the sol (aqueous) and gel phases of ASL, both layers moving together, axially along the airways driven by the actions of the cilia to generate mucociliary clearance. In CF the situation is quite different in that fluid production is curtailed yet fluid absorption, driven by sodium transport, is enhanced. This situation is depicted in the right-hand half of Fig. 5.3. The proposed reduction in the thickness of the ASL layer brings the gel layer in contact with the cilia reducing their effectiveness and the failure of mucociliary clearance. This hypothesis is the so called 'volume hypothesis' (Kilburn, 1968; Boucher, 1994) for mucociliary clearance, and its failure in CF and the subsequent consequences for infection and lung function.

A totally different hypothesis, called the 'concentration hypothesis' (Smith et al., 1996), predicts that it is the concentration of salts in the ASL that changes and is responsible for respiratory failure in CF. Evidence for the concentration hypothesis was obtained as follows. Airway epithelia from both normal and CF airways were grown on permeable supports and exposed to various strains of *P. aeruginosa*. The bacteria were rapidly killed by normal epithelia, but not those derived from CF tissues. Transfection of CF airway monolayers with an adenovirus vector expressing CFTR restored the ability of the monolayers to destroy bacteria. Defensin-like factors are normally present in ASL and are presumably involved in keeping airway surfaces free of aspirated bacteria. CF airways apparently generate defensins in the same manner as do non-CF epithelia so a straightforward explanation for the difference between normal and CF epithelia cannot be simply due to the absence of defensins. Bactericidal activity in ASL is inhibited by increased salt concentration and it was found that dilution of ASL of CF epithelia restored their antimicrobial capability. The concentration hypothesis predicts therefore that ASL in CF has a raised NaCl concentration which inhibits the effects of the defensins. It is not easy to see how raised NaCl concentration could be achieved in ASL in a simple epithelial model with high water permeability. One suggestion made was that normally the passive flow of Cl anions accompanying electrogenic Na absorption crossed the epithelium transcellularly and intercellularly, the former route requiring CFTR. In the absorptive duct of the sweat gland chloride ions accompany the actively transported sodium ions by a transcellular route, explaining why the salt concentration in sweat is elevated in CF. However, the absorptive part of the sweat duct is water impermeable.

Clearly to answer the conundrum of the volume versus the concentration model of ASL in CF depends on measuring the salt composition of the ASL. This seemingly simple measurement is, unfortunately, exceedingly difficult to measure in practice. ASL has been estimated to have a thickness of between 10 and 20 μm (Van As, 1977), which is a depth greater than the length of the cilia of 6 μm (Serafini and Michaelson, 1977). Thus ciliary action takes place in the less viscous sol phase, whereas in CF, if the volume hypothesis is correct, ciliary action will be impaired by entanglement with the gel layer. The ASL is the first contact point for inhaled pathogens and particulates and therefore ciliary clearance is a primary defence mechanism against infection. Sampling ASL by the filter paper or porous membrane technique, or by the use of glass pipettes, poses severe technical problems, caused by evaporative losses and possible stimulation of fluid secretion by the collecting probe. Experimental data from these approaches can be found in the literature supporting both the volume and concentration hypotheses.

There is also the possibility that the ASL may vary in ion composition in different locations. In the submucosal glands ENaCs are found in the duct (Burch et al., 1995) and if this part of the duct is relatively water impermeable electrogenic sodium absorption will leave a ductal secretion hypotonic to plasma in rather the same way as for the absorptive part of the sweat duct. If CFTR is required for counterion transport, as in the sweat gland, then dilution of the primary secretion would not occur in CF, favouring the inhibition of defensin activity as a cause of bacterial infection. However, no evidence for such a mechanism is known at present. Much evidence points to the submucosal glands as crucial in the genesis of CF and doubtless these will be intensively investigated in the future.

Filter paper sampling of the lower airways indicated the ASL was hypotonic (Knowles et al., 1997), the tonicity falling the longer the collection period and this was interpreted to be the result of drawing fluid from the submucosal glands. Similar results were obtained in normal and CF subjects, which could mean that the salt absorption in the duct was not CFTR dependent or that the filter paper technique did not accurately report ASL composition for the ducts. Clearly improved methods for measuring ASL properties are required. Recently a noninvasive method using confocal microscopy and ratio-imaging microscopy, together with fluorescent indicators, has been used to measure thickness, pH, and salt concentration of ASL in cultured monolayers of airway epithelial cells, isolated human bronchi, and mouse trachea through a small transparent window. In mice the results indicate that ASL is nearly isotonic, with no difference between normal and CF mice (Jayaraman et al., 2001). However CF mice do not show the same airway pathophysiology as in humans because of the presence of alternative chloride channels to CFTR (Anderson and Welsh, 1991). These optical techniques have been used, together with fluorescence recovery after photobleaching, to measure viscosity to study secretions issuing from single human tracheobronchial glands, both normal and CF (Joo et al., 2001). Secretion was essentially the same in normal and CF secretions, in terms of thickness, pH, ion composition, and rate of secretion; however, viscosity was elevated in CF. Whether the primary secretion is different in normal and CF glands, but modifications during passage to the surface remove these differences remains to be seen.

In summary, in CF the lack of CFTR increases the susceptibilty to airway infection, particularly by *P. aeruginosa*, possibly by affecting the volume or composition of ASL, resulting in reduced mucociliary clearance, leading eventually to bronchiectasis and the destruction of lung tissue. However, other possible mechanisms may be more causal or are additional to the effects on ASL, namely those connected with the failure of intracellular events which are CFTR-dependent or processes related to other regulatory effects of CFTR. These are considered in the next sections.

5.2 Protein Processing, Bacterial Adherence, and CF

Changes in protein processing are found in CF. These include reduced protein sialylation and increased fucosylation and sulphation. The latter two processes are competitive with sialylation, the enzymes having neutral pH optima, while sialylation requires mildly acid conditions (Paulson et al., 1978; Green et al., 1984). For example, mucus sulphation is increased in CF with a consequent increase in mucus viscosity (Mian et al., 1982). Reduced mucociliary clearance, as discussed earlier, increases the bacterial load in CF, but this alone cannot explain the relationship between *P. aeruginosa* and CF. In primary ciliary dyskinesia mucociliary clearance is reduced but without attracting colonisation with *P. aeruginosa* (Levison et al., 1983). Asialogangliosides have been claimed as putative binding sites for *P. aeruginosa* and other bacterial pathogens (Krivan et al., 1988; Imundo et al., 1995; Bryan et al., 1998).

The changes in protein processing have been attributed to CFTR-dependent changes in organellar pH. Unfortunately no consistent pattern emerges from these studies, and claims for both hypo- and hyperacidification have been made. One study (Barasch et al., 1991) used the uptake of a weak base to estimate pH and showed that the Golgi, trans-Golgi network (TGN), prelysosomes, and endosomes were less acidic in CF cells compared to non-CF from nasal polyps. It was proposed that the action of H^+-ATPase in organellar membranes was limited by the absence of a chloride conductance, due to CFTR, so preventing the movement of the counterion into the organelles (Fig. 5.4). In a second study CF and corrected CF human bronchial epithelial cells were used (Poschet et al., 2001) and the pH measured using a pH-sensitive fluorescent green protein that localised to the TGN. This study showed the TGN was hyperacidified. It was proposed that the absence of CFTR relieved the proton pump from membrane potential buildup by the

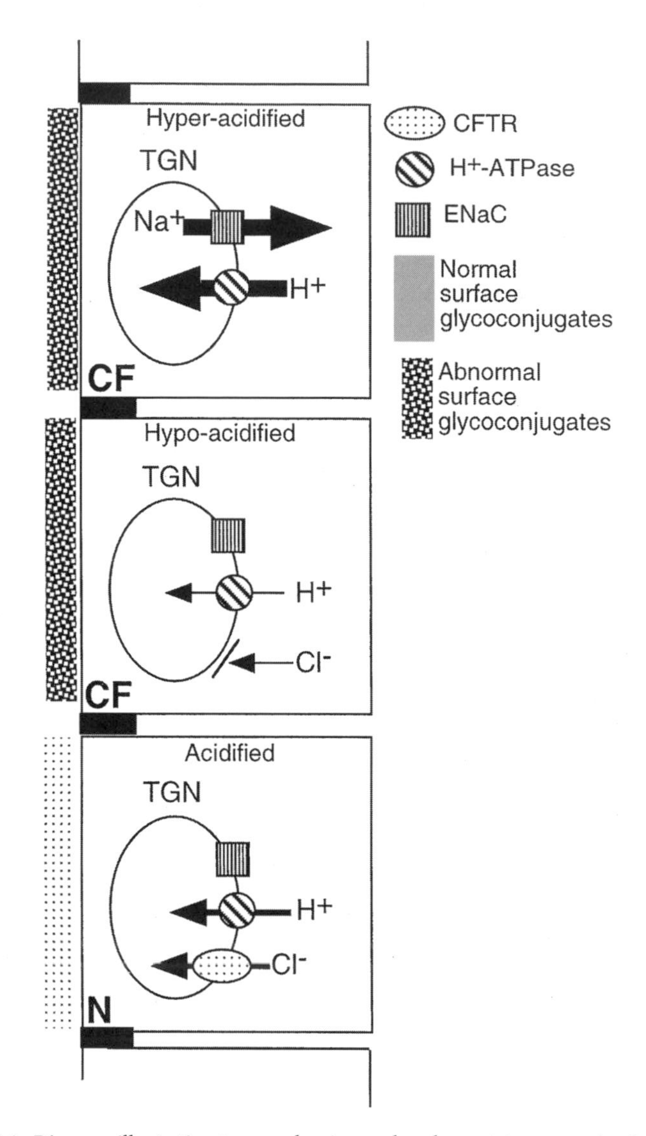

Figure 5.4. Diagram illustrating two mechanisms whereby protein processing in the trans-Golgi network is modified in CF. Proton pumping into the TGN is promoted by a chloride conductance (CFTR) allowing the movement of counterions (bottom). In the absence of CFTR proton pumping is restricted leading to hypoacidification (middle) or is amplified by a deregulated sodium conductance, leading to hyperacidification (top). In both instances the alteration in pH produces abnormal surface glycoconjugate.

efflux of Na^+ through an up-regulated sodium conductance (Fig. 5.4). With both of the foregoing hypotheses the alteration in pH removes sialylating enzymes away from their pH optima.

5.3 Other Regulatory Functions of CFTR and Bacterial Infection

The effect of CFTR in regulating ENaC was discussed in an earlier section. Up-regulation of ENaC is responsible for increased osmotic water flow along with the hyperabsorption of NaCl as a major cause of reduction in ASL thickness and the basis of the 'volume hypothesis'. As mentioned in the introduction CFTR has been shown to regulate a number of other epithelial transporting activities, and in particular the secretion of bicarbonate. For example the alkaline secretion of the human duodenum is abolished in CF (Pratha et al., 2000).

Recently there have been some important developments in the area of bicarbonate secretion that may be relevant to lung infection in CF. In a series of studies Muallem and his colleagues (Lee et al., 1999a, b; Choi et al., 2001) showed that forskolin stimulated luminal Cl^-/HCO_3^- exchange activity in wild-type but not CF epithelia, that the presence of CFTR but not its conductive function was necessary for Cl^-/HCO_3^- exchange activity, and that different CFTR mutations accurately predicted whether the phenotype was of the pancreatic sufficient or pancreatic insufficient kind. This last point is of considerable significance and implies that if a particular CF mutation supports a chloride conductance it may not support Cl^-/HCO_3^- exchange. How are these new data to be interpreted? It may mean that that luminal CFTR, not necessarily wild-type, is necessary for Cl^-/HCO_3^- exchange function or that CFTR is able to operate as a Cl^-/HCO_3^- exchanger in some situations. In a complementary study another group showed in tracheal epithelial cells expressing functional CFTR showed upregulation of mRNA for DRA ('down regulated in adenoma' protein) which itself acts as a Cl^-/HCO_3^- exchanger (Wheat et al., 2000). These processes may be of the utmost importance if they also apply to the submucosal ducts, where inspissation of mucus occurs first in CF. This taken together with the beneficial effects of an alkaline HCO_3 secretion on mucus viscosity (Bansil et al., 1995; Bhaskar et al., 1991; Veerman et al., 1989) and the presence of bacterial-binding sites in mucin (Bryan et al., 1998; Imundo et al., 1995; Krivan et al., 1988; Schenkels et al., 1993; Veerman et al. 1997; Wanke et al., 1990) makes the submucosal duct a point where defence against entrained bacteria may be severely weakened.

6. CFTR AND THE MALARIAL PARASITE

CFTR is found at very low levels in human erythrocytes and has been connected with the deformation-induced release of ATP (Sprague et al., 1998). Channels were also found in mouse erythrocytes with the electrical characteristics of CFTR and hypertonicity induced an additional chloride conductance with strongly inwardly rectifying characteristics that was absent in murine $CF^{-/-}$ erythrocytes. This CFTR-dependent additional conductance was associated with increased ATP release (Verloo et al., 2001). Others (Desai et al., 2000) have described a voltage-dependent channel for nutrient uptake in erythrocytes infected with the malarial parasite – the so-called new permeation pathway (NPP) that is apparently absent in CF erythrocytes (de Jonge HR, personal communication).

Whether the NPP (nutrient uptake) and the CFTR dependent IR pathways (ATP release) are the same and whether CF prevents malarial infection are possibilities on which further research is needed.

7. HETEROZYGOTE ADVANTAGE AGAINST INFECTIOUS DISEASE

The prevalence and persistence of the CF mutation in populations has led to suggestions that there is an heterozygote advantage in carriers. Among Caucasian populations 1 person in 25 has one copy of a mutated CF gene. Among the various advantages that might follow from this situation is the suggestion that heterozygotes have a reduced susceptibility to a number of infectious diseases. The diseases cited involve those in which secretory diarrhoea is the hallmark. These diseases include secretory diarrhoeas due to *Vibrio chlolerae, Escherichia coli*, and *Salmonella typhi*. The loss of fluid in these conditions is severe and the annual death toll, largely in populations where medical care is minimal, reaches many millions. The secretory response in both the intestine and the large bowel is dependent on electrogenic chloride secretion (see Fig. 5.2) in which CFTR provides the exit route for chloride ions from the epithelia. Gut epithelia from persons with CF or homozygous CF mice fail to show chloride secretory responses in response to a variety of agonists which increase cAMP or cGMP in epithelial cells (Berschneider et al., 1988; Cuthbert et al., 1994). The toxins produced by *V. cholerae* and *E. coli* activate adenylate cyclase and guanylate cyclase respectively.

The basis of the heterozygote hypothesis is straightforward. If CFTR is the rate-limiting step in the electrogenic chloride secretion and heterozygotes have only 50% of the CFTR present in homozygotes then secretion will be limited and the effect of the disease less severe (Baxter et al., 1988). As a corollary it is anticipated that persons with CF would not be susceptible to diseases involving secretory diarrhoea.

The availabilty of the CF mouse allowed for the direct testing of the heterozygote advantage hypothesis. The maximal chloride secretory responses to a variety of chloride secretagogues was compared, *in vitro*, in colonic epithelia from wild-type and littermate CF heterozygotes (Cuthbert et al., 1995). In these experiments the mice carried a null mutation, so the CF allele produced no protein. This may differ from the human condition where some mutations produce protein but with much reduced transport capability. Thus CF carriers might be expected to produce reduced responses to chloride secretagogues, but not necessarily reduced by 50%. In the event no differences in the maximal chloride secretory responses were recorded in heterozygote mice in response to cholera toxin, heat-stable enterotoxin (Sta), forskolin, vasoactive intestinal polypeptide, isoprenaline, (these latter three activating adenylate cyclase), guanylin (activates guanylate cyclase), lysyl bradykinin (causes prostaglandin release that activates adenylate cyclase) and carbachol when compared to wild-type controls. Furthermore heterozygote murine colons maintained chloride secretion at normal levels in response to forskolin during a period of 6 hours (Manson et al., 1997). In a further unpublished study, by the author, the chloride secretory current in murine neonate colons in response to forskolin was measured. All mice weighed around 2.5 g and the acute responses to forskolin were 112 ± 16 μA cm^{-2} in wild type and 57 ± 17 μA cm^{-2} in heterozygotes quite different to the result in adult mice.

In another study in which cholera toxin was administered to mice fluid accumulation in the intestine was significantly greater than in heterozygote littermates when collected after 6 hours (Gabriel et al., 1994). Similarly in murine intestinal epithelia studied *in vitro* the total chloride secretory response to cholera toxin in heterozygotes measured over 6 hours was circa 50% of that of wild-type littermates. However the difference in the secretory response appeared only after some 2.5 hours of incubation with cholera toxin, possibly reflecting a reduced ability of heterozygotes to recruit more endosomal CFTR. Alternatively, it has been suggested that the reduced sialylation of GM1 cholera toxin receptors is CFTR-dependent (Fishman et al., 1976), reducing the effectiveness in heterozygotes.

Quite a different role for CFTR was proposed in studies using *S. typhi*. Human epithelial cells, either wild type or expressing the ΔF508 CFTR mutation, were exposed to the organism and the uptake of bacteria was measured. Significantly more were taken up if CFTR was present in the membrane and it was proposed that CFTR acted as a receptor for the uptake process. Monoclonal antibodies directed against the first extracellular domain of CFTR effectively blocked internalisation. The translocation of *S. typhi* into the submucosal intestinal space was further studied in wild-type and CF heterozygote mice carrying the ΔF508 mutation and the latter showed a significant reduction in transfer (Pier et al., 1998). Although florid diarrhoea appears only late in the sequence of symptoms in typhoid fever the proposal from this work involves CFTR not as an effector of chloride secretion but rather as a receptor which allows the organism to gain systemic access, which is reduced in heterozygotes. Incidentally, CFTR has been proposed as a receptor for *P. aeruginosa* in the airways, where its internalisation into the cell leads to its destruction. This is the opposite of the argument made for *S. typhi*, which survives internalisation (Pier, 2000). The heterozygote advantage hypothesis has been investigated also in human CF carriers (Hogenauer et al., 2000). Jejunal secretion was measured using a nonabsorbable marker for two periods of 2 hours in each subject before and after the administration of a prostaglandin E analogue, misoprostol. The stimulated secretion rates of normal and heterozygous individuals did not differ and the increase in jejunal PD was similar in both groups, whereas CF patients had virtually no response.

In summary, although the hetrozygote advantage hypothesis is attractive and often quoted, the evidence to support it is still equivocal. CF mutations are not common in parts of the world where cholera is epidemic and the disease did not appear in Europe until 1832. Although the mouse model for the response to cholera toxin did show a reduction in secretion in heterozygotes, the reduced response did not appear immediately, but only after 2.5 hours. In humans where the prostaglandin agonist was used it was infused for 2 hours with no diminution in secretion found. In the mouse large gut no differences in the acute responses to a variety of agonists were found, even after using forskolin, a powerful stimulant of chloride secretion, for 6 hours. However, in the neonate colon acute responses to forskolin were reduced in heterozygotes. It would seem that if heterozygote advantage exists then the hypothesis in its simplest form is untenable. A more subtle formulation may require reappraisal of whether CFTR is the rate-limiting process in secretion, rates of removal and replenishment of CFTR at the apical surface, and the effects of bacterial toxins on other components of the transport process.

8. SUMMARY

In cystic fibrosis airway disease is the major cause of morbidity and mortality. CF is characterised by chronic airway infection, often by *P. aeruginosa*. The absence of the product of the CF gene, the cystic fibrosis transmembrane conductance regulator (CFTR), is related to the susceptibility to airway infection in a variety of ways. Lack of CFTR affects the thickness and/or the composition of airway surface liquid, resulting in a reduction of mucociliary clearance. Lack of CFTR and/or its regulatory functions affects chloride/bicarbonate exchange processes, resulting in lack of bicarbonate-rich secretion from submucosal glands, with secondary effects on mucus viscosity. Intracellular protein processing is altered in the absence of CFTR to produce secreted or membrane proteins that may act as binding sites for *P. aeruginosa*. Persons with CF are incapable of responding to organisms invading the gastrointestinal tract that normally cause secretory diarrhoea, and heterozygotes may also have a partial advantage.

REFERENCES

Anderson, M. P., Gregory, R. J., Thompson, S., et al. (1991). Demonstration that CFTR is a chloride channel by alteration of its anion selectivity. *Science* **253**, 202–205.

Anderson, M. P. and Welsh, M. J. (1991). Calcium and cAMP activate different chloride channels in the apical membrane of normal and cystic fibrosis epithelia. *Proc. Natl. Acad. Sci. USA* **88**, 6003–6007.

Bansil, R., Stanley, E. and LaMont, J. T. (1995). Mucin biophysics. *Annu. Rev. Physiol.* **57**, 635–657.

Barasch, J., Kiss, B., Prince, A., et al. (1991). Defective acidification of intracellular organelles in cystic fibrosis. *Nature* **352**, 70–74.

Baxter, P. S., Goldhill, J., Hardcastle, J., et al. (1988). Accounting for cystic fibrosis. *Lancet* **333**, 464–466.

Berschneider, H. M., Knowles, M. R., Azizkhan, R. G., et al. (1988). Altered intestinal chloride transport in cystic fibrosis. *FASEB J.* **2**, 2625–2629.

Bhaskar, K. R., Gong, D. H., R., et al. (1991). Profound increase in viscosity and aggregation of pig gastric mucin at low pH. *Am. J. Physiol.* **261**, G827–G832.

Boucher, R. C. (1994). Human airway ion transport. Part one. *Am. J. Respir. Crit. Care Dis.* **150**, 271–281.

Bryan, R., Kube, D., Perez, A., et al. (1998). Overproduction of the CFTR R domain leads to increased levels of asialo GM1 and increased *Pseudomonas aeruginosa* binding by epithelial cells. *Am. J. Respir. Cell Mol. Biol.* **19**, 269–277.

Burch, L. H., Talbot, C. R., Knowles, M. R., et al. (1995). Relative expression of the human epithelial Na^+ channel subunits in normal and cystic fibrosis airways. *Am. J. Physiol.* **269**, C511–C518.

CF Trust's Control of Infection Group. (2001). *Pseudomonas aeruginosa* infection in people with cystic fibrosis. *Cystic Fibrosis Trust*, 1–20.

Choi, J. Y., Muallem, D., Kiselyov, K., et al. (2001). Aberrant CFTR-dependent HCO_3^- transport in mutations associated with cystic fibrosis. *Nature* **410**, 94–97.

Cuthbert, A. W. (1999). Functional role of CFTR chloride channels in airway and gut epithelia. In: *Chloride Channels.* Kozlowski, R. (ed.). Oxford: Isis Medical Media Ltd., pp. 79–95.

Cuthbert, A. W., Halstead, J., Ratcliff, R., et al. (1995). The genetic advantage hypothesis in cystic fibrosis heterozygotes: a murine study. *J. Physiol.* **482**, 449–454.

Cuthbert, A. W., MacVinish, L. J., Hickman, M. E., et al. (1994). Ion-transporting activity in the murine colonic epithelium of normal animals and animals with cystic fibrosis. *Pflugers Arch.* **428**, 508–515.

Desai, S. A., Bezrukov, S. M. and Zimmerberg, J. (2000). A voltage-dependent channel involved in nutrient uptake by red blood cells infected with the malaria parasite. *Nature* **406**, 1001–1005.

Devor, D. C., Singh, A. K., Lambert, L. C., et al. (1999). Bicarbonate and chloride secretion in Calu-3 human airway epithelial cells. *J. Gen. Physiol.* **113**, 743–760.

Engelhardt, J. F., Yankaskas, J. R., Ernst, S. A., et al. (1992). Submucosal glands are the predominant site of CFTR expression in the human bronchus. *Nat. Genet.* **2**, 240–248.

Fishman, P. H., Moss, J. and Vaughan, M. (1976). Uptake and metabolism of gangliosides in transformed mouse fibroblasts. Relationship of ganglioside structure to choleragen response. *J. Biol. Chem.* **251**, 4490–4494.

FitzSimmons, S. (1996). The Cystic Fibrosis Foundation Patient Registry Report, 1996. *Pediatr. Pulmonol.* **21**, 267–275.

Gabriel, S. E., Brigman, K. N., Koller, B. H., et al. (1994). Cystic fibrosis heterozygote resistance to cholera toxin in the cystic fibrosis mouse model. *Science* **266**, 107–109.

Green, E. D., Gruenebaum, J., Bielinska, M., et al. (1984). Sulfation of lutropin oligosaccharides with a cell-free system. *Proc. Natl. Acad. Sci. USA* **81**, 5320–5324.

Henry, R. L., Mellis, C. M. and Petrovic, L. (1992). Mucoid *Pseudomonas aeruginosa* is a marker of poor survival in cystic fibrosis. *Pediatr. Pulmonol.* **12**, 158–161.

Hogenauer, C., Santa Ana, C. A., Porter, J. L., et al. (2000). Active intestinal chloride secretion in human carriers of cystic fibrosis mutations: An evaluation of the hypothesis that heterozygotes have subnormal active intestinal chloride secretion. *Am. J. Hum. Genet.* **67**, 1422–1427.

Hudson, V. L., Wielinski, C. L. and Regelmann, W. E. (1993). Prognostic implications of initial oropharyngeal bacterial flora in patients with cystic fibrosis diagnosed before the age of two years. *J. Pediatr.* **122**, 854–860.

Imundo, L., Barasch, J., Prince, A., et al. (1995). Cystic fibrosis epithelial cells have a receptor for pathogenic bacteria on their apical surface. *Proc. Natl. Acad. Sci. USA* **92**, 3019–3023.

Jayaraman, S., Song, Y., Vetrivel, L., et al. (2001). Non-invasive *in vivo* fluorescence measurement of airway-surface liquid depth, salt concentration, and pH. *J. Clin. Invest.* **107**, 317–324.

Joo, N. S., Krouse, M. E., Wu, J. V., et al. (2001). Transport in relation to mucus secretion from submucosal glands. *J. Pancreas (Online)* **2**(Suppl. 4), 280–284.

Kerem, B., Rommens, J. M., Buchanan, J. A., et al. (1989). Identification of the cystic fibrosis gene: genetic analysis. *Science* **245**, 1073–1080.

Kilburn, K. H. (1968). A hypothesis for pulmonary clearance and its implications. *Am. Rev. Respir. Dis.* **98**, 449–463.

Knowles, M. R., Gatzy, J. and Boucher, R. (1981). Increased bioelectric potential difference across respiratory epithelia in cystic fibrosis. *N. Engl. J.Med.* **305**, 1489–1495.

Knowles, M. R., Robinson, J. M., Wood, R. E., et al. (1997). Ion composition of airway surface liquid of patients with cystic fibrosis as compared with normal and disease-control subjects. *J. Clin. Invest.* **100**, 2588 2595.

Krivan, H. C., Ginsburg, V., and Roberts, D. D. (1988). *Pseudomonas aeruginosa* and *Pseudomonas cepacia* isolated from cystic fibrosis patients bind specifically to gangliotetraosylceramide (asialo GM1) and gangliotriaosylceramide (asialo GM2). *Arch. Biochem. Biophys.* **260**, 493–496.

Lee, M. G., Choi, J. Y., Luo, X., et al. (1999a). Cystic fibrosis conductance regulator regulates luminal Cl^-/HCO_3^- exchange in mouse submandibular and pancreatic ducts. *J. Biol. Chem.* **274**, 14670–14677.

Lee, M. G., Wigley, W. C., Zeng, W., et al. (1999b). Regulation of Cl^-/HCO_3^- exchange by cystic fibrosis transmembrane conductance regulator expressed in NIH 3T3 and HEK 293 cells. *J. Biol. Chem.* **274**, 3414–3421.

Levison, H., Mindorrf, C. M., Chao, J., et al. (1983). Pathophysiology of the ciliary motility syndromes. *Eur. J. Respir. Dis. Suppl.* **127**, 102–117.

Loussouarn, G., Demolombe, S., Mohammad-Panah, R., et al. (1996). Expression of CFTR controls cAMP dependent activation of epithelial K^+ currents. *Am. J. Physiol.* **271**, C1565–C1573.

Mall, M., Hipper, A., Greger, R., et al. (1996). Wild type but not ΔF508 CFTR inhibits Na^+ conductance when coexpressed in *Xenopus* oocytes. *FEBS Lett.* **381**, 47–52.

Manson, A. L., Trezise, A. E. O., MacVinish, L. J., et al. (1997). Complementation of null CF mice with a human CFTR YAC transgene. *EMBO J.* **16**, 4238–4249.

Mian, N., Pope, A. J., Anderson, C. E., et al. (1982). Factors influencing the viscous properties of chicken tracheal mucins. *Biochim. Biophys. Acta.* **717**, 41–48.

Ornoy, A., Arnon, J., Katznelson, D., et al. (1987). Pathological confirmation of cystic fibrosis in the fetus following prenatal diagnosis. *Am. J. Med. Genet.* **28**, 935–947.

Paulson, J. C., Prieels, J.-P., Glasgow, L. R., et al. (1978). Sialyl- and fructosyltransferases in the biosynthesis of asparaginyl-linked oligosaccharides in glycoproteins. Mutually exclusive glycosylation by beta-galactoside alpha2 go to 6 sialyltransferase and *N*-acetylglucosaminidine alpha1 goes to 3 fucosyltransferase. *J. Biol. Chem.* **253**, 5617–5624.

Pier, G. B. (2000). Role of the cystic fibrosis transmembrane conductance regulator in innate immunity to *Pseudomonas aeruginosa* infections. *Proc. Natl. Acad. Sci. USA* **97**, 8822–8828.

Pier, G. B., Grout, M., Zaidi, T., et al. (1998). *Salmonella typhi* uses CFTR to enter intestinal epithelial cells. *Nature* **393**, 79–82.

Pilewski, J. M. and Frizzell, R. A. (1999). Role of CFTR in airway disease. *Physiol. Rev.* **79**, S215–S255.

Poschet, J. F., Boucher, J. C., Tatterson, L., et al. (2001). Molecular basis for defective glycosylation and *Pseudomonas* pathogenesis in cystic fibrosis lung. *Proc. Natl. Acad. Sci. USA* **98**, 13972–13977.

Pratha, V. S., Hogan, D. L., Martensson, B. A., et al. (2000). Identification of transport abnormalities in duodenal mucosa and duodenal enterocytes from patients with cystic fibrosis. *Gastroenterology* **118**, 1051–1060.

Quinton, P. M. (1983). Chloride impermeability in cystic fibrosis. *Nature* **382**, 421–422.

Riordan, J. R., Rommens, J. M., Kerem, B., et al. (1989). Identification of the cystic fibrosis gene: cloning and characterization of complementary DNA. *Science* **245**, 1066–1073.

Rommens, J. M., Iannuzzi, M. C., Kerem, B., et al. (1989). Identification of the cystic fibrosis gene: chromosome walking and jumping. *Science* **245**, 1059–1065.

Schenkels, L. C., Ligtenberg, A. J., Veerman, E. C., et al. (1993). Interaction of the salivary glycoprotein EP-GP with the bacterium *Streptococcus salivarius* HB. *J. Dent. Res.* **72**, 1559–1565.

Schreiber, R., Greger, R., Nitschke, R., et al. (1997). Cystic fibrosis transmembrane conductance regulator activates water conductance in *Xenopus* oocytes. *Eur. J. Physiol.* **434**, 841–847.

Serafini, S. M. and Michaelson, E. D. (1977). Length and distribution of cilia in human and canine airways. *Bull. Eur. Physiopathol. Respir.* **13**, 551–559.

Shumaker, H. and Soleimani, M. (1999). CFTR upregulates the expression of the basolateral Na^{+}-K^{+}-$2Cl^{-}$ cotransporter in cultured pancreatic duct cells. *Am. J. Physiol.* **277**, C1100–C1110.

Smith, J. J., Travis, S. M., Greenberg, E. P., et al. (1996). Cystic fibrosis airway epithelia fail to kill bacteria because of abnormal airway surface fluid. *Cell* **85**, 229–236.

Sprague, R. S., Ellsworth, M. L., Stephenson, A. H., et al. (1998). Deformation-induced ATP release from red blood cells requires CFTR activity. *Am. J. Physiol.* **275**, H1726–H1732.

Stutts, M. J., Rossier, B. C. and Boucher, R. C. (1997). Cystic fibrosis transmembrane conductance regulator inverts protein kinase A-mediated regulation of epithelial sodium channel single channel kinetics. *J. Biol. Chem.* **272**, 14037–14040.

Trapnell, B. C., Chu, C. S., Paakko, P. K., et al. (1991). Expression of the cystic fibrosis transmembrane conductance regulator gene in the respiratory tract of normal individuals and individuals with cystic fibrosis. *Proc. Natl. Acad. Sci. USA* **88**, 6565–6569.

Van As, A. (1977). Pulmonary airway clearance mechanisms: a reappraisal. *Am. Rev. Respir. Dis.* **115**, 721–726.

Veerman, E. C., Bank, C. M., Namavar, F., et al. (1997). Sulfated glycans on oral mucin as receptors for *Helicobacter pylori. Glycobiology* **7**, 737–743.

Veerman, E. C., Valentijn-Benz, M. and Nieuw Amerongen, A. V. (1989). Viscosity of human salivary mucins: effect of pH and ionic strength and role of sialic acid. *J. Biol. Buccale* **17**, 297–306.

Verloo, P., Hogema, B., Tilly, B., et al. (2001). CFTR-dependent ATP release in human erythrocytes: Possible involvement of an inwardly rectifying chloride channel. *Ped. Pulmanol. Suppl.* **22**, 177.

Wanke, C. A., Cronan, S., Goss, C., et al. (1990). Characterization of binding of *Escherichia coli* strains which are enteropathogens to small bowel mucin. *Infect. Immun.* **58**, 794–800.

Wang, X., Moylan, B., Leopold, D. A., et al. (2000). Mutation in the gene responsible for cystic fibrosis and predisposition to chronic rhinosinusitis in the general population. *J. Am. Med. Assoc.* **284**, 1814–1819.

Wei, L., Vankeerberghen, A., Cuppens, H., et al. (1999). Interactions between calcium-activated chloride channels and the cystic fibrosis transmembrane conductance regulator. *Eur. J. Physiol.* **438**, 635–641.

Wheat, V. J., Shumaker, H., Burnham, C., et al. (2000). CFTR induces the expression of DRA along with Cl^-/HCO_3^- exchange activity in tracheal epithelial cells. *Am J. Physiol. Cell Physiol.* **279**, C62–C71.

Willumsen, N. J. and Boucher, R. C. (1991). Transcellular sodium transport in cultured cystic fibrosis human nasal epithelium. *Am. J. Physiol.* **261**, C332–C341.

Willumsen, N. J., Davis, C. W. and Boucher, R. C. (1989). Intracellular Cl^- activity and cellular Cl^- pathways in cultured human airway epithelium. *Am. J. Physiol.* **261**, C1033–C1044.

CHAPTER 6

The influence of inherited traits on malaria infection

David J. Roberts and Tyler Harris

Nuffield Department of Clinical Laboratory Sciences and National Blood Service-Oxford Centre, John Radcliffe Hospital, United Kingdom

Thomas Williams

Wellcome Trust-KEMRI Centre for Geographic Medicine, Kenya; Department of Paediatrics, Faculty of Medicine, Imperial College of Science Technology and Medicine, Exhibition Road United Kingdom

1. INTRODUCTION

The history of genetics and the study of malaria are inextricably linked. The burden of disease due to malaria across much of the world has selected for a series of very visible traits of major medical importance, including the alleles of genes encoding haemoglobin, red cell enzymes, and membrane proteins. Furthermore, it now appears that many other genes may also influence the outcome of infection, including some that modulate the immune responses and others that encode for endothelial proteins as might be expected from the intricate life cycle of the parasite in the human host (Fig. 6.1).

A short chapter cannot hope to be a comprehensive description of such a large field of scientific endeavour. The selection of evidence and the discussion of its significance are inevitably personal. Nevertheless, we will attempt to address some of the major questions that have been at the heart of studies of the genetic influence on malaria infection, including 'What is the overall impact of genetics on malaria infection?', 'What protective traits have been identified already?', 'How can we identify new protective traits?', and 'How do these traits result in protection?' The approaches we can use to study these questions continue to change with new genetic and molecular techniques that allow us to extract and analyse more genetic information, not least of which is the sequence of both the human and parasite genomes. However, we must not be satisfied with a simple catalogue of resistance traits to malaria. We must now strive to relate this information to understanding how protective traits actually modify malaria infection. By doing so, we stand to learn much more about the subtle interaction between parasite and host and how immunity develops in nature. As we shall see, the answers to this crucial question are themselves complex and often elusive.

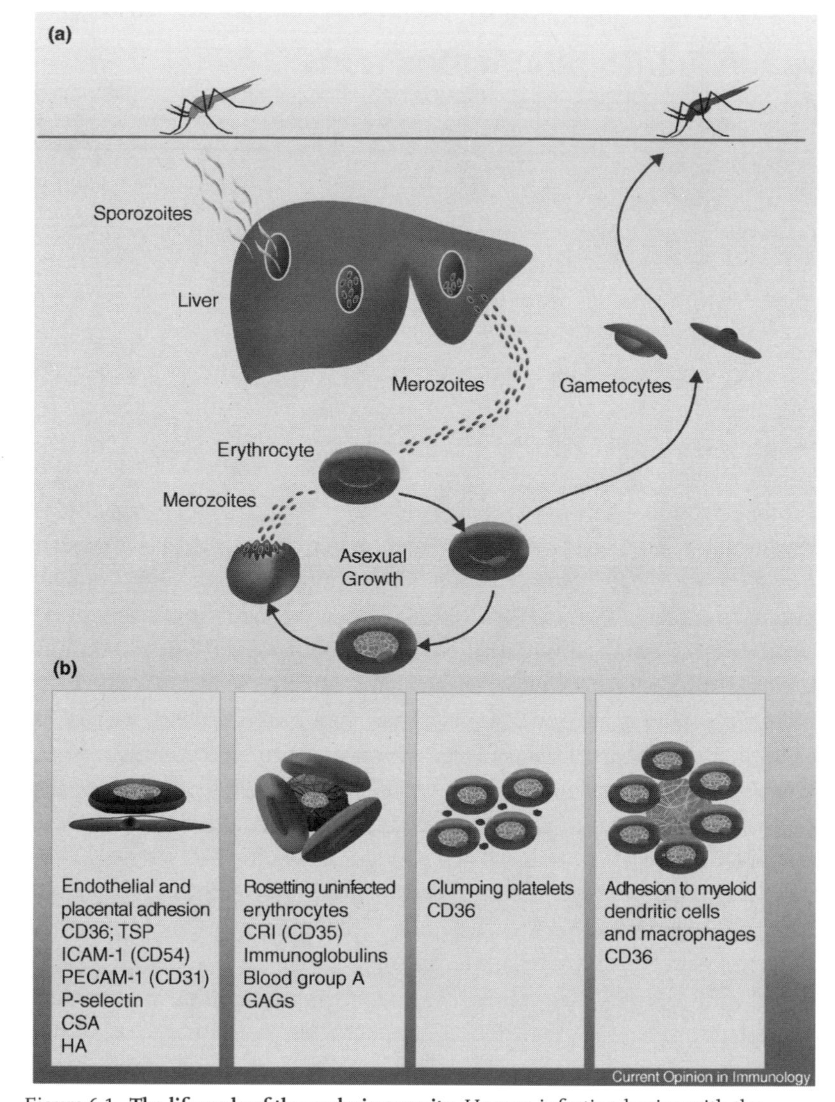

Figure 6.1. **The life cycle of the malaria parasite.** Human infection begins with the injection of sporozoites from female Anopheles mosquitoes taking a blood meal. Within hepatocytes, replication takes place within 10–14 days to generate approximately 10,000 merozoites. After rupture of the hepatocyte, the infective merozoites are released into the circulation and enter and multiply within erythrocytes. The parasites can multiply approximately eightfold every 2 days and so within a short time a high proportion of erythrocytes may be infected, comprising up to several grams of foreign antigens. Asexual parasites may differentiate into sexual forms or gametocytes. The parasite life cycle is

(*cont.*)

2. THE CONTRIBUTION OF GENETIC TRAITS TO RESISTANCE TO MALARIA

The first observations suggesting that infectious diseases are influenced by inherited factors are perhaps the striking species specificity for bacterial pathogens noted by Louis Pasteur, as evidence for the germ theory of disease. In reviewing the epidemiology of malaria in North America, Klebs and Tomassi-Crudeli commented on the apparent resistance of those of African descent to malaria who 'better than any other resist (malaria's) action' (Klebs and Tommasi-Crudeli, 1888). However, more than 100 years passed before A. E. Garrod stated a clear hypothesis that constitutional factors may underlie the striking individual variation in the outcome of infectious disease. Later, support for a contribution of genetic factors to malaria susceptibility has been garnered from a variety of sources. During the era of malaria therapy, when patients with neurosyphyilis were deliberately infected with malaria for treatment purposes, the marked variation in susceptibility observed in non-immune adults was accepted as general evidence for a genetic contribution. However, the first clues about some of the genes that might be contributing to malaria protection were provided by population genetic studies. As we will see in subsequent sections, a number of genes, including those for sickle haemoglobin (HbS), the thalassaemias, and a range of red cell membrane and enzyme abnormalities, share a distribution that is strikingly similar to that of malaria. It has now been confirmed that many of these traits provide substantial protection against the disease. With the list growing all the time, the overall importance of genetic factors has become increasingly obvious. However, other factors also affect the outcome of individual malaria infections, including previous exposure and immune status and determinants of parasite virulence.

Figure 6.1. (*cont.*) completed within the female Anopheles mosquito after the sexual forms of the parasite are ingested in a blood meal. During the second half of the erythrocytic cycle, PfEMP-1 or the variant antigen is expressed on the surface of infected erythrocytes on the surface. This family of proteins (possibly with other adhesive proteins) mediate adhesion of infected erythrocytes to endothelium on postcapillary venules. Infected erythrocytes may also form rosettes with uninfected erythrocytes mediated in different strains by CR1 (CD35), ABO blood group antigens, immunoglobulins, and glycosamaminoglycans. Some 'clumping' strains are able to form large aggregates of infected erythrocytes and platelets, requiring expression of CD36 on platelets. Finally adhesion of infected erythrocytes to leucocytes directly or indirectly via CD36 and CD51 (α_V integrin of the $\alpha_V\beta_3$ or $\alpha_V\beta_5$ heterodimer) may modulate host cell function. (See color plate.)

So what is the relative importance of each of these factors to the overall burden of malaria disease in human populations? Obviously, without a complete list of candidate genes it is hard to come up with a precise estimate. However, Mackinnon and colleagues (2000) have recently addressed this question through pedigree analysis of malaria data collected during longitudinal cohort studies in Sri Lanka (Mackinnon et al., 2000). Compared with many parts of Africa, current levels of malaria transmission and mortality in Sri Lanka are relatively low. Nevertheless, they were able to determine that human genetic factors, housing, and predisposing systemic effects, such as sex, age, or village, each explained about 15% of the variation in the frequency of malaria infection. The genetic component contributing to the severity of the clinical illness was slightly greater and here 20% of the variation in the intensity of disease was explained by repeatable differences between patients and approximately half of this variation was attributable to host genetics. A similar analysis of the total genetic contribution to malaria infection and disease is underway in Africa where the high rate of transmission and increased severity of disease may suggest a greater contribution of human genetic factors to the outcome of disease.

3. THE SEARCH FOR INHERITED TRAITS CONFERRING RESISTANCE TO MALARIA

The early search for malaria-protective genes took two main lines: a population genetic approach and a case control approach. Population genetic evidence is strongest for the polymorphisms of the red blood cells, including the haemoglobinopathies (such as HbS, HbC, and HbE and the thalassaemias), red cell membrane anomalies (including ovalocytosis and elliptocytosis), and enzyme defects (such as G6PD deficiency). This evidence will be discussed in detail in the appropriate sections; however, in general terms, such studies provide only an indication of which genes might be protective and ultimately require the corroboration of more direct evidence.

3.1 Association Studies

Association studies measuring gene frequencies in carefully ascertained cases of malaria and in ethnically matched controls have been the backbone of subsequent work aimed at defining protective traits. These simple methods have the advantage of statistical power, providing a quantitative estimate of protection from malaria by severity and syndrome for any demographic group. However, they do have several drawbacks. First, they may give false-positive or false-negative results if the control groups are not drawn from the same ethnic background as the cases. Gene frequencies are likely to differ even between quite closely related groups due to drift or other selective forces. Many association studies do not report the ethnic composition of cases and controls and their results are consequently of uncertain reliability. Matching for age and/or sex is of secondary importance as autosomal allele frequencies are unlikely to change with age or by sex. Using parental controls and analysing the transmission of alleles may circumvent some of these confounding factors with some loss of statistical power (for recent review see Schulze and McMahon, 2002). Second, case definitions have to be as precise as possible. Only well-defined cases should be included in such studies. In most malaria endemic areas this requires sufficient clinical and laboratory support to exclude the many other phenotypically similar but pathologically distinct diseases that may mimic the various manifestations of severe malaria. These include acute respiratory infection, bacterial or viral meningitis, and simple iron deficiency anaemia. Finally, such studies should be designed to have sufficient power to detect effects that are likely to be statistically significant to avoid false-negative results.

3.2 Studies Comparing Ethnic Groups

Within geographic regions with similar transmission of malaria, different ethnic groups have different malaria-specific morbidity. Careful comparative studies within West Africa suggest the Fulani are more resistant to malaria than the Mossi and Rimaibe ethnic groups living in the same area. These differences cannot be explained by known environmental or genetic factors and it seems possible that novel genetic traits have been selected in the Fulani groups (Modiano et al. 1996). A polymorphism in the IL-4 promoter (IL4-524 T) has been associated with increased levels of antimalarial antibodies (Luoni et al., 2001). However, this association has not been confirmed in other populations, nor has it been associated with the outcome of malaria infection.

3.3 Linkage Studies

Linkage studies identify the cosegregation genetic markers and the presence of disease in individuals within families. In general these methods have been very successful in identifying loci responsible for monogenic diseases. In malaria studies they are attractive because they avoid the problem of matching for ethnic group, which is crucial for association studies. However, it is difficult to assemble sufficient family members who have suffered from severe malaria. Instead segregation and linkage analyses have studied the development of mild malaria or examined intermediate disease phenotypes such as parasite density.

Groups working in Burkina Faso have demonstrated linkage of high parasite densities to chromosome 5q31-q33 (Rihet et al., 1999). This region does contain several promising candidate genes for the control of blood stage parasitaemia, including IL-4, IL-12, IL-3, and granulocyte-macrophage colony stimulating factor and their receptors (CSF1, CSF2, and CSF1R). Furthermore the syntenic regions in the mouse do influence immunity to parasites (Riopel et al., 2001).

Other studies have measured parasite density in pedigrees in Cameroon. A segregation analysis of blood levels of parasites in 42 families was consistent with the effect of a common recessive gene, rendering about 20% of the population susceptible to high levels of infection (Abel et al., 1992). However, the gene(s) responsible for susceptibility to malaria in this population have not been identified.

In other polygenic diseases, where there is strong clustering of disease within families, a genomewide scan of anonymous markers in affected siblings has identified several candidate loci for disease susceptibility. A recent twin study of more than 200 pairs of dizygous Gambian twins monitored the incidence of mild malaria and showed that where both twins developed mild malaria, sharing two MHC haplotypes was more common than would be expected by chance alone (Jepson et al., 1997). No other susceptibility loci were identified in this study. The power of similar studies would be increased if siblings who had suffered from severe disease could be recruited. Recruitment of several hundred affected sibling pairs would require linked medical and family records from several hundred thousand people. This is beyond the scope of present epidemiological surveys.

An alternative approach would be to use a series of association studies in well matched case control studies to determine the role of some of the 1,000,000 polymorphisms now identified in the 30,000 genes of the human genome (Todd, 1999). Beyond the technical and organisational problems of such a major exercise, the outstanding statistical problems are considerable. It would be a great advantage if results could be combined from different studies. However, a formal meta-analysis of genetic epidemiology studies may be difficult in the light of the heterogeneous clinical epidemiology and thus one assumes heterogeneous underlying pathophysiology in different geographical areas and under different transmission conditions. Nevertheless, such large-scale association studies are the most likely way forward in the next 10 years.

4. PROTECTIVE TRAITS

The life cycle of malaria parasites in their human host involves many tissues, specific host–parasite interactions, and several effector mechanisms for removal of the parasite by the immune system (Figs. 6.1 and 6.2). The number of genetic traits that could conceivably modulate the outcome of malaria infection is correspondingly large. While the role of many such candidate genes expressed in the immune system or influencing specific receptor ligand interactions have been explored the most compelling epidemiological evidence for protection from malaria is for the haemoglobinopathies and deficiency of glucose-6-phosphate dehydrogenase (G6PDH).

4.1 Haemoglobinopathies

It was first suggested over half a century ago that the haemoglobinopathies, abnormalities of haemoglobin structure or synthesis, might protect against *Plasmodium falciparum* malaria. First Beet and later Allison suggested that the peculiar distribution of sickle cell trait in African populations might result from a selective advantage against malaria (Beet, 1946; Allison, 1954). However, the 'malaria hypothesis' is generally ascribed to Haldane (1949) who, reflecting on the evolving descriptions of the global distribution of β-thalassaemia (Cooley's anaemia), wrote these landmark words:

> Neel and Valentine believe that the heterozygote is less fit than normal, and think that the mutation rate is above 4×10^{-4} rather than below it. I believe that the possibility that the heterozygote is fitter than normal must be seriously considered. Such increased fitness is found in the case of several lethal and sublethal genes in *Drosophila* and *Zea*. A possible mechanism is as follows. The corpuscles of the anaemic heterozygotes are smaller than normal, and more resistant to hypotonic solutions. It is at least conceivable that they are also more resistant to attacks by the sporozoa which cause malaria, a disease prevalent in Italy, Sicily and Greece, where the gene is frequent.

Since Haldane's first statement of his 'malaria hypothesis', in such clear terms, evidence that many genes provide protection from *P. falciparum* malaria has accumulated relentlessly.

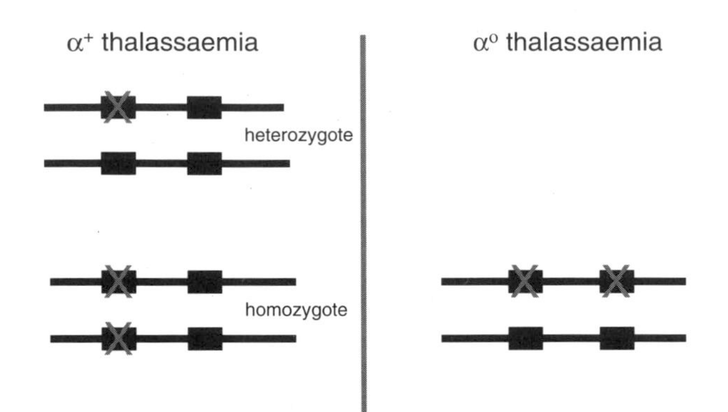

Figure 6.2. **The genetics of thalassaemia.** Normal individuals have four α-globin genes, two per haploid genome ($\alpha\alpha/\alpha\alpha$). These genes, the $\alpha 1$ and the highly homologous $\alpha 2$ genes, are arranged in pairs within the α-globin gene cluster on chromosome 16. β-Globin, on the other hand, is encoded by single copy genes on chromosome 11. The genetic defects underlying the thalassaemias are extremely diverse and are reviewed in Higgs et al. (1984) and Weatherall and Clegg (2002). The α-thalassaemias can be generally classified into two main groups: the α^+-thalassaemias, which affect only one of the paired α-globin genes, and the α^0-thalassaemias, which affect both genes. Thus, homozygotes for α^+-thalassaemia have only two functional α-globin genes ($-\alpha/-\alpha$), which in general terms is clinically indistinguishable from heterozygous α^0-thalassaemia ($-/\alpha\alpha$), both resulting in a mild form of microcytic anaemia (Williams et al., 1996). Most forms of both α^+- and α^0-thalassaemia originated from deletions of genetic material from within the α-globin gene cluster and arose through unequal meiotic crossover events. In general, individuals with two or more functional genes are clinically normal, whereas those with only one functional gene (compound heterozygotes for α^0- and α^+-thalassaemia) have haemoglobin H disease, which is usually associated with clinically significant anaemia. Homozygous α^0-thalassaemia ($-/-$) results in hydrops fetalis and is incompatible with survival. The β-thalassaemias have arisen through mutations that affect β-globin production through a variety of mechanisms causing either reduced (β^+) or no (β^0) globin production. Unlike α-thalassaemia the abnormal gene sequences are point mutations rather than gene deletions. They may affect the promoter or enhancer regions, the coding, or intronic sequence, the latter of which affects mRNA splicing. Many different mutations cause β-thalassaemia in any area but locally particular mutations predominate, for example, β^0 39 and β^+ IVS 1-100 in the western and eastern Mediterranean, respectively. Heterozygous β-thalassaemia is clinically silent, whereas homozygous β-thalassaemia is generally associated with a chronic, transfusion-dependent anaemia and premature death.

4.1.1 Sickle Cell Trait

The most celebrated example of a malaria protective gene is that of sickle haemoglobin. Haemoglobin S (HbS) is a variant form of haemoglobin composed of two normal α-globin molecules in association with two abnormal β-globin molecules (βs) ($\alpha_2\beta s_2$). Production of abnormal βs globin results from a point mutation of the β globin gene such that the codon determining the amino acid at position β^6 is changed from GAG (coding for glutamic acid) to GTG (coding for valine). Homozygotes for the βs mutation suffer from sickle cell disease, a debilitating form of anaemia and organ damage following infarction associated with premature death in most developing countries. Nevertheless, on the basis of surveys conducted during the 1940s and 1950s it became increasingly obvious that the carrier state (sickle cell trait; HbAS) was extremely common in much of sub-Saharan Africa. Allison provided support for this hypothesis of a *balanced polymorphism* by showing that heterozygotes were less likely to have parasites in their blood and were less likely to die from severe malaria than normals (Allison, 1954). Case-control studies have since confirmed that sickle cell trait is 90% protective against severe and complicated malaria (cerebral malaria and severe anaemia) and 60% protective against clinical malaria leading to hospital admission (Willcox et al., 1983; Hill et al., 1991; Marsh, 1992). A recent cohort study has showed sickle trait confers 60% protection against mortality between 2 and 16 months of age in an area of high transmission (Aidoo et al., 2002).

In spite of this dramatic protective effect, clinical epidemiology has yielded few clues regarding the mechanisms involved. Allison's early report (1954) showed a reduced parasite prevalence in heterozygotes (HbAS) (Allison, 1954). However, subsequent surveys have not produced consistent evidence for protection against asymptomatic parasitaemia. Although there is some evidence for protection against mild clinical malaria (Fleming et al., 1979; Marsh et al., 1989; Jakobsen et al., 1991), this is generally manifest as reduced parasite densities rather than protection against parasitaemia *per se*.

4.1.2 The Thalassaemia Syndromes

The thalassaemia syndromes are disorders of haemoglobin production. They fall into two main groups, the α- and β-thalassaemias, characterised by underproduction of α- and β-globin respectively (Fig. 6.2).

4.1.2.1 Population Genetics of the Thalassaemias

The initial evidence supporting malaria protection by both forms of thalassaemia was derived from population genetic studies. Surveys conducted throughout the world have shown that, as a group, the thalassaemias are the commonest single gene disorders so far described (summarised by Weatherall and Clegg, 2002). Overall, gene frequencies of >0.1 are the norm in tropical populations, whereas frequencies are somewhat lower in the subtropics and rare in the temperate zones. In fact, the only tropical populations in which the thalassaemias have not been found at polymorphic frequencies are in the New World, where malaria has only been introduced during the past few hundred years (Dunn, 1965). Moreover, isolated examples of extreme frequencies have been described in particular ethnic groups, particularly in certain tribes in India and Nepal (Brittenham et al., 1980; Kulozik et al., 1988; Labie et al., 1989; Modiano et al., 1991). In one of these tribal groups, the Tharu people of Nepal, the α-thalassaemia gene frequency reaches 0.78 and there is some evidence to suggest that these people are also more resistant to malaria than their non-Tharu neighbours (Terrenato et al., 1988).

Although the global distribution of the thalassaemias provides reasonable evidence for malaria protection, it is the extreme diversity of the molecular origins of these conditions that lends the most compelling support for genetic selection as an explanation for their extraordinary distribution. The thalassaemias have arisen throughout the world, through a wide variety of rare independent genetic events. However, in some populations they have risen to polymorphic frequencies while in others they remain rare, suggesting differential selection according to location. Although a number of diseases follow a tropical distribution, the probability that malaria was responsible for this selection was further supported by microepidemiological data from Italy and the Pacific. In a series of classic studies conducted in the 1960s, Siniscalco and colleagues (1961) demonstrated a cline (or gradient) in the population frequencies of β-thalassaemia in Sardinia that correlated with altitude (Siniscalco et al., 1961). Although malaria was no longer endemic in Sardinia when these studies were conducted, historically, the incidence of malaria was known to have correlated closely with altitude. The authors' favoured hypothesis was that the cline therefore represented selection for the β-globin gene under pressure from malaria. Support for this hypothesis has since come from the Pacific, where Hill and colleagues found evidence for similar clines in the population frequencies of both α- and β-thalassaemia in the island populations of Melanesia. Although the gene frequency for β-thalassaemia followed a similar correlation with altitude in Papua New Guinea (Hill et al., 1988), it was the data for α-thalassaemia collected from communities throughout Melanesia that was perhaps the more dramatic. These data resulted in four important observations (Flint et al., 1986; Hill, 1986; Yenchitsomanus et al., 1986). First, the α-gene frequency was found to be consistently low in areas that are known historically to have been nonmalarious. Second, α-thalassaemia was found in all malaria-exposed populations and at gene frequencies that were proportional to the incidence of malaria based on historical reports: a cline in the α-thalassaemia haplotype frequency was seen both from north to south and with increasing altitude, each of which are paralleled by a cline in the historical incidence of malaria. Third, even within this relatively small area, the genetic deletions responsible were numerous and regionally specific. Finally, the population frequencies of other 'neutral' genetic markers, including the γ-globin haplotypes -γ and -$\gamma\gamma\gamma$ and the haptoglobin polymorphism Hp^1, for which there is no evidence of malaria protection, showed no such correlation. These observations provided some of the strongest circumstantial evidence for malaria protection by the thalassaemias.

4.1.2.2 Epidemiological Studies

Case-control studies have now confirmed that both α-thalassaemia (Allen et al., 1997) and β-thalassaemia (Willcox et al., 1983) provide a high degree of protection against clinical malaria presenting to hospital. However, of particular further interest, Allen and colleagues also found evidence that α-thalassaemia conferred a similar degree of protection against hospital admission with illnesses other than malaria. Genetic traits that directly protect against severe malaria in children or nonimmunes may also provide protection to the newborn from the indirect morbidity of malaria associated with pregnancy. However, few studies of the influence of haemoglobinopathies on the outcome of pregnancy have been published. One study has described an association between heterozygous α-thalassaemia and malarial anaemia in pregnant women (Mockenhaupt et al., 1999). The influence of sickle cell trait on the outcome of pregnancy is unclear.

Somewhat paradoxically, two studies that have examined the relationship between mild clinical malaria and thalassaemia found evidence for a raised rather than a reduced incidence in children with α-thalassaemia (Oppenheimer et al., 1987; Williams et al., 1996). These epidemiological data therefore provide a perplexing picture: on one hand, α-thalassaemia appears to predispose children to mild clinical malaria, whereas on the other, it is associated with high-grade protection against severe malaria and other illnesses. It is possible to reconcile this paradox by arguing that α-thalassaemia protects against severe malaria by a 'vaccine' effect: predisposing to mild malaria in early life such that individuals develop sufficient immunity to produce protection against severe infections later in life (Williams et al., 1996).

4.1.3 Other Haemoglobinopathies

Although the literature is rather less comprehensive, there is also evidence for malaria protection by other haemoglobinopathies. Flatz and colleagues (1965) demonstrated a positive geographical correlation between frequencies of HbE and high malaria endemicity in Thailand and this is supported by recent experimental evidence of reduced invasion of HbAE erythrocytes compared with normal controls containing HbAA (Flatz et al., 1965). However, a protective effect has not been documented and quantified in case-control studies of malaria.

In Mali, Burkina Faso, and northern Ghana the gene for haemoglobin C (HbC) ($beta_6$glu-lys) is common, reaching frequencies of up to 9% in Mali. The high frequency of this gene in these malarious areas of West Africa has strongly hinted that this haemoglobinopathy does protect from malaria. Two case-control studies now support this hypothesis. A case-control study of severe and mild malaria in Mali, HbC gave 80% protection against severe malaria (odds ratio 0.22) (Agarwal et al., 2000). Furthermore, a large case-control study performed in Burkina Faso of over 4000 people, showed HbC was associated with a 29% reduction in risk of clinical malaria in HbAC heterozygotes and of 93% in HbCC homozygotes (Modiano et al., 2001). Those carrying the gene for HbAC and HbCC are largely asymptomatic and it has been suggested that this allele would replace HbS, at least in central West Africa, if malaria continued to cause appreciable mortality.

4.2 Red Cell Enzymes

Glucose-6-phosphate dehydrogenase deficiency is widespread in malarial endemic areas and shows considerable molecular heterogeneity (for review see Mehta et al., 2000). Its frequency is correlated with the historical distribution of malaria in Sardinia (Siniscalco et al., 1961). Taken together these data strongly suggest that G6PDH deficiency protects against malaria. Clinical studies evaluating the degree of protection conferred by this X-linked gene have found apparently conflicting results. Earlier, case-control studies have shown protection for both sexes against severe disease (Gilles et al., 1967), protection of female heterozygotes against nonsevere malaria (Bienzle et al., 1972), and no protection for females or males (Martin et al., 1979). Interpretation of these results is complicated by the continuous distribution of G6PDH activity in erythrocytes within a population of female heterozygotes (due to variable X chromosome inactivation), the different criteria for severe disease used in these studies, and the small size of the study by Martin and his colleagues. However, a larger case-control study has now shown that both female heterozygotes and male hemizygotes enjoy about 50% reduction in the risk of severe malaria (Ruwende et al., 1995). These data provide compelling evidence for a protective role for G6PDH deficiency in malaria. A role of red cell enzymes, involved in redox metabolism, other than G6PDH, in conferring resistance to malaria has not been established.

4.3 Mechanisms of Protection in the Haemoglobinopathies and G6PDH Deficiency

Despite the fact that its now more than 50 years since the protective effects of the haemoglobinopathies and G6PDH were first suspected, there is still no consensus regarding the mechanisms involved. Although it would be tempting to believe that the various types of haemoglobinopathy all act through a common mechanism it is still not possible to draw this conclusion from the available evidence. A number of mechanisms have been proposed, including decreased invasion or growth of malaria parasites in variant red blood cells, decreased virulence through reduced rosetting or cytoadherence, and increased removal through enhanced immunological recognition. The evidence for each of these mechanisms is better for some traits than for others and in all cases lack sufficient strength to draw firm conclusions. It is difficult to summarise this literature succinctly; however, rather than presenting the evidence for each condition separately, it might be more relevant to approach the subject from the perspective of putative mechanism.

4.3.1 Reduced Invasion and Growth

As can be seen from Haldane's first statement of the malaria hypothesis, reduced invasion or growth of parasites in variant red blood cells was proposed at the outset. Since it finally became possible to culture *P. falciparum* in the 1970s (Trager and Jensen, 1976), a number of studies have been conducted that have sought to investigate this hypothesis *in vitro*. Nevertheless, this line of investigation has produced conflicting and at times confusing results.

Convincing evidence supports reduced invasion or growth in the more severe haemoglobinopathies such as homozygous HbE (Nagel et al., 1981; Bunyaratvej et al., 1992), haemoglobin H disease (Ifediba et al., 1985; Brockelman et al., 1987; Bunyaratvej et al., 1992), homozygous HbC (Friedman et al., 1979), or compound states of two different haemoglobinopathies (Friedman et al., 1979; Brockelman et al., 1987; Yuthavong et al., 1987; Udomsangpetch et al., 1993). However, the cellular pathology of these conditions varies substantially and, furthermore, it is not known whether individuals with many of these complex or severe haemoglobinopathies enjoy malaria protection.

In contrast, the majority of studies have found that, *in vitro*, the invasion of red cells from individuals with clinically silent haemoglobinopathies, such as the heterozygous forms of α- or β-thalassaemia, HbS, HbE, or HbC, the phenotypes that are presumed to be protective *in vivo*, is normal (Nagel et al., 1981; Ifediba et al., 1985; Kaminsky et al., 1986; Yuthavong et al., 1988; Luzzi et al., 1990; Bunyaratvej et al., 1992; Williams et al., 2002). However, several groups have reported significantly reduced invasion of red cells from individuals with some of these minor forms of thalassaemia (Brockelman et al., 1987; Udomsangpetch et al., 1993). Moreover, two studies have found evidence for reduced growth or reinvasion when parasites are cultured in variant cells for prolonged periods (Kaminsky et al., 1986; Senok et al., 1997).

A number of investigators have found that when *P. falciparum* is cultured in HbAS red cells *in vitro*, growth is dramatically reduced under conditions of low oxygen tension (Friedman, 1978; Pasvol et al., 1978; Roth et al., 1978). Such cells also appear to be prone to sickle and this phenomenon has been attributed to haemoglobin polymerisation (Luzzatto et al., 1970) or to biochemical changes, including potassium loss or the toxic effects of haem (Orjih et al., 1985; Ginsburg et al., 1986).

A series of *in vitro* studies have suggested several possible mechanisms of protection of G6PDH-deficient red cells. G6PDH-deficient red cells do not support parasite growth in aerobic conditions *in vitro* (Roth et al., 1983; Roth et al., 1983; Usanga and Luzzatto, 1985) or under oxidative stress (Friedman et al., 1979; Golenser et al., 1983). These experimental results are in agreement. However, these *in vitro* studies cannot reproduce the conditions prevailing *in vivo* and there remains the possibility that the survival and/or ability of G6PDH-deficient erythrocytes to support parasite growth is artefactually reduced.

Overall, the consensus is that heterozygous forms of thalassaemia have little or no effect on growth of falciparum parasites *in vitro*, whereas the growth of these parasites in HbAS erythrocytes is reduced at low oxygen tension and the growth of parasites in G6PDH-deficient cells is reduced at high oxygen tension. The relevance of such findings for protection *in vivo* remains to be determined as behaviour and viability of all erythrocytes and particularly variant erythrocytes is quite different *in vitro* and *in vivo*.

4.3.2 Cytoadherence of *P. falciparum*-Infected Variant Red Blood Cells

The pathological effects of *P. falciparum* malaria are mediated by a range of mechanisms that probably include adhesion of infected cells to vascular endothelium, to other uninfected red cells (rosetting), and to other infected red cells (clumping) (see Fig. 6.1 and for recent review see Ho and White, 1999, and Craig and Scherf, 2001). Adhesion of infected cells to specific host ligands is mediated at least in part by a family of parasite-encoded antigens expressed at the surface of infected erythrocytes (for recent review, see Kyes et al., 1997). As a consequence, a number of studies have examined the biological properties of variant red blood cells infected with malaria parasites, focussing on the structural and functional characteristics expressed at the surface of infected erythrocytes.

Two groups have presented data suggesting that rosette formation, a phenomenon that is correlated with an adverse outcome in malaria infection (Carlson, 1993; Rowe et al., 1995), may be reduced in variant red blood cells. Udomsangpetch and colleagues found that, *in vitro*, both infected heterozygous α- and β-thalassaemic red cells were 60% less likely to be involved in rosettes than normal cells (Udomsangpetch et al., 1993). Although Carlson found a somewhat smaller effect (a 10–50% reduction), they also reported that in experiments using red blood cells from individuals with heterozygous α thalassaemia rosettes were smaller (as defined by the number of uninfected cells binding to each parasitised red blood cell) and more easily disrupted than controls (Carlson et al., 1994). The adhesive characteristics of malaria-infected HbAS erythrocytes has received little attention and here a single study has shown that rosetting is not changed in malaria-infected cells from individuals with sickle cell trait (Udomsangpetch et al., 1993).

Three reports have addressed the cytoadherence phenotype of infected α-thalassaemic red cells. Luzzi and Pasvol (1990) found no reduction in cytoadherence to C32 amelanotic melanoma cells of heterozygous thalassaemic red cells infected with ITO strain *P. falciparum* parasites, whereas Udomsangpetch and colleagues found that cytoadherence to human umbilical-vein endothelial cells (HUVEC) by heterozygous α- and β-thalassaemic red cells infected with TM267R strain parasites was reduced by about 60% (Udomsangpetch et al., 1993). Surprisingly, the latter study also reported reduced binding of the human monoclonal antibody MAb 33G2, which reacts with a number of *P. falciparum* antigens, including Pf155, antigen 332, and Pf11.1 not widely thought to be expressed on the surface of malaria-infected erythrocytes. A more recent study has found no difference in the adhesive phenotype of heterozygous thalassaemic erythrocytes infected with malaria compared with control-infected erythrocytes (Williams et al., 2002).

These cytoadherence studies are difficult to interpret not only because the results are variable but also because the causal role of specific adhesive phenotypes in severe disease has not been precisely delineated. However, variant erythrocytes infected with malaria parasites do not have striking differences in their adhesive phenotypes compared with infected erythrocytes from normal individuals.

4.3.3 Phagocytosis

Rapid clearance of infected erythrocytes by host phagocytes is an important component of host defences. There is mounting evidence that variant erythrocytes infected with malaria parasites are indeed cleared more rapidly than normal erythrocytes.

One group have reported an increased susceptibility of *P. falciparum*-infected thalassaemic red blood cells to phagocytosis (Bunyaratvej et al., 1986; Yuthavong et al., 1988). These observations held for infected red cells from both heterozygous and homozygous HbE (Bunyaratvej et al., 1986) and various forms of α-thalassaemia with or without HbCS (Yuthavong et al., 1988). In these studies, uninfected variant red cells were also more susceptible to phagocytosis, consistent with the observations of others in human thalassaemia and a murine model of the disease (Knyszynski et al., 1979; Rachmilewitz et al., 1980).

However, it is less clear how phagocytes may recognise malaria-infected variant cells. Luzzi and colleagues (1991) found that infected α- and β-thalassaemic red cells bound respectively 1.69 and 1.23 times as much antibody per unit area from immune serum as control cells (Luzzi et al., 1991). Udomsangpetch and colleagues (1993), on the other hand, found no significant difference in binding to either variant cell type (Udomsangpetch et al., 1993). These disparate results may be explained by methodological differences: although the former group used a radiometric method to assess binding to live cells the latter used a flow cytometric method to quantify binding to gluteraldehyde-fixed cells. A further study has shown that malaria-infected thalassaemic cells have increased binding of immunoglobulins from immune sera compared to controls using live cells and flow cytometry. This increased binding of immunoglobulins is not associated with increased expression and/or recognition of the variant antigen Pf-EMP-1 (Williams et al., 2002).

There is also indirect evidence from an animal model for enhanced clearance of infected cells from individuals with sickle cell trait. Mice expressing high levels of a HbS but not HbA transgene were protected from the rodent malaria *P. chabaudi adami*. However, the protective effect was abrogated by splenectomy of the HbS transgenic mice (Shear et al., 1993). These data suggest that expression of HbS may result in enhanced splenic clearance of parasitised red cells.

Finally, there is good evidence that malaria infected erythrocytes from G6PDH-deficient individuals are more likely to be phagocytosed by macrophages than malaria-infected erythrocytes from those with normal enzymes levels. Moreover, the relative enhancement of phagocytosis of G6PDH-deficient-infected erythrocytes is much greater for early ring stage parasites than the more mature trophozoites (Cappadoro et al., 1998). Here during the early (ring) stage of malaria infection the G6PDH-deficient erythrocytes bind increased autologous IgG and complement C3 fragments compared to normal erythrocytes.

4.3.4 Immune-Mediated Mechanisms

There are a series of intriguing and unexplained observations that suggest the protection of people with sickle cell trait may be mediated, at least in part, by enhancing the acquired immune response to malaria. Such an explicit immune-mediated mechanism(s) would explain the observations of Gugenmoos–Holzmann and colleagues who showed that the effect of AS on parasite density was strongly age-dependent beginning after only 2 years (Guggenmoos-Holzmann et al., 1981). Later Bayoumi formally proposed that the immune response to malaria parasites is enhanced in those with sickle cell trait (Bayoumi, 1987, 1997).

Some data support the hypothesis of an enhanced immunological response in HbAS. Two studies (Edozien et al., 1960; Cornille-Brogger et al., 1979) have found significantly raised gamma globulin levels in young AS children. More specifically, children with the HbAS phenotype had higher levels of antisporozoite antibodies and agglutinating antibodies to variant surface antigens (Marsh et al., 1989) and significantly increased lymphoproliferative responses to malaria antigens (Abu-Zeid et al., 1992). The clinical significance of these changes in the antibody and cellular responses to malaria antigens is uncertain. Nevertheless this data is at least consistent with the hypothesis of an enhanced acquired immune response in sickle cell trait. It is possible that an interaction between sickle cell trait and the acquisition of an effective immune response to malaria is less evident at higher levels of transmission (Aidoo et al., 2002).

There is no similar data for the other haemoglobinopathies, indeed a small study in Gambian children with thalassaemia trait showed no differences in their immune responses to malaria compared to controls (Allen et al., 1993).

The mechanism of an enhanced immunological response has not been established either *in vitro* or by *ex vivo* clinical studies. Infected variant erythrocytes may be able to stimulate immune responses in some way. Alternatively,

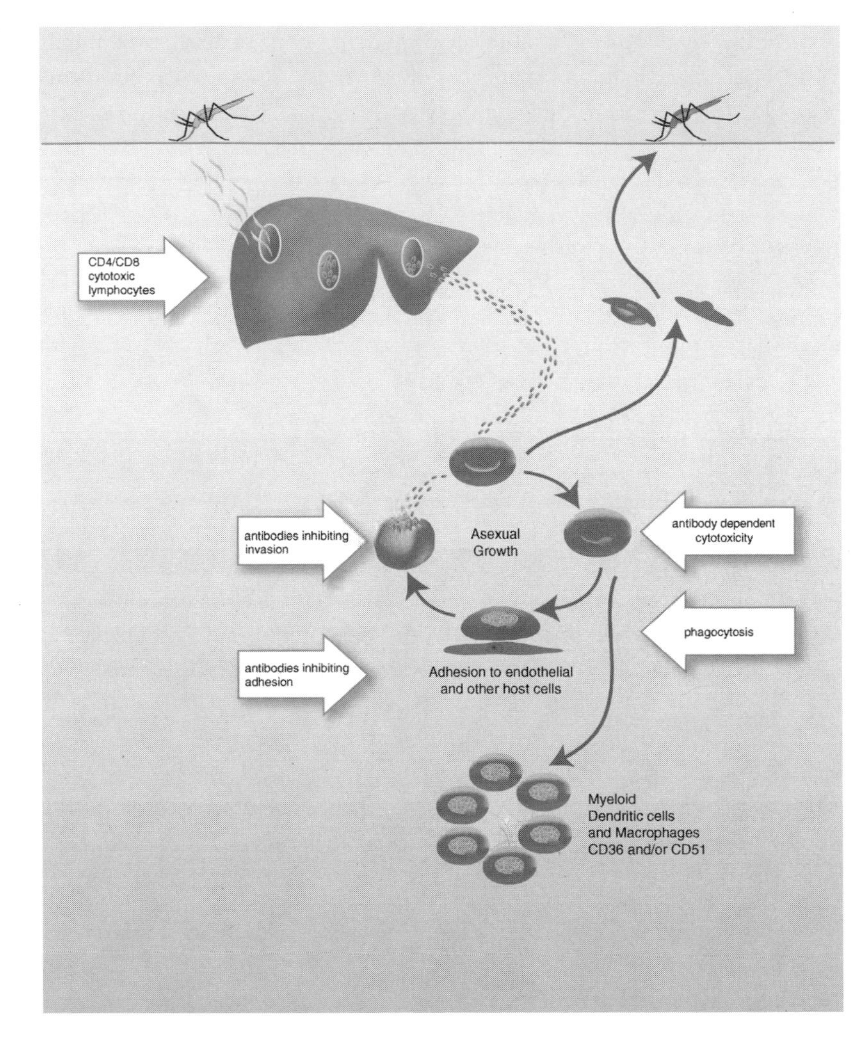

Figure 6.3. **Immune responses to falciparum malaria.** Immune responses against the plasmodium parasites are distinct for liver-stage and blood-stage parasites. Cytotoxic lymphocytes can recognise and destroy intrahepatic parasites and secretion of IFN-γ by both CD4+ and CD8+ T cells is associated with protective immune responses. Specific immunity against blood-stage antigens includes antibodies directed against parasite molecules required for invasion of erythrocytes and antibodies recognising the adhesive and variant proteins (PfEMP-1) expressed on the surface of erythrocytes. Other antibodies are able to stimulate antibody-dependent cytotoxicity and splenic macrophages phagocytose infected erythrocytes. Other effector mechanisms are poorly defined. Immune responses may be subverted during the blood stage of infection. Adhesion of

(*cont.*)

early clearance of infected erythrocytes at the ring-stage of development may enhance or prevent the inhibition of development of the acquired immune response against malaria parasites. First, removal of ring-stage parasites would reduce production of malaria pigment formed by the digestion of haemoglobin in the host erythrocytes and subsequent dysregulation of monocyte function (Schwarzer et al., 1993). Second, removal of ring-stage parasites would reduce adhesion of infected erythrocytes to myeloid dendritic cell and macrophages and so decrease down-regulation of innate and acquired immune responses (Urban et al., 1999; McGilvray et al., 2000) (Fig. 6.3 and for review see Urban and Roberts, 2002).

4.3.5 Conclusion

Therefore there is some agreement that enhanced phagocytosis of variant erythrocytes may occur *in vitro* and in animal models, although evidence for the mechanisms for rapid phagocytosis is fragmentary. At present there is no direct data from clinical studies to suggest that enhanced phagocytosis occurs during malaria infection in those carrying sickle cell trait, thalassaemia traits or G6PDH deficiency. On balance, however, the available data suggest these traits permit enhanced immunological clearance of infected erythrocytes rather than grant simple resistance of the variant red cells to invasion or parasite growth. Clearance of ring-stage parasites would alter the outcome of infection by reducing sequestration of infected erythrocytes in peripheral tissues and subsequent organ dysfunction. Blocking the development of mature trophozoites may also alter the immune response to the parasite. A substantial amount of experimental, clinical, and epidemiological evidence is still required to support this scheme.

4.4 Membrane Proteins

4.4.1 Duffy Blood Group System

The most complete story that links a genetic trait affording protection from malaria with well-defined cellular and molecular interaction between host and parasite is the Duffy blood group and erythrocyte invasion. Early

Figure 6.3. (*cont.*) infected erythrocytes to dendritic cells via CD36 and CD51 (α_V integrin of the $\alpha_V\beta_3$ or $\alpha_V\beta_5$ heterodimer) may modulate DC function. These interactions reduce the ability of myeloid DCs to stimulate primary and secondary CD4+ T-cell responses. Adhesion of infected erythrocytes to macrophages may reduce secretion of inflammatory cytokines, including TNF-α. (See color plate.)

comparative studies showed the invasion of erythrocytes by merozoites was dependent on specific determinants on the red cell membrane (Hadley and Miller, 1988). The first merozoite receptor identified on human erythrocytes was the Duffy blood group antigen. Red cells lacking these determinants [Fy(a-)Fy(b-) phenotype and FyFy genotype] are resistant to invasion by *Plasmodium knowlesi in vitro* (Miller et al., 1975), although *P. falciparum* merozoites can invade FyFy cells (Miller et al., 1977). These studies suggested that the Duffy blood group antigen could be a receptor for *Plasmodium vivax*. The essential role of the Duffy blood group antigen for erythrocyte invasion was obtained in epidemiological studies of *P. vivax* infection (Miller et al., 1976) and more recently in Papua New Guinea (Zimmerman et al., 1999). It is now known that the Duffy blood group antigen is a receptor for IL-8 and melanoma growth stimulatory activity and invasion of *P. knowlesi* blood stage merozoites into erythrocytes carrying the Duffy antigen can be inhibited by MGSA and IL-8 (Horuk et al., 1993).

In Africa the Fy(a-)Fy(b-) phenotypes reach 100% in much of sub-Saharan Africa. It seems that the selective advantage of the FyFy genotype has driven this genotype nearly to fixation. This is certainly plausible but somewhat puzzling in the light of the low virulence of *P. vivax* under present epidemiological conditions and given appropriate medical treatment. We would have to conclude that mortality from *P. vivax* was much higher in the past in the absence of suitable drug treatment and possibly with a more virulent parasite.

These studies led to the detailed characterisation of the Duffy binding proteins in *P. vivax* and the simian malaria *P. knowlesi*. Chetan Chitnis and Lou Miller (1994) identified one protein from *P. vivax* and three proteins from *P. knowlesi* that bound to the Duffy blood group antigen. The proteins shared sequence homology and Miller and Chitnis identified a cysteine-rich domain within *P. vivax*, Duffy binding ligand, mediating adhesion to Duffy blood group-positive but not Duffy blood group-negative human erythrocytes. The homologous domain of the proteins from the simian malaria parasite *P. knowlesi* also bound erythrocytes but had different specificities (Chitnis et al., 1994).

The erythrocyte binding domains in these proteins, namely the Duffy binding domains, also showed sequence conservation with the domain for erythrocyte binding in the *P. falciparum* protein, erythrocyte-binding antigen-175, which bound sialic acid on human erythrocytes. The conservation of this binding domain between distant malaria species strongly suggested that these structures were functionally significant and potential vaccine candidates. These suggestions were confirmed by hypervariability of the erythrocyte-binding domain in the *P. vivax* Duffy binding protein (Tsuboi et al., 1994). The binding domains within these erythrocyte-binding proteins (PvDBPs) lie within a conserved N-terminal cysteine-rich region of 330 amino acids (Ranjan and Chitnis, 1999). The major challenge in using this information is to use correctly folded recombinant proteins that stimulate blocking antibodies. A recombinant form of the *P. vivax* binding protein that is both functionally active and capable of eliciting antibodies that block erythrocyte invasion has now been achieved (Singh et al., 2001). These techniques also used for the *P. falciparum* EBA-175 (Pandey et al., 2002). These proteins are now leading candidate vaccine antigens for an invasion blocking the vaccine (for a review see Chitnis, 2001).

4.4.2 Other Red Blood Cell Surface Antigens

Malaria parasites use several receptors on the surface of erythrocytes during invasion and during the rosetting of unaffected erythrocytes by infected erythrocytes. It has therefore seemed possible that some of the many polymorphic blood group systems may be linked to protection from malaria. However, somewhat surprisingly the epidemiological and laboratory evidence for significant protection against falciparum malaria by blood group polymorphisms is limited and somewhat inconsistent.

The ABO(H) blood group system is most accessible for study and two studies have suggested that group A is associated with severe malaria (Fischer and Boone, 1998; Lell et al., 1999). These observations are consistent with low frequency of blood group A in many malaria endemic areas. A possible mechanism for such a protective effect is the modulation of rosetting exhibited by some *P. falciparum* strains by the ABO(H) group of erythrocytes *ex vivo* and by ABO(H) blood group substances *in vitro* (Carlson and Wahlgren, 1992; Barragan et al., 2000). Clinical evidence from Thailand and East Africa has supported the role for the influence of rosetting by ABO blood group type. In these studies the frequency of rosetting parasites was less frequently isolated from group O patients than from patients with groups A and B (Rowe et al., 1995) or A and AB (Udomsangpetch et al., 1993; Chotivanich et al., 1998).

The role of blood group antigens in modulating the invasion of malaria parasites is somewhat incomplete. The capacity of malaria parasites to invade cells with modified O-linked saccharides is reduced. Removal of sialic acid (*N*-acetyl neuraminic acid) reduced invasion by malaria parasites (Miller et al., 1977; Perkins, 1981; Breuer et al., 1983; Friedman et al., 1984). Tn+ erythrocytes (constitutively deficient in sialic acid and galactose in O-linked oligosaccharides) and Cad+ erythrocytes (with an extra *N*-acetyl galactosamine residue next to sialic acid residues) are resistant to invasion (Pasvol et al., 1982; Cartron et al., 1983). However, there is no evidence that the blood groups Tn+, Cad+, or En- (sialic acid deficient) or other variants of glycophorin A reach polymorphic frequencies in those living in endemic areas for malaria (Mourant et al., 1976; Brindle et al., 1995). This is somewhat puzzling as a ligand for glycophorin A has been identified, namely EBA-175. Furthermore, the coding sequences of the gene encoding this protein appear to have undergone immune selection. One would have to conclude that the parasite could use alternate invasion pathways, so minimising any selective advantage of mutation in glycophorin A.

The story surrounding glycophorin B is also tantalisingly incomplete. There is some experimental evidence that S-s-U- cells deficient in glycophorin B are relatively resistant to invasion. The use of glycophorin B may, however, be specific to parasite strains as these findings have not, however, been confirmed by others (Hadley et al., 1988). Nevertheless, the blood group S-s-U- is found in 2–5% of Africans, suggesting malaria may have provided a selective force for this polymorphism (Mourant, 1968). However, a parasite ligand for glycophorin B has not been identified.

On the other hand, a parasite receptor has now been identified for the polymorphic glycophorin C that determines the Gerbich (Ge) blood group system. Deletion of exon 3 in the glycophorin C gene reaches a high frequency (46.5%) in coastal areas of Papua New Guinea, where malaria is hyperendemic. The receptor for *P. falciparum* erythrocyte-binding antigen 140 (EBA140, also known as BAEBL) is glycophorin C and that this interaction mediates a principal *P. falciparum* invasion pathway into human erythrocytes (Maier et al., 2002). These findings strongly suggest that Ge negativity has arisen in Melanesian populations through natural selection by severe malaria, although the published studies have not shown that glycophorin C deficiency is associated with differences in either *P. falciparum* or *P vivax* infection (Patel et al., 2001). Further case-control studies are awaited with interest.

4.5 Red Blood Cell Cytoskeletal Proteins

Polymorphisms of the genes encoding the red cell cytoskeletal proteins are far less common than globin variants or G6PDH deficiency (for review see McMullin, 1999, and Bennett and Baines, 2001).

4.5.1 Southeast Asian Ovalocytosis

In Southeast Asian ovalocytosis (SAO), an autosomal dominant trait reaches a frequency over 10% in some areas of Papua New Guinea (Amato and Booth, 1977). The genetic defect lies in band 3, the erythrocyte anion transporter, where two linked mutations (lys_{56} – glu and deletion of residues

400–408) have been identified in all the cases of SAO examined so far (Liu et al., 1990; Jarolim et al., 1991; Schofield et al., 1992). The cells from affected individuals show reduced anion transport suggesting the abnormal protein has little or no transport activity (Schofield et al., 1992).

The distribution of SAO within PNG is restricted to malaria-endemic regions. Although parasite rates are similar in ovalocytics and nonovalocytics (Serjeantson et al., 1977), two case-control studies have recently confirmed that it is highly protective against severe malaria (Genton et al., 1995; Allen et al., 1999). Somewhat suprisingly, this protection appears to be highly specific to cerebral malaria (Allen et al., 1999).

The mechanism of protection is unclear. Initially, it was thought that protection was due to resistance to invasion of such cells by malaria parasites (Kidson et al., 1981) and a number of hypotheses have been proposed to explain the mechanism of a protective effect. These include reduced deformability of the red cell membrane (Mohandas and Evans, 1984); depression of the blood group antigens U, Wr(b), and En(a) implicated in invasion (Booth et al., 1977); high intracellular sodium concentrations (Honig et al., 1971); and finally reduced anion transport (Schofield et al., 1992). The relative importance of these factors has not been determined. Moreover, the hypothesis that protection in SAO is mediated by induced invasion is somewhat at odds with the observation that parasitaemia is not reduced in this population.

4.5.2 Other Red Cell Cytoskeletal Proteins

Other forms of elliptocytosis have more restricted distribution. Defects may involve the α- and less commonly β-chain of spectrin, protein 4.1, or absence of glycophorin C (and Gerbich blood group antigens) (reviewed by Palek and Lambert, 1990). The α-spectrin variants $\alpha 1/65$ and $\alpha 1/46$ in West African populations (named after the pattern of peptides produced after tryptic digestion) reached polymorphic frequencies in two West African populations (Morle et al., 1989). In addition, red cells from those with the α-spectrin variants $\alpha 1/74$ and the $\alpha 1/65$ were resistant to invasion by *P. falciparum* merozoites (Facer, 1989, 1995). In these studies the reduction in invasion correlated with the percentage of spectrin dimers present within the membrane of variant cells. Invasion and growth of *P. falciparum* is also reduced in elliptocytic red blood cells with a combined deficiency of protein 4.1, glycophorin C, and p55 (Chishti et al., 1996). In summary, this evidence is consistent with the role for protection from malaria by spectrin variants but falls short of detailed descriptions obtained for other red cell defects.

4.6 Immune Response Genes

The control of the growth of malaria parasites is stage-specific and various effector pathways of the innate and acquired immune system control parasite growth (Fig. 6.3) (for recent review see Plebanski and Hill, 2000, and Malaguarnera and Musumeci, 2002). Both CD8+ and CD4+ lymphocytes are able to kill the developing parasites within hepatocytes. The growth of blood stage parasites may be inhibited by antibodies that reduce the invasion of red blood cells, that activate antibody dependent cytotoxicity, or that recognise and clear the parasite-determined neoantigens on the surface of infected red blood cells. The production of high-affinity antibodies requires help from CD4+ T cells. Crucially, infected red blood cells are cleared from the circulation by macrophages, whereas the role of NK cells and neutrophils is less well defined. This range of protective immune responses suggested that many genes expressed in the cells of the immune system influence the outcome of severe malaria. The central role of HLA class I and II molecules in the initiation of specific acquired immune responses made the respective genes with the MHC locus an obvious starting point to identify protective alleles influencing the immune response to malaria.

4.6.1 HLA Class I and Class II

In a landmark case-control study the class I allele HLA-Bw53 was associated with resistance to cerebral malaria and to severe malarial anaemia and the class II haplotype HLA-DRB1*1302-DQB1*0501 with resistance to severe malarial anaemia (Hill et al., 1991). This study suggested that in Gambia the protection afforded by these two alleles was comparable to that provided by sickle cell trait. The study was of widespread interest for two reasons. First, the data supported the widely held and attractive hypothesis that the astonishing diversity of HLA class I and class II alleles was driven by the selective forces exerted by infectious diseases. Second, the association suggested a direct route from which to identify epitopes in liver stage malarial antigens that could be used as candidate antigens for vaccine development.

Some evidence is available that specific peptides on liver stage antigen 1 (LSA-1) may be presented preferentially in HLA B53-positive individuals (Hill et al., 1992). As a result of these observations, it is now generally considered important that vaccine preparations that include antigens expressed at liver stage should include epitopes that are recognised by a wide range of class I molecules (Hoffman et al., 1997).

The associations of the MHC class I allele HLA-Bw53 and the class II haplotype HLA-DRB1*1302-DQB1*0501 with protection from malaria have not been replicated elsewhere in Africa. A possible explanation for these results must include frequency-dependent selection where the selective advantage of different alleles varies over time and place. Thus as the association of the respective class I and II alleles with protection would vary with the frequency of parasite genotype, so would the corresponding class I and/or class II restricting elements wax and wane. It is also possible that these results represent, at least in part, a true association to linked alleles within the MHC locus.

Further analysis of the association of HLA class I alleles with parasite strains defined by polymorphisms of the putative T-cell epitopes presented by the respective class I molecules provided evidence for antagonism of T-cell responses by different parasite strains. In Gambia there was a nonrandom association of parasite types in those carrying the HLA-B*35 allele (Gilbert et al., 1998). The parasite strains defined by a HLA class I-restricted epitope in the major coat protein of the sporozoites were found more commonly in those carrying certain HLA class I alleles. Experimental evidence suggests that the mutually restricted strains can antagonise CD8 T-cell responses (Plebanski et al., 1999). Although these association have not to date been found elsewhere they do suggest how future study of immune response genes and parasite genotypes in the same sample set may yield important insights into the ecology of parasite infections.

4.6.2 Tumour Necrosis Factor and Other Cytokines

Cytokines have powerful pleiomorphic effects on protective and harmful immune responses. The genes encoding the cytokines known to be raised in severe malaria: TNF-α, IL-6, IL-1, and IFN-γ and their respective receptors have therefore attracted attention as candidate genes for susceptibility traits for malaria.

Tumour necrosis factor-α was the first cytokine shown to be elevated in severe disease and polymorphisms of the noncoding region of this gene have been extensively investigated. Gambian children who are homozygous for the *TNF-308A* allele (or *TNF2*) allele have an increased susceptibility to cerebral malaria (McGuire et al., 1994). The *TNF2* allele has subsequently been associated with severe disease in Sri Lanka (Wattavidanage et al., 1999) but not in Thailand (Hananantachai et al., 2001). This allele is also associated with severe malaria, leishmaniasis, scarring trachoma, and lepromatous leprosy (reviewed by Knight and Kwiatkowski, 1999). Furthermore, a study in western Kenya has shown that homozygotes for this allele are predisposed to high-density *P. falciparum* parasitemia (Aidoo et al., 2001). These data strongly suggest that the *TNF-308A* allele may be functionally significant and reporter gene assays have demonstrated up to a fivefold differences in transcription between *TNF1* and *TNF2* allelic constructs when the TNF 3′UTR is present (Wilson et al., 1997). Thus, the majority of the data support a direct role for the *TNF2* allele in the elevated TNF levels observed in *TNF2* homozygotes and so in the predisposition to cerebral malaria (reviewed by Abraham and Kroeger, 1999, and Knight et al., 1999).

A second allele, *TNF-376A*, has been associated with susceptibility to cerebral malaria in both East and West Africans. This association is independent of the association of *TNF-308A* and linked HLA alleles with severe malaria. The polymorphism causes the helix-turn-helix transcription factor OCT-1 to bind to the promoter and so alters gene expression in human monocytes. The OCT-1-binding genotype, found in approximately 5% of Africans, is associated with fourfold increased susceptibility to cerebral malaria (Knight et al., 1999).

Finally, a third polymorphism, the *TNF-238A* allele, has been associated with severe malarial anaemia in Gambia, but not in Kenya (McGuire et al., 1999). It has been suggested that this allele may facilitate chronic low-level transcription of TNF-α and thus predispose to malarial anaemia through inhibition of erythropoiesis and/or increased erythropoiesis. If this is true it remains puzzling why some, but not all, polymorphisms at this locus, apparently enhancing TNF-α production are associated with cerebral malaria but not with malarial anaemia. A more detailed understanding of the pathogenesis of severe malarial anaemia may illuminate these relationships.

Interferon-γ (IFN-γ) plays an important role in the immune response to malaria. Koch and Kwaitowski and colleagues examined the relationship between polymorphisms in the promoter region of the gene encoding IFN-γ receptor 1 (*IFNGR1*) and susceptibility to malaria in African children. They identified four polymorphisms and six haplotypes after sequencing the region between −1400 and +100 nt of the translational start site and analysis of nuclear families. In a large case-control study of 562 Gambian children with severe malaria they showed that in Mandinka, the major Gambian ethnic group, heterozygotes for the *IFNGR1-56* polymorphism were protected against cerebral malaria (odds ratio 0.54) and against death resulting from cerebral malaria (odds ratio 0.22) (Koch et al., 2002).

Few other association studies of cytokines have been published, although the influence of IL-10 and IL-12 polymorphisms are of critical importance in the control of the immune response. Intriguingly, recent studies do show associations of IL-12 promoter and 3′UTR polymorphisms with malaria in Tanzania but not in Kenya (Morahan et al., 2002) and further genetic studies of functional polymorphisms of these genes are awaited with interest.

4.6.3 Nitric Oxide Synthase

The role of nitric oxide in malaria is controversial and this messenger has been proposed both as a mediator of coma in children with malaria and as a mediator of parasite destruction. In these circumstances, the genetic epidemiology and appropriate functional studies may help resolve this debate and so further understanding of the role of these effector molecules in malaria.

Several studies have examined the association of the *NOS2-G954C* polymorphism in promoter region of the inducible nitric oxide synthase gene (iNOS) with malaria. Heterozygotes for this iNOS point mutation were strongly protected from severe malaria in Gabon (Kun et al., 2001).

A second *NOS2* promoter polymorphism, a CCTTT microsatellite repeat, was associated with fatal cerebral malaria in Gambia (Burgner et al., 1998). Genotypes with longer forms of the CCTTT repeat (alleles of $\geq$15 repeats) were significantly associated with severe malaria (odds ratio 2.1) in Thailand (Ohashi et al., 2002).

However, a study from Tanzania suggested that these associations might not define casual relationships among the respective polymorphism, NO production, and the outcome of infection. Here, there was no significant correlation of either the *NOS2-G954C* or the CCTTT repeat with disease severity or with measures of NO production and *NOS2* expression (Levesque et al., 1999).

Finally, a new *NOS2* promoter polymorphism *NOS2-C1173T* associated with increased nitric oxide production and strongly associated with protection from severe malarial anaemia (RR 0.25) (Hobbs et al., 2002). This polymorphism may be functional, as this association was independent of the previously recognised *NOS2-G954C* polymorphism. Furthermore, the (CCTTT) repeat polymorphism was not associated with severe malaria in this study.

Taken together these studies encapsulate the promise and the problems of using genetic epidemiology to understand the immune response to malaria. Different genetic studies in different areas have given widely contrasting results. The most simple explanation for these apparently inconsistent results (other than the trivial explanation that some of the positive associations were due to chance) is that the associations between a genetic polymorphism and direct or indirect measurements of gene expression do not represent causal relationships. Thus, the respective polymorphisms may be linked to other, as-yet-unidentified or partly characterised functional polymorphisms. Further genetic studies analysing the extended haplotypes at this locus with the outcome of malaria infection and a molecular analysis of relations between polymorphisms and gene expression are clearly required for further progress. These considerations illustrate how difficult it is to interpret the results of single studies of association of genetic polymorphisms with susceptibility or resistance to malaria with additional functional or genetic data

4.6.4 Other Immune Response Genes

In the *Plasmodium berghei anka* model of murine malaria, CD40–CD40L interaction leads to the breakdown of the blood–brain barrier, macrophage sequestration, and platelet consumption (Piguet et al., 2001). In a case-control study of severe malaria, males hemizygous for the *CD40L-726C* polymorphism have a reduced risk of severe malaria (Sabeti et al., 2002) and are consistent for the role of CD40–CD40L interaction in human disease.

There is additional evidence for recent selection at the CD40L locus by analysing the long-range haplotypes. The age of each core haplotype can be measured by the decay of its association to alleles at various distances from the locus, as measured by extended haplotype homozygosity (EHH). Haplotypes with high levels of surrounding homozygosity (and so a high EHH) and a high population frequency suggest the recent selection of a mutation. This approach was validated for G6PDH locus and the same techniques also show significant evidence of selection at the locus for CD40 ligand (Sabeti et al., 2002). This method may represent a useful, and much needed, method of scanning the entire genome for evidence of recent positive selection.

The role of the Fc-γ receptors has not been defined in malaria infection although it seems likely that they are crucial for several effector mechanisms. The inhibition of *P. falciparum* blood-stage parasite growth by antibody-dependent cellular inhibition *in vitro* is mediated by IgG1 and IgG3, but not IgG2, malaria-specific antibodies and monocytes via the Fc-γ receptor II (Fc-γRII). It has been suggested that the Fc-γRIIa-Arg/Arg131 genotype, which does not bind to IgG2, may confer protection from malaria. A single study from Kenya has shown that the Fc-γRIIa-Arg/Arg131 genotype is associated with a lower risk for high-density falciparum infection (Shi et al., 2001). Further evaluation of the role of this polymorphism at this locus would be interesting.

4.7 Host Receptors for Infected Erythrocytes on Endothelium and Uninfected Erythrocytes

Sequestration of infected erythrocytes has been implicated in the pathogenesis of severe malaria in a series of histopathological studies. The host receptors for the clonally variant antigen or Pf-EMP-1 (*P. falciparum* membrane protein-1) encoded by the *var* gene expressed at the surface of the infected erythrocyte have been well characterised and include ICAM-1, CD36, and CD31. Furthermore, some falciparum strains demonstrate rosetting of uninfected erythrocytes by infected erythrocytes, and this interaction involves the variant antigen and complement receptor 1 (CR1) or CD35 on the surface of infected erythrocytes (Fig. 6.1 and for review see Ho et al., 1999, and Craig et al., 2001). Association studies using large case-control studies have examined the association of polymorphisms of all those genes with severe malaria.

A high-frequency conservative amino-acid substitution in the first domain of ICAM-1 (*ICAM-1Kilifi*) has been associated with severe malaria in Kilifi, Kenya (Fernandez-Reyes et al., 1997) but not in Gambia (Bellamy et al., 1998). In Gabon this mutation was associated with protection from malaria (Kun et al., 1999). A soluble construct of ICAM-1Kilifi has a number of different binding affinities from the wild-type protein; ICAM-1Kilifi has reduced avidity for LFA-1, does not bind fibrinogen, and has reduced binding to ICAM-1-binding parasite strains (Craig et al., 2000). The clinical significance of these functional differences is unclear and once again we would expect additional genetic and functional studies might elucidate the association.

Most parasite strains adhere to CD36 and one may have expected selective forces to select for mutations or deletions of the CD36 gene. Two studies of the relationship between a common stop codon in CD36 and severe malaria have been published. The first study showed a significant association between the CD36 stop codon and susceptibility to malaria (Aitman et al., 2000). A second study of 700 patients and controls showed an association of the mutation with protection from severe malarial anaemia, hypoglycaemia, and respiratory distress, but no protection from cerebral malaria (Pain et al., 2001). These two studies are not necessarily contradictory. An association of this mutation with susceptibility or no protection from cerebral malaria may reflect the complex role of interactions involving CD36 in malaria. CD36 is expressed on endothelium and platelets and thus high levels of CD36 expression would be expected to increase sequestration and clumping of infected erythrocytes with platelets. On the other hand, adhesion of infected erythrocytes to CD36 may down-regulate inflammatory and immune responses (Fig. 6.3) (see Urban et al., 2001, and review by Urban and Roberts, 2002). These two important roles for CD36 in malaria, one augmenting and one ameliorating disease pathology, suggest it is quite conceivable that polymorphisms at this locus have different effects in the different syndromes of severe disease. Such speculation can be answered only with further studies.

Two studies have failed to show an association of a functional polymorphism in CD31 codon 125 leu-val with severe disease (Casals-Pascual et al.,

2001). However, a study from Thailand shows an association of this polymorphic allele with susceptibility to severe malaria (Kikuchi et al., 2001). Additional studies of the extended haplotype at this locus would clearly be informative.

Rosetting of uninfected erythrocytes by infected erythrocytes is mediated at least in part by complement receptor 1 (CD35) (Rowe et al., 1997). The locus encoding this protein is highly polymorphic in African populations and the variation in the expression level and number of repeated domains and a series of single amino acid polymorphisms coexist. An earlier study failed to show an association of a polymorphism known to influence expression level of CR1 in Caucasians with malaria in Gambian children (Bellamy et al., 1998). However, this polymorphism is not linked to changes in expression level of this protein in an African population (Rowe et al., 2002). The relationships between the CR1 allotypes and genotypes in the African population have recently been established (Moulds et al., 2001) but further studies of the role of this CR1 phenotype with disease must await a full description of the relationship between the expression level phenotype and genotype or a 'back to basics' study of red cell phenotype and disease.

Surveying these studies it is somewhat surprising that no well-defined, consistent associations have been established for polymorphisms of the host receptors for infected erythrocytes. Although we can only reemphasise the need for further studies, it is possible here those functional polymorphisms of these receptors have opposing or differential effects in the various syndromes of severe malaria. Case mix and transmission (and its geographical location) may be capable of unexplained confounding factors when comparing different studies.

5. NEGATIVE ASSOCIATION STUDIES

We must conclude this section with a mention of those studies where no significant association between functional gene polymorphisms and malaria have been established. These include the mannose binding protein and interleukin 1 receptor antagonist (IL-IRA) (Bellamy et al., 1998a, 1998b). What can we conclude here? Strictly speaking absence of an association does not prove there is no effect, but simply defines confidence intervals for the relative risk or odds ratio of the association. Furthermore, false-negative studies may be obtained by chance or by different relationships between phenotype and genotype in different populations.

Given these provisos, such negative studies, like the 'dog that didn't bark in the night' in the Sherlock Holmes short story may be useful evidence when trying to understand the role of particular host parasite interactions in severe malaria. They may also lead further lines of enquiry to understand the selective forces driving high-frequency expression of mutant alleles in apparently strategic immune response genes. Finally, if negative association studies do not enter the public domain there is a consistent publication bias to positive association studies. It is obviously important that negative studies are published.

6. CONCLUSION

The search for genetic traits that protect from malaria began to prove Haldane and Beet's hypothesis that the major haemoglobinopathies with a global distribution arose from the selective forces imposed by the malaria parasite. More recently, genetic association studies have extended these general findings and are consistent with the notion that the diversity at the HLA class I and class II loci arise through natural selection for survival against infectious pathogens, including malaria. These association studies have now been extended to examine numerous other immune responses and other host molecules that may be involved in the pathogenesis of severe malaria. As the catalogue of these traits has extended the focus has shifted to try to understand in some detail how these traits confer protection. With a few exceptions, notably the identification of the Duffy binding proteins of *P. vivax* and other malarias, this search has proved both difficult and been poorly rewarded. It may be worth considering exactly why this should be so.

The biology of the malaria parasite provides the answers. The blood stage life cycle involves multiple rounds of growth, before symptoms appear. It is also clear that the parasite uses alternative parallel pathways for invasion of new red cells and adhesion of infected red blood cells to host endothelium. The difficulty faced by experimental biologists is therefore twofold. First, relatively small changes in any one parameter may have a significant effect over multiple growth cycles. In the laboratory it is difficult to measure differences in growth or adhesion of the order of 10%. However, such differences may provide significant biological effects over several rounds of growth. Second, the use of multiple receptors for a single function may mask significant host–parasite interactions in an experiment designed only to look at a single pathway. Examining the effect of host polymorphisms coding for immune response genes faces additional difficulties. Frequently, multiple cell types may express a cytokine or surface molecule and the multiplicity of functions may make the prediction of the outcome of infection difficult from single assays.

We must ask ourselves whether we actually understand more about the pathogenesis of malaria through malaria resistance traits? It is tempting to answer that malaria has taught us more about genetics than genetics has about malaria. The hurdles for a more useful application of genetic epidemiology to understanding pathogenesis have been discussed and include the uncertainty of the significance of protective genetic traits identified in single studies. There is also considerable ignorance regarding the cellular mechanisms of pathology.

In the future, it seems the most constructive approach would be, on one hand, a more sophisticated approach to the study of the genetic epidemiology, including the use of halpotype analysis and comparison of polymorphisms in multiple case-control studies and, on the other hand, close integration of genetic epidemiology and laboratory and clinical studies. With these ideals we may well achieve a truly useful application of genetics to malaria.

ACKNOWLEDGMENTS

D. J. R. and T. H. are supported by the National Blood Service UK and the Howard Hughes Medical Institute. T.W. is supported by the Wellcome Trust UK. We thank Dr. A. Norton for comments on the chapter.

REFERENCES

Abel, L., Cot, M., Mulder, L. et al. (1992). Segregation analysis detects a major gene controlling blood infection levels in human malaria. *Am. J. Hum. Genet.* **50**, 1308–1317.

Abraham, L. J. and Kroeger, K. M. (1999). Impact of the -308 TNF promoter polymorphism on the transcriptional regulation of the TNF gene: Relevance to disease. *J. Leukoc. Biol.* **66**, 562–566.

Abu-Zeid, Y. A., Abdulhadi, N. H., Theander, T. G., et al. (1992). Seasonal changes in cell mediated immune responses to soluble *Plasmodium falciparum* antigens in children with haemoglobin AA and haemoglobin AS. *Trans. R. Soc. Trop. Med. Hyg.* **86**, 20–22.

Agarwal, A., Guindo, A., Cissoko, Y., et al. (2000). Hemoglobin C associated with protection from severe malaria in the Dogon of Mali, a West African population with a low prevalence of hemoglobin S. *Blood* **96**, 2358–2363.

Aidoo, M., McElroy, P. D., Kolczak, M. S., et al. (2001). Tumor necrosis factor-alpha promoter variant 2 (TNF2) is associated with pre-term delivery, infant mortality, and malaria morbidity in western Kenya: Asembo Bay Cohort Project IX. *Genet. Epidemiol.* **21**, 201–211.

Aidoo, M., Terlouw, D. J., Kolczak, M. S., et al. (2002). Protective effects of the sickle cell gene against malaria morbidity and mortality. *Lancet* **359**, 1311–1312.

Aitman, T. J., Cooper, L. D., Norsworthy, P. J., et al. (2000). Malaria susceptibility and CD36 mutation. *Nature* **405**, 1015–1016.

Allen, S. J., O'Donnell, A., Alexander, N. D., et al. (1997). Alpha+-Thalassemia protects children against disease caused by other infections as well as malaria. *Proc. Natl. Acad. Sci. USA* **94**, 14736–14741.

Allen, S. J., O'Donnell, A., Alexander, N. D., et al. (1999). Prevention of cerebral malaria in children in Papua New Guinea by Southeast Asian ovalocytosis band 3. *Am. J. Trop. Med. Hyg.* **60**, 1056–1060.

Allen, S. J., Rowe, P., Allsopp, C. E., et al. (1993). A prospective study of the influence of alpha thalassaemia on morbidity from malaria and immune responses to defined *Plasmodium falciparum* antigens in Gambian children. *Trans. R. Soc. Trop. Med. Hyg.* **87**, 282–285.

Allison, A. C. (1954). Protection afforded by sickle cell trait against subtertian malarial infection. *Br. Med. J.* **1**, 290–295.

Amato, D. and Booth, P. B. (1977). Hereditary ovalocytosis in Melanesians. *P. N. G. Med. J.* **20**, 26–32.

Barragan, A., Kremsner, P. G., Wahlgren, M., et al. (2000). Blood group A antigen is a coreceptor in *Plasmodium falciparum* rosetting. *Infect. Immun.* **68**, 2971–2975.

Bayoumi, R. A. (1987). The sickle-cell trait modifies the intensity and specificity of the immune response against *P. falciparum* malaria and leads to acquired protective immunity. *Med. Hypotheses* **22**, 287–298.

Bayoumi, R. A. (1997). Does the mechanism of protection from falciparum malaria by red cell genetic disorders involve a switch to a balanced TH1/TH2 cytokine production mode? *Med. Hypotheses* **48**, 11–17.

Beet, E. A. (1946). Sickle cell disease in the Balovale district of Northern Rhodesia. *E. African Med. J.* **23**, 75–86.

Bellamy, R., Kwiatkowski, D. and Hill, A. V. (1998a). Absence of an association between intercellular adhesion molecule 1, complement receptor 1 and interleukin 1 receptor antagonist gene polymorphisms and severe malaria in a West African population. *Trans. R. Soc. Trop. Med. Hyg.* **92**, 312–316.

Bellamy, R., Ruwende, C., McAdam, K. P., et al. (1998b). Mannose binding protein deficiency is not associated with malaria, hepatitis B carriage nor tuberculosis in Africans. *Q. J. Med.* **91**, 13–18.

Bennett, V. and Baines, A. J. (2001). Spectrin and ankyrin-based pathways: Metazoan inventions for integrating cells into tissues. *Physiol. Rev.* **81**, 1353–1392.

Bienzle, U., Ayeni, O., Lucas, A. O., et al. (1972). Glucose-6-phosphate dehydrogenase and malaria: greater resistance of females heterozygous for enzyme deficiency and of males with non-deficient variant. *Lancet* **1**(7742), 107–10.

Booth, P. B., Serjeantson, S., Woodfield, D. G., et al. (1977). Selective depression of blood group antigens associated with hereditary ovalocytosis among melanesians. *Vox Sang* **32**(2), 99–110.

Breuer, W. V., Kahane, I., Baruch, D., et al. (1983). Role of internal domains of glycophorin in *Plasmodium falciparum* invasion of human erythrocytes. *Infect Immun* **42**(1), 133–40.

Brindle, P. M., Maitland, K., Williams, T. N., et al. (1995). A survey for the rare blood group antigen variants, En(a-), Gerbich negative and Duffy negative on Espiritu Santo, Vanuatu in the South Pacific. *Hum. Hered.* **45**, 211–214.

Brittenham, G., Lozoff, B., Harris, J. W., et al. (1980). Thalassemia in southern India. Interaction of genes for beta$^+$-, beta 0-, and delta 0 beta 0-thalassemia. *Acta. Haematol.* **63**, 44–48.

Brockelman, C. R., Wongsattayanont, B., Tan-ariya, P., et al. (1987). Thalassemic erythrocytes inhibit in vitro growth of *Plasmodium falciparum*. *J. Clin. Microbiol.* **25**, 56–60.

Bunyaratvej, A., Butthep, P., Sae-Ung, N., et al. (1992). Reduced deformability of thalassemic erythrocytes and erythrocytes with abnormal hemoglobins and relation with susceptibility to *Plasmodium falciparum* invasion. *Blood* **79**, 2460–2463.

Bunyaratvej, A., Butthep, P., Yuthavong, Y., et al. (1986). Increased phagocytosis of *Plasmodium falciparum*-infected erythrocytes with haemoglobin E by peripheral blood monocytes. *Acta. Haematol.* **76**, 155–158.

Burgner, D., Xu, W., Rockett, K., et al. (1998). Inducible nitric oxide synthase polymorphism and fatal cerebral malaria. *Lancet* **352**, 1193–1194.

Cappadoro, M., Giribaldi, G., O'Brien, E., et al. (1998). Early phagocytosis of glucose-6-phosphate dehydrogenase (G6PD)-deficient erythrocytes parasitized by *Plasmodium falciparum* may explain malaria protection in G6PD deficiency. *Blood* **92**, 2527–2534.

Carlson, J. (1993). Erythrocyte rosetting in *Plasmodium falciparum* malaria – with special reference to the pathogenesis of cerebral malaria. *Scand. J. Infect. Dis. Suppl.* **86**, 1–79.

Carlson, J., Nash, G. B., Gabutti, V., et al. (1994). Natural protection against severe *Plasmodium falciparum* malaria due to impaired rosette formation. *Blood* **84**, 3909–3914.

Carlson, J. and Wahlgren, M. (1992). *Plasmodium falciparum* erythrocyte rosetting is mediated by promiscuous lectin-like interactions. *J. Exp. Med.* **176**, 1311–1317.

Cartron, J. P., Prou, O., Luilier, M., et al. (1983). Susceptibility to invasion by *Plasmodium falciparum* of some human erythrocytes carrying rare blood group antigens. *Br. J. Haematol.* **55**, 639–647.

Casals-Pascual, C., Allen, S., Allen, A., et al. (2001). Short report: codon 125 polymorphism of CD31 and susceptibility to malaria. *Am. J. Trop. Med. Hyg.* **65**, 736–737.

Chishti, A. H., Palek, J., Fisher, D., et al. (1996). Reduced invasion and growth of *Plasmodium falciparum* into elliptocytic red blood cells with a combined deficiency of protein 4.1, glycophorin C, and p55. *Blood* **87**, 3462–3469.

Chitnis, C. E. (2001). Molecular insights into receptors used by malaria parasites for erythrocyte invasion. *Curr. Opin. Hematol.* **8**, 85–91.

Chitnis, C. E. and Miller, L. H. (1994). Identification of the erythrocyte binding domains of *Plasmodium vivax* and *Plasmodium knowlesi* proteins involved in erythrocyte invasion. *J. Exp. Med.* **180**, 497–506.

Chotivanich, K. T., Udomsangpetch, R., Pipitaporn, B., et al. (1998). Rosetting characteristics of uninfected erythrocytes from healthy individuals and malaria patients. *Ann. Trop. Med. Parasitol.* **92**, 45–56.

Cornille-Brogger, R., Fleming, A. F., Kagan, I., et al. (1979). Abnormal haemoglobins in the Sudan savanna of Nigeria. II. Immunological response to malaria in normals and subjects with sickle cell trait. *Ann. Trop. Med. Parasitol.* **73**, 173–183.

Craig, A., Fernandez-Reyes, D., Mesri, M., et al. (2000). A functional analysis of a natural variant of intercellular adhesion molecule-1 (ICAM-1Kilifi). *Hum. Mol. Genet.* **9**, 525–530.

Craig, A. and Scherf, A. (2001). Molecules on the surface of the *Plasmodium falciparum* infected erythrocyte and their role in malaria pathogenesis and immune evasion. *Mol. Biochem. Parasitol.* **115**, 129–143.

Dunn, F. L. (1965). On the antiquity of malaria in the western hemisphere. *Hum. Biol.* **37**, 385–393.

Edozien, J. C., Boyo, A. E., and Morley, D. C. (1960). The relationship of serum gamma-globulin concentration to malaria and sickling. *J. Clin. Pathol.* **13**, 118–123.

Facer, C. A. (1989). Malaria, hereditary elliptocytosis, and pyropoikilocytosis. *Lancet* **1**, 897.

Facer, C. A. (1995). Erythrocytes carrying mutations in spectrin and protein 4.1 show differing sensitivities to invasion by *Plasmodium falciparum*. *Parasitol. Res.* **81**, 52–57.

Fernandez-Reyes, D., Craig, A. G., Kyes, S. A., et al. (1997). A high frequency African coding polymorphism in the N-terminal domain of ICAM-1 predisposing to cerebral malaria in Kenya. *Hum. Mol. Genet.* **6**, 1357–1360.

Fischer, P. R. and Boone, P. (1998). Short report: severe malaria associated with blood group. *Am. J. Trop. Med. Hyg.* **58**, 122–123.

Flatz, G., Pik, C. and Sringam, S. (1965). Haemoglobin E and beta-thalassaemia: their distribution in Thailand. *Ann. Hum. Genet.* **29**, 151–170.

Fleming, A. F., Storey, J., Molineaux, L., et al. (1979). Abnormal haemoglobins in the Sudan savanna of Nigeria. I. Prevalence of haemoglobins and relationships between sickle cell trait, malaria and survival. *Ann. Trop. Med. Parasitol.* **73**, 161–172.

Flint, J., Hill, A. V., Bowden, D. K., et al. (1986). High frequencies of alpha-thalassaemia are the result of natural selection by malaria. *Nature* **321**, 744–750.

Friedman, M. J. (1978). Erythrocytic mechanism of sickle cell resistance to malaria. *Proc. Natl. Acad. Sci. USA* **75**, 1994–1997.

Friedman, M. J., Blankenberg, T., Sensabaugh, G., et al. (1984). Recognition and invasion of human erythrocytes by malarial parasites: contribution of sialoglycoproteins to attachment and host specificity. *J. Cell Biol.* **98**, 1672–1677.

Friedman, M. J., Roth, E. F., Nagel, R. L., et al. (1979). The role of hemoglobins C, S, and Nbalt in the inhibition of malaria parasite development in vitro. *Am. J. Trop. Med. Hyg.* **28**, 777–780.

Genton, B., al-Yaman, F., Mgone, C. S., et al. (1995). Ovalocytosis and cerebral malaria. *Nature* **378**, 564–565.

Gilbert, S. C., Plebanski, M., Gupta, S., et al. (1998). Association of malaria parasite population structure, HLA, and immunological antagonism. *Science* **279**, 1173–1177.

Gilles, H. M., Fletcher, K. A., Hendrickse, R. G., et al. (1967). Glucose-6-phosphate-dehydrogenase deficiency, sickling, and malaria in African children in South Western Nigeria. *Lancet* **1**, 138–140.

Ginsburg, H., Handeli, S., Friedman, S., et al. (1986). Effects of red blood cell potassium and hypertonicity on the growth of *Plasmodium falciparum* in culture. *Z. Parasitenkd* **72**, 185–199.

Golenser, J., Miller, J., Spira, D. T., et al. (1983). Inhibitory effect of a fava bean component on the in vitro development of *Plasmodium falciparum* in normal and glucose-6-phosphate dehydrogenase deficient erythrocytes. *Blood* **61**, 507–510.

Guggenmoos-Holzmann, I., Bienzle, U. and Luzzatto, L. (1981). *Plasmodium falciparum* malaria and human red cells. II. Red cell genetic traits and resistance against malaria. *Int. J. Epidemiol.* **10**, 16–22.

Hadley, T. J. and Miller, L. H. (1988). Invasion of erythrocytes by malaria parasites: erythrocyte ligands and parasite receptors. *Prog. Allergy* **41**, 49–71.

Haldane, J. B. S. (1949). Disease and evolution. *La Ricerca Scientifica* **19**(Suppl.), 68–75.

Hananantachai, H., Patarapotikul, J., Looareesuwan, S., et al. (2001). Lack of association of −308A/G TNFA promoter and 196R/M TNFR2 polymorphisms with disease severity in Thai adult malaria patients. *Am. J. Med. Genet.* **102**, 391–392.

Higgs, D. R., Hill, A. V., Bowden, D. K., et al. (1984). Independent recombination events between the duplicated human alpha globin genes: implications for their concerted evolution. *Nucleic Acids Res.* **12**, 6965–6977.

Hill, A. V. (1986). The population genetics of alpha-thalassemia and the malaria hypothesis. *Cold Spring Harb. Symp. Quant. Biol.* **51**(Pt 1), 489–498.

Hill, A. V., Allsopp, C. E., Kwiatkowski, D., et al. (1991). Common west African HLA antigens are associated with protection from severe malaria. *Nature* **352**, 595–600.

Hill, A. V., Bowden, D. K., O'Shaughnessy, D. F., et al. (1988). Beta thalassemia in Melanesia: association with malaria and characterization of a common variant (IVS-1 nt 5 G – C). *Blood* **72**, 9–14.

Hill, A. V., Elvin, J., Willis, A. C., et al. (1992). Molecular analysis of the association of HLA-B53 and resistance to severe malaria. *Nature* **360**, 434–439.

Ho, M. and White, N. J. (1999). Molecular mechanisms of cytoadherence in malaria. *Am. J. Physiol.* **276**, C1231–1242.

Hobbs, M. R., Udhayakumar, V., Levesque, M. C., et al. (2002). A new NOS2 promoter polymorphism associated with increased nitric oxide production and protection from severe malaria in Tanzanian and Kenyan children. *Lancet* **360**, 1468–1475.

Hoffman, S. L., Doolan, D. L., Sedegah, M., et al. (1997). Toward clinical trials of DNA vaccines against malaria. *Immunol. Cell. Biol.* **75**, 376–381.

Honig, G. R., Lacson, P. S. and Maurer, H. S. (1971). A new familial disorder with abnormal erythrocyte morphology associated with increased permeability of erythrocytes to sodium and potassium. *Pediatr. Res.* **5**, 159–166.

Horuk, R., Chitnis, C. E., Darbonne, W. C., et al. (1993). A receptor for the malarial parasite *Plasmodium vivax*: the erythrocyte chemokine receptor. *Science* **261**, 1182–1184.

Ifediba, T. C., Stern, A., Ibrahim, A., et al. (1985). *Plasmodium falciparum in vitro*: diminished growth in hemoglobin H disease erythrocytes. *Blood* **65**, 452–455.

Jakobsen, P. H., Riley, E. M., Allen, S. J., et al. (1991). Differential antibody response of Gambian donors to soluble *Plasmodium falciparum* antigens. *Trans. R. Soc. Trop. Med. Hyg.* **85**, 26–32.

Jarolim, P., Palek, J., Amato, D., et al. (1991). Deletion in erythrocyte band 3 gene in malaria-resistant Southeast Asian ovalocytosis. *Proc. Natl. Acad. Sci. USA* **88**, 11022–11026.

Jepson, A., Sisay-Joof, F., Banya, W., et al. (1997). Genetic linkage of mild malaria to the major histocompatibility complex in Gambian children: study of affected sibling pairs. *Br. Med. J.* **315**, 96–97.

Kaminsky, R., Kruger, N., Hempelmann, E., et al. (1986). Reduced development of *Plasmodium falciparum* in beta-thalassaemic erythrocytes. *Z. Parasitenkd.* **72**, 553–556.

Kidson, C., Lamont, G., Saul, A., et al. (1981). Ovalocytic erythrocytes from Melanesians are resistant to invasion by malaria parasites in culture. *Proc. Natl. Acad. Sci. USA* **78**, 5829–5832.

Kikuchi, M., Looareesuwan, S., Ubalee, R., et al. (2001). Association of adhesion molecule PECAM-1/CD31 polymorphism with susceptibility to cerebral malaria in Thais. *Parasitol. Int.* **50**, 235–239.

Klebs, E. and Tomassi-Crudeli, C. (1888). 'On the nature of malaria'. Translated from Reale Academia dei Lincei, Roma.

Knight, J. C. and Kwiatkowski, D. (1999). Inherited variability of tumor necrosis factor production and susceptibility to infectious disease. *Proc. Assoc. Am. Physicians* **111**, 290–298.

Knight, J. C., Udalova, I., Hill, A. V., et al. (1999). A polymorphism that affects OCT-1 binding to the TNF promoter region is associated with severe malaria. *Nat. Genet.* **22**, 145–150.

Knyszynski, A., Danon, D., Kahane, I., et al. (1979). Phagocytosis of nucleated and mature beta thalassaemic red blood cells by mouse macrophages in vitro. *Br. J. Haematol.* **43**, 251–255.

Koch, O., Awomoyi, A., Usen, S., et al. (2002). IFNGR1 gene promoter polymorphisms and susceptibility to cerebral malaria. *J. Infect. Dis.* **185**, 1684–1687.

Kulozik, A. E., Kar, B. C., Serjeant, G. R., et al. (1988). The molecular basis of alpha thalassemia in India: its interaction with the sickle cell gene. *Blood* **71**, 467–472.

Kun, J. F., Klabunde, J., Lell, B., et al. (1999). Association of the ICAM-1Kilifi mutation with protection against severe malaria in Lambarene, Gabon. *Am. J. Trop. Med. Hyg.* **61**, 776–779.

Kun, J. F., Mordmuller, B., Perkins, D. J., et al. (2001). Nitric oxide synthase 2 (Lambarene) (G-954C), increased nitric oxide production, and protection against malaria. *J. Infect. Dis.* **184**, 330–336.

Kyes, S., Taylor, H., Craig, A., et al. (1997). Genomic representation of *var* gene sequences in *Plasmodium falciparum* field isolates from different geographic regions. *Mol. Biochem. Parasitol.* **87**, 235–238.

Labie, D., Srinivas, R., Dunda, O., et al. (1989). Haplotypes in tribal Indians bearing the sickle gene: evidence for the unicentric origin of the beta S mutation and the unicentric origin of the tribal populations of India. *Hum. Biol.* **61**, 479–491.

Lell, B., May, J., Schmidt-Ott, R. J., et al. (1999). The role of red blood cell polymorphisms in resistance and susceptibility to malaria. *Clin. Infect. Dis.* **28**, 794–799.

Levesque, M. C., Hobbs, M. R., Anstey, N. M., et al. (1999). Nitric oxide synthase type 2 promoter polymorphisms, nitric oxide production, and disease severity in Tanzanian children with malaria. *J. Infect. Dis.* **180**, 1994–2002.

Liu, S. C., Zhai, S., Palek, J., et al. (1990). Molecular defect of the band 3 protein in southeast Asian ovalocytosis. *N. Engl. J. Med.* **323**, 1530–1538.

Luoni, G., Verra, F., Arca, B., et al. (2001). Antimalarial antibody levels and IL4 polymorphism in the Fulani of West Africa. *Genes Immun.* **2**, 411–414.

Luzzatto, L., Nwachuku-Jarrett, E. S. and Reddy, S. (1970). Increased sickling of parasitised erythrocytes as mechanism of resistance against malaria in the sickle-cell trait. *Lancet* **1**, 319–321.

Luzzi, G. A., Merry, A. H., Newbold, C. I., et al. (1991). Surface antigen expression on *Plasmodium falciparum*-infected erythrocytes is modified in alpha- and beta-thalassemia. *J. Exp. Med.* **173**, 785–791.

Luzzi, G. A. and Pasvol, G. (1990). Cytoadherence of *Plasmodium falciparum*-infected alpha-thalassaemic red cells. *Ann. Trop. Med. Parasitol.* **84**, 413–414.

Luzzi, G. A., Torii, M., Aikawa, M., et al. (1990). Unrestricted growth of *Plasmodium falciparum* in microcytic erythrocytes in iron deficiency and thalassaemia. *Br. J. Haematol.* **74**, 519–524.

Mackinnon, M. J., Gunawardena, D. M., Rajakaruna, J., et al. (2000). Quantifying genetic and nongenetic contributions to malarial infection in a Sri Lankan population. *Proc. Natl. Acad. Sci. USA* **97**, 12661–12666.

Maier, A. G., Duraisingh, M. T., Reeder, J. C., et al. (2002). *Plasmodium falciparum* erythrocyte invasion through glycophorin C and selection for Gerbich negativity in human populations. *Nat. Med.* **9**, 9.

Malaguarnera, L. and Musumeci, S. (2002). The immune response to *Plasmodium falciparum* malaria. *Lancet Infect. Dis.* **2**, 472–478.

Marsh, K. (1992). Malaria – a neglected disease? *Parasitology* **104**(Suppl.), S53–S69.

Marsh, K., Otoo, L., Hayes, R. J., et al. (1989). Antibodies to blood stage antigens of *Plasmodium falciparum* in rural Gambians and their relation to protection against infection. *Trans. R. Soc. Trop. Med. Hyg.* **83**, 293–303.

Martin, S. K., Miller, L. H., Alling, D., et al. (1979). Severe malaria and glucose-6-phosphate-dehydrogenase deficiency: A reappraisal of the malaria/G-6-P. D. hypothesis. *Lancet* **1**, 524–526.

McGilvray, I. D., Serghides, L., Kapus, A., et al. (2000). Nonopsonic monocyte/macrophage phagocytosis of *Plasmodium falciparum*-parasitized erythrocytes: A role for CD36 in malarial clearance. *Blood* **96**, 3231–3240.

McGuire, W., Hill, A. V., Allsopp, C. E., et al. (1994). Variation in the TNF-alpha promoter region associated with susceptibility to cerebral malaria. *Nature* **371**, 508–510.

McGuire, W., Knight, J. C., Hill, A. V., et al. (1999). Severe malarial anemia and cerebral malaria are associated with different tumor necrosis factor promoter alleles. *J. Infect. Dis.* **179**, 287–290.

McMullin, M. F. (1999). The molecular basis of disorders of the red cell membrane. *J. Clin. Pathol.* **52**, 245–248.

Mehta, A., Mason, P. J. and Vulliamy, T. J. (2000). Glucose-6-phosphate dehydrogenase deficiency. *Baillieres Best Pract. Res. Clin. Haematol.* **13**.

Miller, L. H., Haynes, J. D., McAuliffe, F. M., et al. (1977). Evidence for differences in erythrocyte surface receptors for the malarial parasites, *Plasmodium falciparum* and *Plasmodium knowlesi*. *J. Exp. Med.* **146**, 277–281.

Miller, L. H., Mason, S. J., Clyde, D. F., et al. (1976). The resistance factor to *Plasmodium vivax* in blacks. The Duffy-blood-group genotype, FyFy. *N. Engl. J. Med.* **295**, 302–304.

Miller, L. H., Mason, S. J., Dvorak, J. A., et al. (1975). Erythrocyte receptors for (*Plasmodium knowlesi*) malaria: Duffy blood group determinants. *Science* **189**, 561–563.

Mockenhaupt, F. P., Bienzle, U., May, J., et al. (1999). *Plasmodium falciparum* infection: Influence on hemoglobin levels in alpha-thalassemia and microcytosis. *J. Infect. Dis.* **180**, 925–928.

Modiano, D., Luoni, G., Sirima, B. S., et al. (2001). Haemoglobin C protects against clinical *Plasmodium falciparum* malaria. *Nature* **414**, 305–308.

Modiano, D., Petrarca, V., Sirima, B. S., et al. (1996). Different response to *Plasmodium falciparum* malaria in west African sympatric ethnic groups. *Proc. Natl. Acad. Sci. USA* **93**, 13206–13211.

Modiano, G., Morpurgo, G., Terrenato, L., et al. (1991). Protection against malaria morbidity: near-fixation of the alpha-thalassemia gene in a Nepalese population. *Am. J. Hum. Genet.* **48**, 390–397.

Mohandas, N. and Evans, E. (1984). Adherence of sickle erythrocytes to vascular endothelial cells: Requirement for both cell membrane changes and plasma factors. *Blood* **64**, 282–287.

Morahan, G., Boutlis, C. S., Huang, D., et al. (2002). A promoter polymorphism in the gene encoding interleukin-12 p40 (IL12B) is associated with mortality from cerebral malaria and with reduced nitric oxide production. *Genes Immun.* **3**, 414–418.

Morle, L., Morle, F., Roux, A. F., et al. (1989). Spectrin Tunis (Sp alpha I/78), an elliptocytogenic variant, is due to the CGG – TGG codon change (Arg – Trp) at position 35 of the alpha I domain. *Blood* **74**, 828–832.

Moulds, J. M., Zimmerman, P. A., Doumbo, O. K., et al. (2001). Molecular identification of Knops blood group polymorphisms found in long homologous region D of complement receptor 1. *Blood* **97**, 2879–2885.

Mourant, A. E. (1968). Genetical polymorphisms and the incidence of disease. *Proc. R. Soc. Med.* **61**, 163.

Mourant, A. E., Kopek, A. C. and Domaniewska-Sobczak, K. A. (1976). *The Distribution of the Human Blood Groups.* London/New York/Toronto: Oxford University Press.

Nagel, R. L., Raventos-Suarez, C., Fabry, M. E., et al. (1981). Impairment of the growth of *Plasmodium falciparum* in HbEE erythrocytes. *J. Clin. Invest.* **68**, 303–305.

Ohashi, J., Naka, I., Patarapotikul, J., et al. (2002). Significant association of longer forms of CCTTT microsatellite repeat in the inducible nitric oxide synthase promoter with severe malaria in Thailand. *J. Infect. Dis.* **186**, 578–581.

Oppenheimer, S. J., Hill, A. V., Gibson, F. D., et al. (1987). The interaction of alpha thalassaemia with malaria. *Trans. R. Soc. Trop. Med. Hyg.* **81**, 322–326.

Orjih, A. U., Chevli, R. and Fitch, C. D. (1985). Toxic heme in sickle cells: an explanation for death of malaria parasites. *Am. J. Trop. Med. Hyg.* **34**, 223–227.

Pain, A., Urban, B. C., Kai, O., et al. (2001). A non-sense mutation in Cd36 gene is associated with protection from severe malaria. *Lancet* **357**, 1502–1503.

Palek, J. and Lambert, S. (1990). Genetics of the red cell membrane skeleton. *Semin. Hematol.* **27**, 290–332.

Pandey, K. C., Singh, S., Pattnaik, P., et al. (2002). Bacterially expressed and refolded receptor binding domain of *Plasmodium falciparum* EBA-175 elicits invasion inhibitory antibodies. *Mol. Biochem. Parasitol.* **123**, 23–33.

Pasvol, G., Wainscoat, J. S. and Weatherall, D. J. (1982). Erythrocytes deficienct in glycophorin resist invasion by the malarial parasite *Plasmodium falciparum.* *Nature* **297**, 64–66.

Pasvol, G., Weatherall, D. J. and Wilson, R. J. (1978). Cellular mechanism for the protective effect of haemoglobin S against *P. falciparum* malaria. *Nature* **274**, 701–703.

Patel, S. S., Mehlotra, R. K., Kastens, W., et al. (2001). The association of the glycophorin C exon 3 deletion with ovalocytosis and malaria susceptibility in the Wosera, Papua New Guinea. *Blood* **98**, 3489–3491.

Perkins, M. (1981). Inhibitory effects of erythrocyte membrane proteins on the *in vitro* invasion of the human malarial parasite (*Plasmodium falciparum*) into its host cell. *J. Cell Biol.* **90**, 563–567.

Piguet, P. F., Kan, C. D., Vesin, C., et al. (2001). Role of CD40-CVD40L in mouse severe malaria. *Am. J. Pathol.* **159**, 733–742.

Plebanski, M., Flanagan, K. L., Lee, E. A., et al. (1999). Interleukin 10-mediated immunosuppression by a variant CD4 T cell epitope of *Plasmodium falciparum. Immunity* **10**, 651–660.

Plebanski, M. and Hill, A. V. (2000). The immunology of malaria infection. *Curr. Opin. Immunol.* **12**, 437–441.

Plebanski, M., Lee, E. A., Hannan, C. M., et al. (1999). Altered peptide ligands narrow the repertoire of cellular immune responses by interfering with T-cell priming. *Nat. Med.* **5**, 565–571.

Rachmilewitz, E. A., Treves, A. and Treves, A. J. (1980). Susceptibility of thalassemic red blood cells to phagocytosis by human macrophages in vitro. *Ann. N. Y. Acad. Sci.* **344**, 314–322.

Ranjan, A. and Chitnis, C. E. (1999). Mapping regions containing binding residues within functional domains of *Plasmodium vivax* and *Plasmodium knowlesi* erythrocyte-binding proteins. *Proc. Natl. Acad. Sci. USA* **96**, 14067–14072.

Rihet, P., Traore, Y., Aucan, C., et al. (1999). Genetic dissection of *Plasmodium falciparum* blood infection levels and other complex traits related to human malaria infection. *Parasitologia* **41**, 83–87.

Riopel, J., Tam, M., Mohan, K., et al. (2001). Granulocyte-macrophage colony-stimulating factor-deficient mice have impaired resistance to blood-stage malaria. *Infect. Immun.* **69**, 129–136.

Roth, E. F., Jr., Friedman, M., Ueda, Y., et al. (1978). Sickling rates of human AS red cells infected in vitro with *Plasmodium falciparum* malaria. *Science* **202**, 650–652.

Roth, E. F., Jr., Raventos Suarez, C., Rinaldi, A., et al. (1983). The effect of X chromosome inactivation on the inhibition of *Plasmodium falciparum* malaria growth by glucose-6-phosphate-dehydrogenase-deficient red cells. *Blood* **62**, 866–868.

Roth, E. F., Jr., Raventos-Suarez, C., Rinaldi, A., et al. (1983). Glucose-6-phosphate dehydrogenase deficiency inhibits in vitro growth of *Plasmodium falciparum. Proc. Natl. Acad. Sci. USA* **80**, 298–299.

Rowe, A., Obeiro, J., Newbold, C. I., et al. (1995). *Plasmodium falciparum* rosetting is associated with malaria severity in Kenya. *Infect. Immun.* **63**, 2323–2326.

Rowe, J. A., Moulds, J. M., Newbold, C. I., et al. (1997). *P. falciparum* rosetting mediated by a parasite-variant erythrocyte membrane protein and complement-receptor 1. *Nature* **388**, 292–295.

Rowe, J. A., Raza, A., Diallo, D. A., et al. (2002). Erythrocyte CR1 expression level does not correlate with a *Hind* III restriction fragment length polymorphism in Africans: Implications for studies on malaria susceptibility. *Genes Immun.* **3**, 497–500.

Ruwende, C., Khoo, S. C., Snow, R. W., et al. (1995). Natural selection of hemi- and heterozygotes for G6PD deficiency in Africa by resistance to severe malaria. *Nature* **376**, 246–249.

Sabeti, P., Usen, S., Farhadian, S., et al. (2002). CD40L association with protection from severe malaria. *Genes Immun.* **3**, 286–291.

Sabeti, P. C., Reich, D. E., Higgins, J. M., et al. (2002). Detecting recent positive selection in the human genome from haplotype structure. *Nature* **419**, 832–837.

Schofield, A. E., Reardon, D. M. and Tanner, M. J. (1992). Defective anion transport activity of the abnormal band 3 in hereditary ovalocytic red blood cells. *Nature* **355**, 836–838.

Schofield, A. E., Tanner, M. J., Pinder, J. C., et al. (1992). Basis of unique red cell membrane properties in hereditary ovalocytosis. *J. Mol. Biol.* **223**, 949–958.

Schulze, T. G. and McMahon, F. J. (2002). Genetic association mapping at the crossroads: which test and why? Overview and practical guidelines. *Am. J. Med. Genet.* **114**, 1–11.

Schwarzer, E., Turrini, F., Giribaldi, G., et al. (1993). Phagocytosis of *P. falciparum* malarial pigment hemozoin by human monocytes inactivates monocyte protein kinase C. *Biochim. Biophys. Acta.* **1181**, 51–54.

Senok, A. C., Li, K., Nelson, E. A., et al. (1997). Invasion and growth of *Plasmodium falciparum* is inhibited in fractionated thalassaemic erythrocytes. *Trans. R. Soc. Trop. Med. Hyg.* **91**, 138–143.

Serjeantson, S., Bryson, K., Amato, D., et al. (1977). Malaria and hereditary ovalocytosis. *Hum. Genet.* **37**, 161–167.

Shear, H. L., Roth, E. F., Jr., Fabry, M. E., et al. (1993). Transgenic mice expressing human sickle hemoglobin are partially resistant to rodent malaria. *Blood* **81**, 222–226.

Shi, Y. P., Nahlen, B. L., Kariuki, S., et al. (2001). Fcgamma receptor IIa (CD32) polymorphism is associated with protection of infants against high-density *Plasmodium falciparum* infection. VII. Asembo Bay Cohort Project. *J. Infect. Dis.* **184**, 107–111.

Singh, S., Pandey, K., Chattopadhayay, R., et al. (2001). Biochemical, biophysical, and functional characterization of bacterially expressed and refolded receptor binding domain of Plasmodium vivax duffy-binding protein. *J. Biol. Chem.* **276**, 17111–17116.

Siniscalco, M., Bernini, L., Latte, B., et al. (1961). Favism and thalassaemia in Sardinia and their relationship to malaria. *Nature* **190**, 1179–1180.

Su, Z. and Stevenson, M. M. (2002). IL-12 is required for antibody-mediated protective immunity against blood-stage *Plasmodium chabaudi* AS malaria infection in mice. *J. Immunol.* **168**, 1348–1355.

Terrenato, L., Shrestha, S., Dixit, K. A., et al. (1988). Decreased malaria morbidity in the Tharu people compared to sympatric populations in Nepal. *Ann. Trop. Med. Parasitol.* **82**, 1–11.

Todd, J. A. (1999). From genome to aetiology in a multifactorial disease, type 1 diabetes. *Bioessays* **21**, 164–174.

Trager, W. and Jensen, J. B. (1976). Human malaria parasites in continuous culture. *Science* **193**, 673–675.

Tsuboi, T., Kappe, S. H., al-Yaman, F., et al. (1994). Natural variation within the principal adhesion domain of the *Plasmodium vivax* duffy binding protein. *Infect. Immun.* **62**, 5581–5586.

Udomsangpetch, R., Sueblinvong, T., Pattanapanyasat, K., et al. (1993). Alteration in cytoadherence and rosetting of *Plasmodium falciparum*-infected thalassemic red blood cells. *Blood* **82**, 3752–3759.

Udomsangpetch, R., Todd, J., Carlson, J., et al. (1993). The effects of hemoglobin genotype and ABO blood group on the formation of rosettes by *Plasmodium falciparum*-infected red blood cells. *Am. J. Trop. Med. Hyg.* **48**, 149–153.

Urban, B. C., Ferguson, D. J., Pain, A., et al. (1999). *Plasmodium falciparum*-infected erythrocytes modulate the maturation of dendritic cells. *Nature* **400**, 73–77.

Urban, B. C. and Roberts, D. J. (2002). Malaria, monocytes, macrophages and myeloid dendritic cells: sticking of infected erythrocytes switches off host cells. *Curr. Opin. Immunol.* **14**, 458–465.

Urban, B. C., Willcox, N. and Roberts, D. J. (2001). A role for CD36 in the regulation of dendritic cell function. *Proc. Natl. Acad. Sci. USA* **98**, 8750–8755.

Usanga, E. A. and Luzzatto, L. (1985). Adaptation of *Plasmodium falciparum* to glucose 6-phosphate dehydrogenase-deficient host red cells by production of parasite-encoded enzyme. *Nature* **313**, 793–795.

Wattavidanage, J., Carter, R., Perera, K. L., et al. (1999). TNFalpha*2 marks high risk of severe disease during *Plasmodium falciparum* malaria and other infections in Sri Lankans. *Clin. Exp. Immunol.* **115**, 350–355.

Weatherall, D. J. and Clegg, J. (2002). *The Thalassaemias*. Oxford: Oxford University Press.

Willcox, M., Bjorkman, A., Brohult, J., et al. (1983). A case-control study in northern Liberia of *Plasmodium falciparum* malaria in haemoglobin S and beta-thalassaemia traits. *Ann. Trop. Med. Parasitol.* **77**, 239–246.

Williams, T. N., Maitland, K., Bennett, S., et al. (1996). High incidence of malaria in alpha-thalassaemic children. *Nature* **383**, 522–525.

Williams, T. N., Maitland, K., Ganczakowski, M., et al. (1996). Red blood cell phenotypes in the alpha + thalassaemias from early childhood to maturity. *Br. J. Haematol.* **95**, 266–272.

Williams, T. N., Weatherall, D. J. and Newbold, C. I. (2002). The membrane characteristics of *Plasmodium falciparum*-infected and -uninfected heterozygous alpha0-thalassaemic erythrocytes. *Br. J. Haematol.* **118**, 663–670.

Wilson, A. G., Symons, J. A., McDowell, T. L., et al. (1997). Effects of a polymorphism in the human tumor necrosis factor alpha promoter on transcriptional activation. *Proc. Natl. Acad. Sci. USA* **94**, 3195–3199.

Yenchitsomanus, P., Summers, K. M., Board, P. G., et al. (1986). Alpha-thalassemia in Papua New Guinea. *Hum. Genet.* **74**, 432–437.

Yuthavong, Y., Butthep, P., Bunyaratvej, A., et al. (1987). Inhibitory effect of beta zero-thalassaemia/haemoglobin E erythrocytes on *Plasmodium falciparum* growth in vitro. *Trans. R. Soc. Trop. Med. Hyg.* **81**, 903–906.

Yuthavong, Y., Butthep, P., Bunyaratvej, A., et al. (1988). Impaired parasite growth and increased susceptibility to phagocytosis of *Plasmodium falciparum* infected alpha-thalassemia or hemoglobin Constant Spring red blood cells. *Am. J. Clin. Pathol.* **89**, 521–525.

Zimmerman, P. A., Woolley, I., Masinde, G. L., et al. (1999). Emergence of FY*A(null) in a Plasmodium vivax-endemic region of Papua New Guinea. *Proc. Natl. Acad. Sci. USA* **96**, 13973–13977.

CHAPTER 7

Polymorphic chemokine receptor and ligand genes in HIV infection

Jianming (James) Tang
Division of Geographic Medicine, Department of Medicine, School of Medicine

Richard A. Kaslow
Department of Epidemiology and International Health, School of Public Health, University of Alabama at Birmingham

Human immunogenetic studies beginning in 1996 have produced clear evidence that initial acquisition of HIV-1 infection can be effectively blocked by homozygosity for a 32-bp deletion (Δ32) in the open reading frame of the beta (C-C motif) chemokine receptor 5 (*CCR5*) and further inhibited by the Δ32 heterozygous genotype or by Δ32 in combination with another mutation that also introduces a premature stop codon in *CCR5*. Conversely, homozygosity for the *CCR2-CCR5* HHE haplotype defined by several single-nucleotide polymorphisms (SNPs) appears to enhance HIV-1 acquisition. The two closely related *CCR2-CCR5* haplotypes HHE and HHG*2 (=*CCR5*-Δ32) and probably others [e.g., HHF*2 (=CCR2-64I)] are also associated with varying rates of HIV-1 disease progression against certain ethnic backgrounds. Additional but less consistent associations with both HIV-1 infection and disease progression have been documented for *SDF-1*, RANTES (*SCYA5*), *CX3CR1*, and *MIP-1α* polymorphisms within the chemokine receptor and ligand system. Both chance association and population heterogeneity probably account for some of the inconsistencies. More recent recognition of *CCR2-CCR5* haplotype-mediated effects on HIV-1 RNA concentration implies that CCR polymorphisms are important early determinants of the virus–host equilibrium. Evolving usage of chemokine receptors by HIV-1 may cloud the interpretation of newly acquired data and impede translation of this research into improvements in clinical care. The functional complexity of the chemokine system and its interactions with other host and viral factors calls for a comprehensive analytic approach to the elucidation of immunogenetic influences on HIV/AIDS and vigilance for effects of viral adaptation.

Family	CLs	CRs	
the α (C-X-C motif) family	CXCL1 (GROα)	CXCR2	→ DARC
	CXCL2 (GROβ)	CXCR2	
	CXCL3 (GROγ)	CXCR2	
	CXCL4/SCYB4	Heparin	
	CXCL5 (ENA78)	CXCR2	→ DARC
	CXCL6 (GCP2)	CXCR1, 2	
	CXCL7 (NAP-2)	CXCR2	→ DARC
	CXCL8 (IL-8)	CXCR1, 2	→ DARC
	CXCL9 (Mig)	CXCR3	→ CCR3
	CXCL10 (IP10)	CXCR3	→ CCR3
	CXCL11 (IP-9, I-TAC)	CXCR3	→ CCR3
	CXCL12 (SDF-1)	CXCR4 (Fusin)	
	CXCL13 (BLR-1, BCA-1)	CXCR5	
	SCYB14 (BRAK)	?	
	CXCL16	CXCR6	
the β (C-C motif) family	CCL1 (I309)	CCR8	
	CCL2 (MCP-1)	CCR2, 10	→ DARC
	CCL3 (MIP-1α, LD78α)	CCR5	
	CCL3L1/SCYA3L1 (LD78β)	CCR5	
	CCL4 (MIP-1β)	CCR5	
	CCL5 (RANTES)	CCR5	→ DARC
	CCL7 (MCP-3)	CCR1, 2, 3	→ CCR5 → DARC
	CCL8 (MCP-2)	CCR1, 2, 3	
	CCL11 (Eotaxin-1)	CCR3	→ CCR2
	CCL13 (MCP-4)	CCR2	
	CCL14, 15 (HCC-1, 2)	CCR1	
	CCL16 (LEC)	CCR1	
	CCL17	CCR4	
	SCYA18 (PARC)	?	
	CCL19 (ELC)	CCR7	
	CCL20 (LARC, MIP-3α)	CCR6	
	CCL21 (SLC)	CCR7	
	CCL22 (MDC)	CCR4	
	CCL23 (CKβ8)	CCR1	→ CCR2
	CCL24 (Eotaxin-2)	CCR3	
	CCL25 (TECK)	CCR9 (D6)	→ CCR2
	CCL26 (Eotaxin-3)	CCR3	
	CCL27 (ILC, CTACK)	CCR10	
	CCL28	CCR3, 10	
Others (γ, δ)	XCL1/SCYC1 (Lymphotactin)	XCR1	
	XCL2/SCYC2 (SCM-1β)	XCR2	
	CX_3CL1/SCYD1 (Fractalkine)	CX_3CR1	

Abbreviations:

BRAK, chemokine from breast and kidney cells
ELC, EBI1-ligand chemokine
GCP-2, granulocyte chemotactic protein 2
HCC, hemofiltrate CC chemokine
ILC (CTACK), IL11RA-locus chemokine
LARC, liver and activation-regulated chemokine
MDC, monocyte-derived chemokine
MIP, macrophage inflammatory protein
PARC, pulmonary and activation-regulated chemokine
SDF-1, stromal cell-derived factor 1
SLC, secondary lymphoid tissue chemokine
BLR1, Burkitt lymphoma receptor-1
DARC, Duffy antigen receptor of chemokines
ENA, epithelial cell-derived neutrophil attractant
GRO, growth-related oncogen
IL-8, interleukin 8
IP, interferon γ-inducible protein (cytokine)
MCP, monocyte chemotactic protein
Mig, monokine induced by interferon
NAP, neutrophil-activation protein
RANTES, regulated on activation, normal T-cell expressed and secreted
TECK, thymus-expressed chemokine

1. CLASSIFICATION, EVOLUTION, AND FUNCTION OF CHEMOKINE RECEPTORS AND THEIR LIGANDS

G-protein-coupled, seven-transmembrane-domain chemokine receptors (GPCRs) and their small polypeptide ligands (chemokines) are key regulators of leukocyte trafficking and T-cell differentiation during inflammatory and immune responses (Baggiolini, 1998; Luther and Cyster, 2001; Thelen, 2001). Two major and two minor classes of chemokines can be defined according to the C-X-C (α), C-C (β), C (γ), and C-X_3-C (δ) motifs of their conserved cysteine residues near the N-terminus (Fig. 7.1). The 50 or so structurally and functionally related chemokine ligands (CLs) pair with about 20 chemokine receptors (CRs) either on a strict one-to-one basis or with varying degrees of promiscuity (Fig. 7.1).

Most of the human C-X-C and C-C chemokine genes and several pseudogenes are tandemly arranged at chromosomes 4q12-q21 and 17q11, most likely as a result of multiple gene duplication (Rollins et al., 1991; Naruse et al., 1996; Modi and Chen, 1998; Maho et al., 1999b; Nomiyama et al., 1999; Erdel et al., 2001; Nomiyama et al., 2001) and gene fusion (Tasaki et al., 1999). Likewise, chromosomes 2q34-q35 and 3p21 harbor many of the receptor genes (Ahuja et al., 1992; Combadiere et al., 1995; Raport et al., 1996; Samson et al., 1996; Daugherty and Springer, 1997; Maho et al., 1999a). As the search for novel and orphan chemokine receptor and ligand genes on other chromosomes continues, more members are being identified and mapped (Yoshie et al., 1997; Choe et al., 1998; Fan et al., 1998; Loetscher et al., 1998a; Nomiyama et al., 1998; Guo et al., 1999; Unutmaz et al., 2000; Nelson et al., 2001). Silent (decoy) receptors such as the Duffy antigen and receptor for chemokines (DARC) add to the multiplicity of interactive pathways underlying the receptor–ligand function (Mantovani et al., 2001). Moreover, chemokine- and chemokine receptor-like genes encoded in human

Figure 7.1. Classification of human chemokine receptors (CRs) and their corresponding chemokine ligands (CLs) based on the structure of conserved cysteine (encircled C) motifs. Chemokines CXCL (SCYB) 15, CCL (SCYA) 6, 10, and 12 have been described in the mouse. Population studies have revealed genetic variations at several loci (underlined) often under their common names (in parentheses) instead of systematic names (Zlotnik and Yoshie, 2000). Complex formation between a number of ligands and alternative receptors (indicated by arrows on dotted lines), including CXCR3, CCR2, CCR5, and the decoy receptor DARC, is associated with antagonist effects characterized by the lack of signal transduction (Fig. 8.2). CCR9 (D6) with promiscuous ligand binding in humans is probably another decoy receptor (Nibbs et al., 1997), as is clearly shown in mice (Mantovani et al., 2001).

poxviruses and herpesviruses can further serve as agonists and antagonists (Gompels et al., 1995; Arvanitakis et al., 1997; Endres et al., 1999; Shan et al., 2000; Zhou et al., 2000). The International Union of Pharmacology is leading the effort to catalogue chemokine receptor and ligand genes (Murphy et al., 2000), but the simultaneous use of systematic and common names prevails in current literature (Fig. 7.1).

Phylogenetic analyses indicate that C-X-C (small inducible cytokine B family or SCYB) and C-C chemokine (SCYA) groups and their respective receptors diverged from each other before the emergence of placental mammals (Hughes and Yeager, 1999). The receptors for the MIP (macrophage inflammatory protein) and MCP (monocyte chemotactic protein) C-C chemokine subfamilies, on the other hand, can be either clustered together or scattered on different branches, suggesting limited receptor and ligand coevolution within the C-C subfamilies (Hughes and Yeager, 1999). The extracellular domains EC3 and EC4 in the C-C chemokine receptors (CCR) may have converged independently to produce the promiscuous receptor–ligand pairing of MCP with both CCR2 and CCR3 (Alkhatib et al., 1997).

The expression of CRs and CLs is often characterised by cell-specific regulation, alternative splicing, and posttranslational modification (Mummidi et al., 1997; Youn et al., 1998; Baird et al., 1999; Khoja et al., 2000; Mummidi et al., 2000; Nelson et al., 2001). For example, CCR5 and CCR3 are mainly expressed in Th1 and Th2 cell lineages, respectively (Sallusto et al., 1997; Bonecchi et al., 1998a, 1998b; Loetscher et al., 1998b; Sallusto et al., 1998), whereas CXCR4 is produced by Th2 as well as Th1 cells (Romagnani et al., 2000). As in the regulation of Th1 and Th2 cytokines, histone acetylation and chromatin remodeling often precede the activation of chemokine receptors and their ligand genes (Finzer et al., 2000; Scotet et al., 2001). In addition, many cytokines are involved in the regulation of CRs and CLs. A short list would include interleukin-2 (IL-2) (Wang et al., 1999), IL-4 (Bonecchi et al., 1998b; Zoeteweij et al., 1998), IL-10 (Houle et al., 1999; Mantovani et al., 2001), interferons (Bonecchi et al., 1998b; Penton-Rol et al., 1998; Zella et al., 1998; Zoeteweij et al., 1998; Lee et al., 1999), and TNFα (Hornung et al., 2000). However, these actions also depend heavily on cell populations and cell lineages involved, probably implying the agonist and antagonist cytokine effects frequently associated with the Th1 and Th2 paradigm.

Posttranslational modification of chemokines is mediated by the cell surface dipeptidyl peptidase (DPP) IV (=CD26) and its homologues like DPP8 and FAP. These enzymes hydrolyse N-terminal Ala-Pro, Arg-Pro, and Gly-Pro residues commonly seen in mature chemokines (Abbott et al., 2000). Compared with the full-length LD78β (MIP-1α or CCL3-L1) and RANTES (regulated on activation, normal T-cell expressed and secreted), LD78β

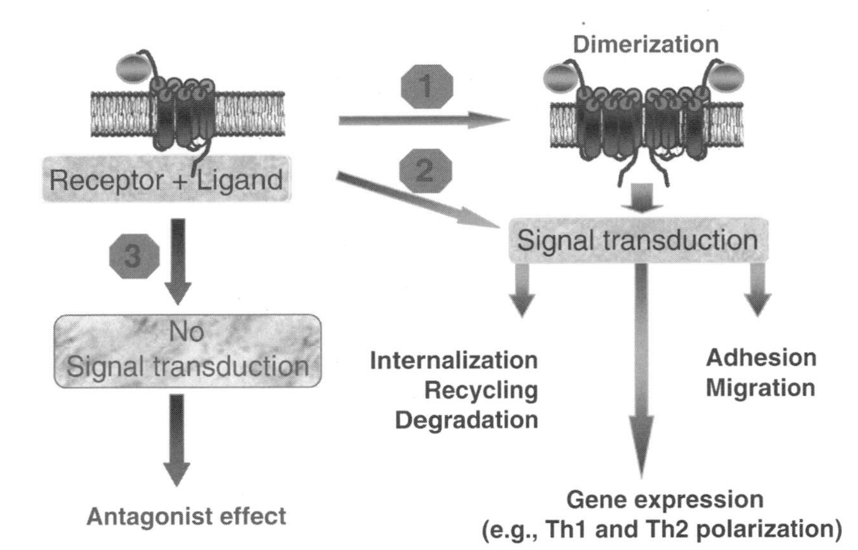

Figure 7.2. Major interactive pathways involving chemokine receptors and their ligands. Dimerisation of receptor–ligand complex has been described for CCR2, CCR5, and CXCR4. Signal transduction leading to the agonist (paths 1 and 2) effects is likely mediated by G proteins as well as Janus kinases (JAK) (Mellado et al., 2001a, 2001b). The antagonist (path 3) effect is driven by the pairing of several functional and decoy receptors with their respective ligands (Fig. 7.1). (See color plate.)

without the NH_2-terminal Ala-Pro dipeptide has a 10-fold higher efficiency in binding chemokine receptors CCR5 and CCR1 (Struyf et al., 2001). In contrast, DPP IV-mediated removal of the NH_2-terminal dipeptide from RANTES reduces its chemotactic potency for monocytes and eosinophils (Struyf et al., 1998; Proost et al., 2000). Introduction of NH_2-terminal pyroglutamic acid (pGlu) protects against degradation by DPP IV, as shown for pGlu-modified MCP-2, which retains its chemotactic activity (Van Coillie et al., 1998).

Depending on the gradient of ligand concentration, receptor–ligand complex formation may lead to hetero- and homodimerisation (Rodriguez-Frade et al., 1999, 2001; Mellado et al., 2001a, 2001b) and internal signal transduction mediated by both G proteins and Janus kinases (Fig. 7.2). The major outcomes include (1) cell adhesion, (2) cell migration into sites of inflammation, (3) internalisation of receptors for recycling and/or degradation, and (4) gene expression in downstream pathways often accompanied by polarisation of helper T-lymphotypes (Th) into the Th1 and Th2 lineages. These outcomes are effectively blocked when antagonist ligands (Blanpain et al., 1999; Loetscher and Clark-Lewis, 2001; Ogilvie et al., 2001) or decoy receptors (Mantovani et al., 2001) are involved in the complex formation (Fig. 7.2).

2. CHEMOKINE SYSTEM AND HIV-1 INFECTION

Recognition of chemokine receptors CCR5 and CXCR4 as the major HIV-1 coreceptors on $CD4^+$ cells (Alkhatib et al., 1996; Feng et al., 1996) has propelled intense research on the interactions between viral products and host molecules in the chemokine system (Deng et al., 1996; Doranz et al., 1996; Liu et al., 1996; Graziosi and Pantaleo, 1997; Scalatti et al., 1997; Zaitseva et al., 1997; Ward et al., 1998). For initiation of infection, the formation of an HIV-1 gp120 and receptor (CD4) coreceptor complex induces conformational changes in gp41, triggering membrane fusion and viral entry into host cells (Wu et al., 1996; Farber and Berger, 2002). Chemokine receptor usage defines three major viral phenotypes, with R5, X4, and dual tropic R5X4 HIV-1 using CCR5, CXCR4, and both CCR5 and CXCR4, respectively (Berger et al., 1998). Despite the availability of multiple coreceptors and especially CXCR4, the dominant viruses derived from primary HIV-1 infection display the R5 phenotype. Active transfer of R5 and not X4 viruses by epithelial cells in the upper gastrointestinal tract correlates with the preferential transmission of R5 HIV-1 via the intestinal mucosa (Meng et al., 2002). Fusion-independent HIV-1 receptors, including dendritic cell-specific ICAM-3-grabbing nonintegrin 1 (DC-SIGN1) and DCSIGN2, may further contribute to the heavy use of CCR5 when these molecules present and promote HIV-1 infection (Geijtenbeek et al., 2000; Pohlmann et al., 2001).

The exact roles of chemokine receptors and their ligands in HIV-1 infection are complicated. High levels of intracellular MIP-1β, a ligand for CCR5, have been linked to low $CD4^+$ cell counts and high viral RNA concentration (viral load) during HIV-1 infection (Tartakovsky et al., 1999), whereas high levels of secreted chemokines may delay clinical AIDS (Garzino-Demo et al., 1999). Patients with rapidly progressive infection typically carry viruses with expanded usage of coreceptors like CCR5, CCR3, CCR2b, and CXCR4 (Connor et al., 1997). Overall, differential distribution of HIV-1 coreceptors on macrophages, activated T cells, and memory T cells correlates with the depletion of specific cell populations mostly in the intestine (Veazey et al., 1998, 2001) and lymphoid tissues (Glushakova et al., 1997, 1998; Grivel and Margolis, 1999; Penn et al., 1999), but not in the lymph nodes or the peripheral blood as originally thought. Studies based on SIV models in macaques have also revealed the gastrointestinal tract as a major site of $CD4^+$ T-cell depletion and viral replication (Veazey et al., 1998), although SIV clearly differs from HIV in coreceptor usage (Chen et al., 1997; Edinger et al., 1997).

It is possible to saturate HIV-1 coreceptors with their natural ligands, thereby blocking HIV-1 transmission. The CXCR4 ligand SDF-1 can prevent $CD4^+$ T cells from being infected by T-cell-line-adapted HIV-1 (Oberlin et al.,

1996), whereas the CCR5 ligands RANTES, MIP-1α (macrophase inflammatory protein 1-α), and MIP-1β can often block infection with the R5 viruses (Margolis et al., 1998) and delay disease progression (Cocchi et al., 1995; Grivel and Margolis, 1999). Several groups have proposed effective manipulations for interfering with the normal function of HIV-1 coreceptors (Schols et al., 1997; Bai et al., 1998; Goila and Banerjea, 1998; Gonzalez et al., 1998; Howard et al., 1998a; Howard et al., 1998b; Bai et al., 2000; Basu et al., 2000; De Clercq, 2000; Steinberger et al., 2000) as a new line of defense against HIV-1 infection in a genotype/subtype-independent manner (Trkola et al., 1998). Although any of these strategies may work well *in vitro*, several obstacles may limit their *in vivo* applications. First and foremost, circulating HIV-1 quasispecies clearly change coreceptor usage. Eliminating a single coreceptor may simply force the virus to seek others. Indeed, in patients with reduced CCR5 expression, viruses seem capable of converting to usage of other coreceptors and the more pathogenic syncytium-inducing (SI) phenotype (D'Aquila et al., 1998). Second, no reliable model predicts how viruses respond to changes in coreceptor availability and competition from ligand or antagonist molecules. For the latter, natural or antagonist ligands bound to coreceptors can either fail to block all sites showing affinity for HIV-1 (Atchison et al., 1996; Rucker et al., 1996; Doranz et al., 1997; Efremov et al., 1998; Hill et al., 1998) or actually enhance HIV-1 infection, as is already seen with RANTES (Chang et al., 2002). As an example, initial success in using a small molecule CCR5 antagonist known as SCH-C (SCH 351125 by Schering-Plough Research Institute) to block HIV-1 entry is now threatened by a major setback with the emergence of escape viruses (Strizki et al., 2001; Trkola et al., 2002). Third, prolonged intervention may disturb the natural chemokine networks and compromise the ability of the host to control other pathogens and autoimmune responses. It is probably more realistic to block regions of HIV-1 gp120 responsible for coreceptor binding (Rizzuto et al., 1998; Salzwedel et al., 2000). The new category of anti-HIV-1 drugs represented by T-20 has already shown the vulnerability of HIV-1 gp41 and gp120 as readily accessible intervention targets (Kilby et al., 1998; Derdeyn et al., 2000).

3. CHEMOKINE RECEPTOR AND LIGAND POLYMORPHISMS: REGIONAL AND GLOBAL DISTRIBUTION AND DELINEATION OF GENOTYPE–PHENOTYPE RELATIONSHIPS

Genotypic variations at the chemokine receptor *DARC* (Duffy antigen and receptor for chemokines) locus clearly illustrate how promoter and coding sequence variations determine phenotypes; how codominant alleles interact to produce a phenotype; how linkage disequilibrium (LD) defines haplotypes;

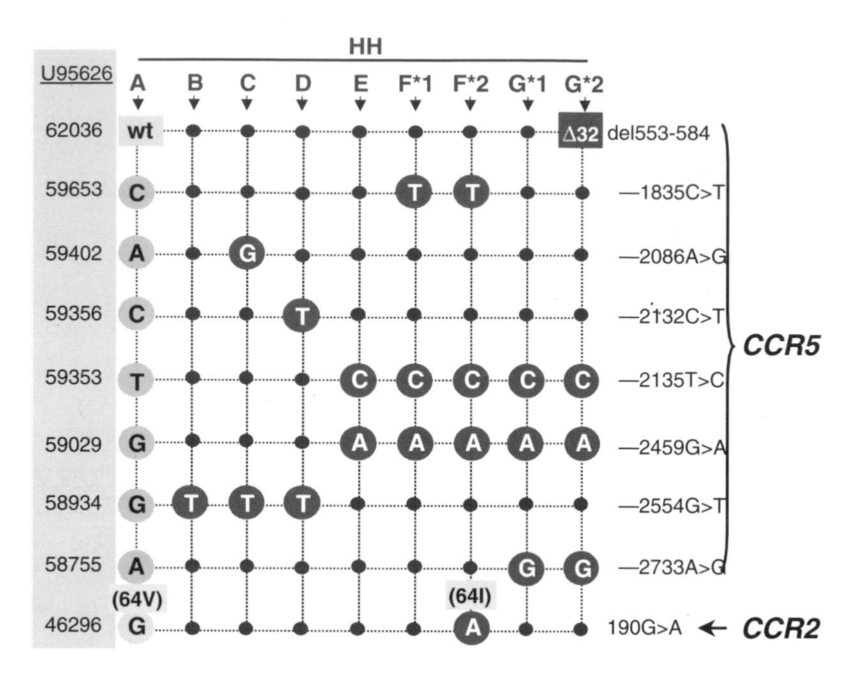

Figure 7.3. Delineation of *CCR2-CCR5* haplotypes on chromosome 3. Eight single-nucleotide polymorphisms (SNPs) and a 32-bp deletion (Δ32) define nine major human haplotypes/haplogroups (HH) designated as HHA through HHG*2 (top). Sequences identical to these shown in HHA are indicated by dots. The five-digit numbers (left) are based on GenBank sequence U95626, whereas three- or four-digit numbers (right) refer to positions relative to the ATG translation start site in the transcribed *CCR2* and *CCR5* sequences. The *CCR2* 190G > A SNP has been commonly known as 64V/I, corresponding to the predicted amino acid substitution. (See color plate.)

and, more importantly, how strong selection by an infectious disease can enrich favorable alleles in certain populations. Research on *DARC* variants has been largely driven by its importance in both malaria and transfusion medicine (Daniel et al., 1998; Pogo and Chaudhuri, 2000).

The wide range of chemokine receptor and ligand production among individuals reflects substantial genetic determination (Paxton et al., 1996, 1998, 2001). At the *CCR2* and *CCR5* loci on chromosome 3, combinations of common single nucleotide polymorphisms (SNPs) and a 32-basepair deletion (Δ32) form nine stable haplotypes (designated HHA through HHG*2) (Gonzalez et al., 1999) as a result of tight linkage disequilibrium (LD) (Fig. 7.3). Of these haplotypes, HHG*2, defining the exclusively linked *CCR5*-Δ32 mutation, is of most recent origin (Mummidi et al., 1998, 2000), having been introduced into the Caucasian populations about 700–2000 years ago

(Stephens et al., 1998). Other estimates also indicate that the *CCR5-Δ32* mutation originated from a single event in northern Europeans and expanded quickly from north to south in less than 1000 years (Libert et al., 1998; Maayan et al., 2000). The spread of Δ32 among Caucasians and certain Jewish populations cannot be solely explained by random genetic drift or migration and therefore may have resulted from strong selection (Libert et al., 1998; Maayan et al., 2000; Klitz et al., 2001). The distributions of other *CCR2-CCR5* haplotypes also differ by race and by geography (Gonzalez et al., 1999, 2001; Tang et al., 2002a). In particular, haplotype HHD is almost exclusively restricted to persons of African ancestry (Martin et al., 1998; Kostrikis et al., 1999; Tang et al., 1999b; Martinson et al., 2000; Mummidi et al., 2000), whereas the *CCR5-Δ32*-carrying HHG*2 haplotype is rather rare outside of Caucasian populations (Dean et al., 1996; Huang et al., 1996; Martinson et al., 1997; Gonzalez et al., 1999, 2001; Tang et al., 2002a).

CCR5-Δ32 introduces a null allelic product that is not expressed on cell surface and can further interfere with the function of the wild-type molecule through the formation of defective heterodimers (Benkirane et al., 1997). As a result, homozygosity with *CCR5-Δ32* renders cells free of surface CCR5 (Liu et al., 1996), whereas *CCR5-Δ32* heterozygosity may reduce CCR5 expression (Paxton et al., 1999). A common nonsynonymous SNP at the *CCR2* locus leads to an amino acid substitution of 64Val to Ile (64V/I). The CCR2-64I, equivalent to *CCR2-CCR5* haplotype HHF*2, is frequent in all major human populations (Smith et al., 1997; Tang et al., 1999b; Martinson et al., 2000; Gonzalez et al., 2001; Tang et al., 2002a). CCR2 carrying either 64V or 64I may cause heterologous desensitisation of CCR5 and CXCR4 (Lee et al., 1998), whereas only CCR-64I appears to facilitate heterodimer formation between CCR2 and CCR5 (Mellado et al., 1999). The noncoding sequence variants in the *CCR5* promoter/regulatory region are located beyond known consensus motifs responsible for promoter activities (Guignard et al., 1998; Liu et al., 1998), but a few of them have been associated with different promoter activity (McDermott et al., 1998) and binding to nuclear transcription factors (Bream et al., 1999; Mummidi et al., 2000). Therefore, promoter sequence polymorphisms may in part explain the existence of multiple CCR5 transcripts (Mummidi et al., 1997, 2000). Several less frequent variants in *CCR5* open reading frame (Ansari-Lari et al., 1997; Carrington et al., 1997) have been shown or predicted to modify or abolish CCR5 expression and function (Quillent et al., 1998; Howard et al., 1999; Blanpain et al., 2000). The two known as del893C (Ansari-Lari et al., 1997) and m303 (303T to A switch) (Quillent et al., 1998) also lead to truncated products, and the former is relatively common (allele frequency = 0.04) in Asians.

Common polymorphisms at other loci encoding chemokine receptor and ligand molecules are being defined and characterised in parallel. A study of Japanese and European Caucasians revealed three SNPs (T51C, G824A, and T971C) in *CCR3* (Fukunaga et al., 2001). The nonsynonymous *CCR3* T51C and G824A SNPs, along with T240C and T1052C, were detected in samples derived from the United States (Zimmermann et al., 1998). Two tightly linked *CX3CR1* variants encoding 249Ile and 280Met appear to affect fractalkine binding (Faure et al., 2000), whereas systematic screening of genetic variation in the entire coding regions of *CXCR1* (*IL8RA*), *CXCR2* (*IL8RB*), and *CXCR3* confirmed two previously reported polymorphisms in *CXCR1* (827G > C) and in *CXCR2* (786C > T) and revealed seven additional SNPs (Kato et al., 2000). Two common SNPs in *CXCR1* and three in *CXCR2* define three *CXCR1* and four *CXCR2* alleles.

At the various chemokine ligand loci, two SNPs (−28C/G and −403G/A) in the *SCYA5* (*RANTES*) promoter (Liu et al., 1999b; McDermott et al., 2000a) form three haplotypes, GC, AC, and AG, counting from the −403 site. The most ancestral AC haplotype has a much lower prevalence in Caucasians than in Africans and Asians (Liu et al., 1999a; Gonzalez et al., 2001). The −28G allele, on the AG haplotype and relatively common in Asians, has been associated with elevated *SCYA5* transcription. *SCYA3 (MIP-1α)* genotypes involving two SNPs (+113C/T and +459C/T) also show ethnic specificity, with the 113C-459T haplotype being mostly confined to Africans (Gonzalez et al., 2001). Four other SNPs within *MIP-1α* have been detected in Japanese: one in exon 2, one in intron 2, and two in exon 3 (Xin et al., 2001). A G-to-A nucleotide change at position 801 relative to the ATG translation start site of *SDF1* (*SDF-1, CXCL4*) defines the 3′ UTR variant commonly referred to as SDF-1–3′A (Winkler et al., 1998), which is more common in Caucasians than in Africans and has an increasing frequency from northeast to southeast Asia (Su et al., 1999). The coding and promoter sequences of eotaxin-1 gene (*SCYA11*) are also polymorphic: a G-to-A SNP at position 67 results in a nonconservative amino acid change of Ala at position 23 to Thr (Ala23Thr) within the signal peptide; two additional SNPs (−426C > T and −384A > G) are found in the 5′-flanking regions (Miyamasu et al., 2001).

The functional significance of many of these genetic variations identified in population studies remains to be established. Systematic pursuit of their role in the pathogenesis of HIV infection and other conditions will require especially meticulous attention to detail because, except for *CCR2* and *CCR5* variants (Carrington et al., 1999a; Mummidi et al., 2000), the nomenclature of most polymorphisms cited above does not comply with the system now proposed for human sequence variations (den Dunnen and Antonarakis, 2000, 2001).

4. POLYMORPHISMS OF CHEMOKINE RECEPTOR AND LIGAND GENES IN RELATION TO HIV-1 INFECTION, HOST–VIRUS EQUILIBRATION, AND DISEASE PROGRESSION

CCR polymorphisms, namely *CCR5*-Δ32, *CCR2*-64I, and *CCR5* promoter variants, have been the subject of intense investigation in HIV-1 infection and disease progression (Kaslow and McNicholl, 1999; O'Brien and Moore, 2000; Ioannidis et al., 2001). Epidemiological findings since 1996 have firmly established that Δ32 homozygosity protects against HIV-1 infection (Dean et al., 1996; Huang et al., 1996; Zimmerman et al., 1997) with rare exceptions (Balotta et al., 1997; Biti et al., 1997; Michael et al., 1998). *CCR5*-Δ32 heterozygosity has also been shown to retard seroconversion among individuals highly exposed to HIV-1 (Mangano et al., 1998; Marmor et al., 2001; Tang et al., 2002a). On the other hand, HIV-1 transmission has occurred more readily in individuals homozygous for a *CCR5* promoter variant (59356T) exclusively found on the *CCR2-CCR5* HHD haplotype (Kostrikis et al., 1999) and in those with the *CCR2-CCR5* HHE/HHE genotype (Mangano et al., 2001; Tang et al., 2002a). *CCR2-CCR5* haplotypes HHG*2 (Δ32) and HHF*2 (64I) both have dominant effects on HIV-1 disease progression, delaying the onset of AIDS for about 2 years in individuals with known dates of seroconversion (Smith et al., 1997; Ioannidis et al., 2001; Mangano et al., 2001; Tang et al., 2002a). The haplotype pair HHE/HHE and perhaps HHC/HHE have been associated with more rapid disease progression (Gonzalez et al., 1999; Mangano et al., 2001; Tang et al., 2002a). In other studies, several *CCR2-CCR5* haplotype variants have been reported to mediate HIV-1 pathogenesis independent of HHF*2 (*CCR2*-64I) and HHG*2 (*CCR5*-Δ32), with CCR5P1/P1 (Martin et al., 1998) or CCR5P1 alone (An et al., 2000) causing a 1-year disease acceleration and 59029G/G (McDermott et al., 1998) retarding clinical AIDS by another 3.8 years. The genotypic effects of CCR5P1/P1 and 59029G/G are largely reciprocal and highly predictable based on the structure of CCR5 promoter alleles (Tang et al., 1999b).

Data pooled from 38 studies (over 13,000 individuals) demonstrated that the majority of HIV-1-infected individuals untreated with potent antiretroviral therapy develop AIDS within the first 2–12 years of acquiring infection, with a median time to AIDS of approximately 9 years (Collaborative Group on AIDS Incubation and HIV Survival, including the CASCADE EU

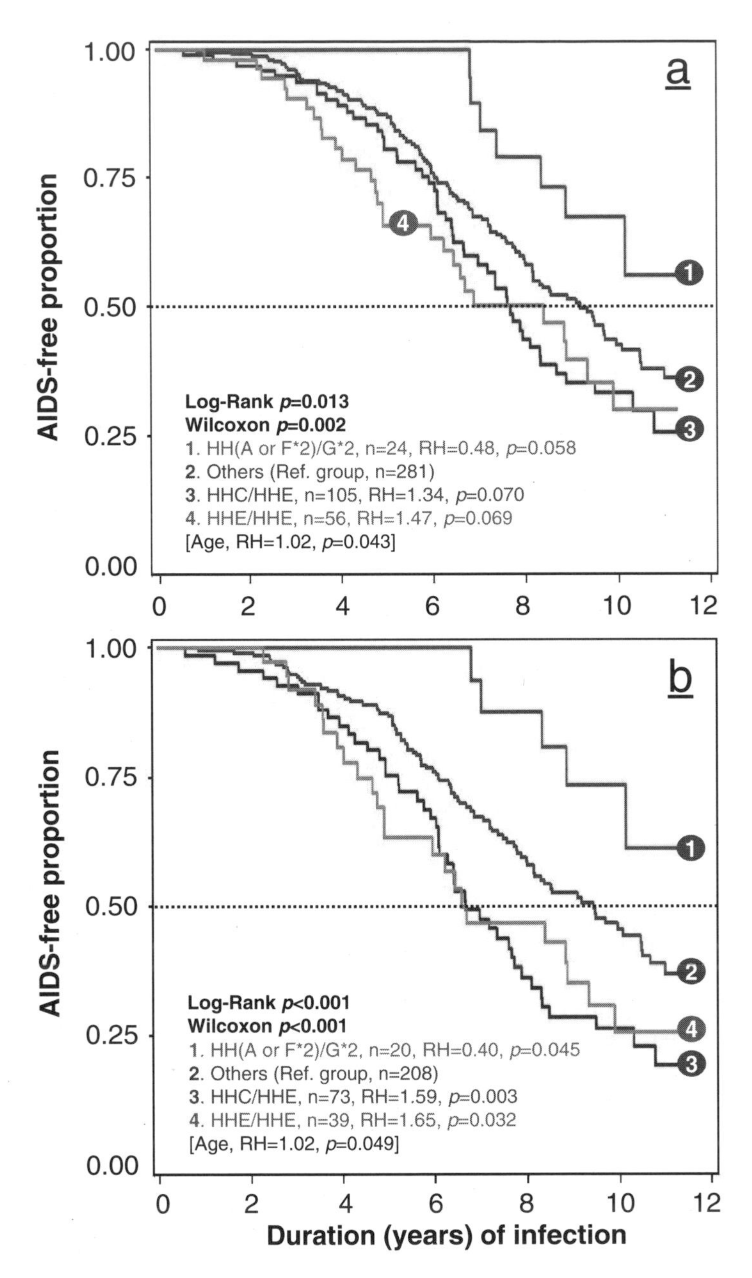

Figure 7.4. Rates of progression to AIDS (CDC 1987) during 11.5 years of follow-up among 470 (a) and 340 (b) HIV-1 seroconverted Caucasians not receiving antiretroviral therapy. Data based on the Multicenter AIDS Cohort Study are derived from Tang et al.

(cont.)

Concerted Action, 2000). Epidemiologic effects attributable to the common *CCR2* and *CCR5* variants must be interpreted with this range in mind, and with special attention to common genetic phenomena such as linkage disequilibrium (LD), population- or race-specific distribution, and allelic/haplotypic interactions. Systematic analyses of fully resolved haplotypes at the neighboring *CCR2* and *CCR5* loci now suggest that several *CCR2-CCR5* haplotypes/genotypes independently influence HIV-1 seroconversion and early HIV-1 RNA levels (viral load) as well as variability in progression to AIDS (Tang et al., 2002a). The effects of four *CCR2-CCR5* variants (the Δ32-linked G*2 haplotype, the 64I-linked HHF*2 haplotype, genotypic combination of HHF*2 and HHG*2, and the HHE/HHE genotype) on HIV-1 viral load are readily discernible within the first 42 months after seroconversion (Tang et al., 2002a). In the Multicenter AIDS Cohort Study, the strong influence of HHA/HHG*2 and HHF*2/HHG*2 on both viral RNA level (Tang et al., 2002a) and disease progression (Fig. 7.4) accounted for much of the effect commonly associated with the HHG*2 and HHF*2 haplotypes and their combination (Buseyne et al., 1998; Rizzardi et al., 1998; Easterbrook et al., 1999; Valdez et al., 1999; Barroga et al., 2000; Guerin et al., 2000; O'Brien et al., 2000b; Ioannidis et al., 2001). Multivariable analysis in this large cohort of HIV-1 seroconverters ($n = 470$) demonstrate that quantification of relative hazards (RH) of AIDS for individual haplotypes and genotypes can be sensitive to the precision of estimated seroconversion window, the number of variants being assessed, separation/aggregation of genetic variants, and age at seroconversion (Fig. 7.4). When seroconverters are restricted to those with observed intervals of ≤12 months (Fig. 7.4b), RH values for *CCR2-CCR5* genotypes HHC/HHE and HHE/HHE differ from those in the analysis allowing broader seroconversion intervals (Fig. 7.4a). The more uncertain the duration of infection in study subjects, the more cautiously estimates of genetic effects from such a study should be treated.

Associations with CCR variants can be both time-dependent and outcome- (endpoint) specific. For example, although haplotype HHE and especially the genotype HHC/HHE had no apparent negative impact on the early viral RNA levels (Tang et al., 2002a), they were associated with accelerated

Figure 7.4. (*cont.*) (2002a) with permission; subjects with ambiguous *CCR2-CCR5* genotypes (paired haplotypes) have been retyped or excluded from this reanalysis. The relative hazards (RH) of AIDS and *P* values in a multivariable model are further adjusted for age at time of seroconversion. Time 0 is the midpoint between last seronegative and first seropositive visits, either within 12 months apart ($n = 340$) or longer (130 additional subjects). (See color plate.)

disease progression in 474 Caucasian seroconverters (Fig. 7.4), as seen earlier in Argentinean children (Mangano et al., 2001) and North American Caucasians (Gonzalez et al., 1999). In contrast, association of 59029G/G with delayed onset of AIDS was not preceded by a substantial decrease in RNA levels (Tang et al., 2002a). The independent and consistent associations of the *CCR2-CCR5* HHE/HHE genotype with several related outcomes (Mangano et al., 2001; Tang et al., 2002a) may be more meaningful than the less consistent findings, especially since the HHE/HHE genotype is relatively common in all major ethnic groups (Gonzalez et al., 1999; Mummidi et al., 2000; Mangano et al., 2001). Furthermore, unlike other variants, including *CCR2*-64I, *CCR5*-Δ32 (Gonzalez et al., 1999; Tang et al., 1999b, 2002a) and *SCYA5* alleles (Liu et al., 1999b; McDermott et al., 2000a), the HHE/HHE genotype seems to exert a comparable effect in populations of both Caucasian and African ancestry.

Influence of coreceptor polymorphisms on viral dynamics (Barroga et al., 2000; Guerin et al., 2000; Kasten et al., 2000; O'Brien et al., 2000a; Tang et al., 2002a) is of particular interest because plasma HIV-1 viral load closely reflects host–virus equilibration (Mellors et al., 1996; Katzenstein et al., 1997; Meyer et al., 1997; Easterbrook et al., 1999; Lockett et al., 1999; Cunningham et al., 2000), which in turn is highly predictive of both subsequent disease progression (O'Shea et al., 1991; Mellors et al., 1996; Shearer et al., 1997; Bratt et al., 1998; Childs et al., 1999; Ioannidis et al., 1999) and HIV-1 transmissibility from infected to uninfected individuals by any of several modes (Operskalski et al., 1997; Garcia et al., 1999; Pedraza et al., 1999; Semba et al., 1999; Quinn et al., 2000; Fideli et al., 2001). Moreover, association of host genetic polymorphisms with early virus concentration can be readily tested in cohorts with adequate virologic measurement in the absence of lengthy follow-up. Assessment of both genetic determinants and viral RNA as simultaneous, independent predictors of long-term outcome will require models specifically developed to account for those factors taken separately and jointly. Exploratory work on such models has been undertaken (Taylor et al., 2000). More comprehensive models of host–virus relationships during the natural course of HIV-1 infection may also need to consider currently circulating HIV-1 genotypes and phenotypes (Dyer et al., 1997; Michael et al., 1997; Alexander et al., 2000), along with other disease-modifying host genetic factors under evaluation. Patients receiving highly effective antiretroviral therapy may be subject to the same genetic influences as untreated individuals and offer opportunities to examine genetic contributions under different circumstances of virologic and immunologic control. Better virologic response to therapy has been reported for the *CCR5* promoter 59029G/G genotype (O'Brien et al.,

2000b). Although this genotype is in exclusive LD with 59353T/T (Fig. 7.3), experimental separation of the linked sequences at 59029 and 59353 suggests reduced promoter function in the 59029G/G construct independent of the 59353T/C SNP (McDermott et al., 1998).

5. APPLICATIONS OF IMMUNOGENETIC FINDINGS

Research on the immunogenetics of HIV/AIDS in general and CCR in particular should yield rich benefits. It can guide functional studies aimed at dissecting the intrinsic mechanisms in host–virus interactions. As examples, alternative classification of genetic variants according to individual sites (McDermott et al., 1998), known function (Gao et al., 2001; MacDonald et al., 2001), and shared motifs (Flores-Villanueva et al., 2001; Kaslow et al., 2001; O'Brien et al., 2001) may reveal novel effects. In the realm of clinical decision making, there are already early indications that *CCR5* genotyping data might be informative: both *CCR5*-Δ32 and *CCR5* 59029G/G have been associated with better virologic and immunologic response to potent antiretroviral therapy (Valdez et al., 1999; Barroga et al., 2000; Guerin et al., 2000; O'Brien et al., 2000b). On the other hand, studies on human major histocompatibility complex (MHC = HLA) (Saah et al., 1998; Carrington et al., 1999b; Hendel et al., 1999; Keet et al., 1999; Magierowska et al., 1999; Tang et al., 1999a) *SCYA5* (Liu et al., 1999b), *IL-4* (Nakayama et al., 2000), *IL-10* (Shin et al., 2000), *CX3CR1* (Faure et al., 2000; McDermott et al., 2000b), *SDF-1* (Mummidi et al., 1998; van Rij et al., 1998; Winkler et al., 1998), and *MIP-1α* (Gonzalez et al., 2001) emphasize the polygenic origin of host influences on HIV/AIDS. Thus, although CCR variants can be shown to modulate HIV-1 transmission, viral RNA level, and disease progression at the population level, effective translation of those population effects into prognostically useful information at the individual level will depend on capturing the multiplicity of genetic effects. CCR genotyping data will likely become part of a comprehensive genetic algorithm that incorporates the effects of numerous determinants, especially those in the cytokine networks and the HLA complex essential for humoral and cell-mediated immunity.

Various algorithmic approaches have been suggested in several studies (Kaslow et al., 1996; Keet et al., 1999; Magierowska et al., 1999; Carrington et al., 2001) dealing with complex polymorphisms when the effect of any single genetic factor could easily be obscured by the effects of others. The population effects of the *CCR2-CCR5* and the HLA systems are largely independent (Keet et al., 1999), yet incorporation of *CCR2* and *CCR5* genotyping data into the analysis of HLA effects improves prediction of late clinical outcomes

(Carrington et al., 2001). Genetic contributions summarised by an algorithm provides a robust assessment of host genetic effects in settings where stratifications and precision of hazard estimates are limited by sample size. Even in large cohorts, conventional stratifications for different host variables are not always feasible. Summary algorithms can also be individualised to distinguish between genetic effects operating early or later in the pathogenetic process, as may be true of CCR markers, whose influence on early viral RNA may not translate directly into significant impact on later disease outcomes or vice versa (Tang et al., 2002a). Of course, any of these analyses should also account for important nongenetic host factors including age, gender, and history of antiretroviral therapy.

Finally, defining regional and global distribution of immunogenetic markers may aid in modeling the dynamics of HIV-1 infection (Hsu Schmitz, 2000; Gonzalez et al., 2001) and the concomitant changes in population genetics (Schliekelman et al., 2001) in association with HIV/AIDS. One recent model shows that *CCR5*-Δ32 can serve as an important determinant of the per-partnership HIV-1 transmission rates in rapid and slow progressors in the San Francisco gay population (Hsu Schmitz, 2000). In a second model, natural selection for 100 years in Africa was projected to produce an increase in the favorable *CCR2-CCR5* haplotypes by approximately 10% and a correspondingly modest (1 year) delay in the average time to AIDS (Schliekelman et al., 2001). These models may be limited by the necessity for proper assessment of demographic characteristics, socioeconomic structures, and reproductive behaviours that are rarely disclosed in immunogenetic analyses. Overall, proper technical resolution and careful statistical analysis of major genetic determinants along with comprehensive documentation of their regional and global distribution will be essential for successful application of host genetic research to the immense challenge of HIV/AIDS.

6. SUMMARY

Description of heritable SNP and other sequence variations in the chemokine receptors and their ligand genes will undoubtedly continue. Techniques for simultaneous analysis of high-resolution genotyping data are essential for accurate assessment of their relative contributions to the HIV/AIDS pandemic. Robust genetic scoring algorithms should be increasingly useful in evaluating the joint effects of multiple markers coexisting in a single individual, along with nongenetic disease-modifying factors. The potential value of genetic data in predicting both disease progression and response to antiretroviral therapy has been enhanced by the recognition that

CCR2-CCR5 variants mediate host-HIV-1 equilibration, as reflected by viral load measured during the early stage of infection when treatment is most likely effective. Although involvement of the chemokine receptor and ligand polymorphisms in HIV-1 infection and disease progression may further guide the development of novel therapeutics, precise clinical applications may prove difficult because the individual effects attributable to *CCR5*, *CCR2*, RANTES (*SCYA5*), and several other CR and CL loci appear to be modest, quantitative, and often time dependent. Over time, changes in population genetics profiles as a result of HIV/AIDS are virtually inevitable in areas with a high prevalence of HIV-1 infection and without effective intervention. Finally, elucidation of the functional signficance of polymorphisms in genes for chemokine receptors and ligands as molecules with broad pathophysiologic relevance should extend benefit to research on infectious diseases other than HIV/AIDS.

ACKNOWLEDGMENTS

We thank Dr. Philip M. Murphy for critical reading of earlier versions of our chapter. As always (Tang et al., 2002a, 2002b), we are also grateful to investigators, staff, and participants from several cohort studies as well as members of the Program in Epidemiology of Infection and Immunity at UAB for their collaboration and valuable contributions to the collection and analyses of clinical and genetic data. Both J.T. and R.A.K. have been supported by NIH Grants R01 AI41951, AI40951, IU01 AI41530, and U01-HD32842.

REFERENCES

Abbott, C. A., Yu, D. M., Woollatt, E., et al. (2000). Cloning, expression and chromosomal localization of a novel human dipeptidyl peptidase (DPP) IV homolog, DPP8. *Eur. J. Biochem.* **267**, 6140–6150.

Ahuja, S. K., Ozcelik, T., Milatovitch, A., et al. (1992). Molecular evolution of the human interleukin-8 receptor gene cluster. *Nat. Genet.* **2**, 31–36.

Alexander, L., Weiskopf, E., Greenough, T. C., et al. (2000). Unusual polymorphisms in human immunodeficiency virus type 1 associated with nonprogressive infection. *J. Virol.* **74**, 4361–4376.

Alkhatib, G., Ahuja, S. S., Light, D., et al. (1997). CC chemokine receptor 5-mediated signaling and HIV-1 co-receptor activity share common structural determinants. Critical residues in the third extracellular loop support HIV-1 fusion. *J. Biol. Chem.* **272**, 19771–19776.

Alkhatib, G., Combadiere, C., Broder, C. C., et al. (1996). CC CKR5: A RANTES, MIP-1alpha, MIP-1beta receptor as a fusion cofactor for macrophage-tropic HIV-1. *Science* **272**, 1955–1958.

An, P., Martin, M. P., Nelson, G. W., et al. (2000). Influence of *CCR5* promoter haplotypes on AIDS progression in African-Americans. *AIDS* **14**, 2117–2122.

Ansari-Lari, M. A., Liu, X.-M., Metzker, M. L., et al. (1997). The extent of genetic variation in the *CCR5* gene. *Nat. Genet.* **16**, 221–222.

Arvanitakis, L., Geras-Raaka, E., Varma, A., et al. (1997). Human herpesvirus KSHV encodes a constitutively active G-protein-coupled receptor linked to cell proliferation. *Nature* **385**, 347–350.

Atchison, R. E., Gosling, J., Monteclaro, F. S., et al. (1996). Multiple extracellular elements of CCR5 and HIV-1 entry: Dissociation from response to chemokines. *Science* **274**, 1924–1926.

Baggiolini, M. (1998). Chemokines and leukocyte traffic. *Nature* **392**, 565–568.

Bai, J., Gorantla, S., Banda, N., et al. (2000). Characterisation of anti-CCR5 ribozyme-transduced CD34+ hematopoietic progenitor cells in vitro and in a SCID-hu mouse model in vivo. *Mol. Therapy* **1**, 244–254.

Bai, X., Chen, J. D., Yang, A. G., et al. (1998). Genetic co-inactivation of macrophage- and T-tropic HIV-1 chemokine coreceptors CCR-5 and CXCR-4 by intrakines. *Gene Therapy* **5**, 984–994.

Baird, J. W., Nibbs, R. J., Komai-Koma, M., et al. (1999). ESkine, a novel beta-chemokine, is differentially spliced to produce secretable and nuclear targeted isoforms. *J. Biol. Chem.* **274**, 33496–33503.

Balotta, C., Bagnarelli, P., Violin, M., et al. (1997). Homozygous delta 32 deletion of the CCR-5 chemokine receptor gene in an HIV-1-infected patient. *AIDS* **11**, F67–71.

Barroga, C. F., Raskino, C., Fangon, M. C., et al. (2000). The CCR5Δ32 allele slows disease progression of human immunodeficiency virus-1-infected children receiving antiretroviral treatment. *J. Infect. Dis.* **182**, 413–419.

Basu, S., Sriram, B., Goila, R., et al. (2000). Targeted cleavage of HIV-1 coreceptor-CXCR-4 by RNA-cleaving DNA-enzyme: inhibition of coreceptor function. *Antiviral Res.* **46**, 125–134.

Benkirane, M., Jin, D. Y., Chun, R. F., et al. (1997). Mechanism of transdominant inhibition of CCR5-mediated HIV-1 infection by CCR5delta32. *J. Biol. Chem.* **272**, 30603–30606.

Berger, E. A., Doms, R. W., Fenyo, E. M., et al. (1998). A new classification for HIV-1. *Nature* **391**, 240.

Biti, R., French, R., Young, J., et al. (1997). HIV-1 infection in an individual homozygous for the *CCR5* deletion allele. *Nat. Med.* **3**, 252–253.

Blanpain, C., Lee, B., Tackoen, M., et al. (2000). Multiple nonfunctional alleles of CCR5 are frequent in various human populations. *Blood* **96**, 1638–1645.

Blanpain, C., Migeotte, I., Lee, B., et al. (1999). CCR5 binds multiple CC-chemokines: MCP-3 acts as a natural antagonist. *Blood* **94**, 1899–1905.

Bonecchi, R., Bianchi, G., Bordignon, P. P., et al. (1998a). Differential expression of chemokine receptors and chemotactic responsiveness of type 1 T helper cells (Th1s) and Th2s. *J. Exp. Med.* **187**, 129–134.

Bonecchi, R., Sozzani, S., Stine, J. T., et al. (1998b). Divergent effects of interleukin-4 and interferon-gamma on macrophage- derived chemokine production: an amplification circuit of polarized T helper 2 responses. *Blood* **92**, 2668–2671.

Bratt, G., Karlsson, A., Leandersson, A. C., et al. (1998). Treatment history and baseline viral load, but not viral tropism or CCR-5 genotype, influence prolonged antiviral efficacy of highly active antiretroviral treatment. *AIDS* **12**, 2193–2202.

Bream, J. H., Young, H. A., Rice, N., et al. (1999). CCR5 promoter alleles and specific DNA binding factors. *Science* **284**, 223a.

Buseyne, P., Janvier, G., Teglas, J. P., et al. (1998). Impact of heterozygosity for the chemokine receptor CCR5 32-bp-deleted allele on plasma virus load and CD4 T lymphocytes in perinatally human immunodeficiency virus-infected children at 8 years of age. *J. Infect. Dis.* **178**, 1019–1023.

Carrington, C., Kissner, T., Gerrard, B., et al. (1997). Novel alleles of the chemokine receptor gene CCR5. *Am. J. Hum. Genet.* **61**, 1261–1267.

Carrington, M., Dean, M., Martin, M. P., et al. (1999a). Genetics of HIV-1 infection: chemokine receptor CCR5 polymorphism and its consequences. *Hum. Mol. Genet.* **8**, 1939–1945.

Carrington, M., Nelson, G. and O'Brien, S. J. (2001). Considering genetic profiles in functional studies of immune responsiveness to HIV-1. *Immunol. Lett.* **79**, 131–140.

Carrington, M., Nelson, G. W., Martin, M. P., et al. (1999b). HLA and HIV-1: heterozygosity advantage and B*35-C*04 disadvantage. *Science* **283**, 1748–1752.

Chang, T. L., Gordon, C. J., Roscic-Mrkic, B., et al. (2002). Interaction of the CC-chemokine RANTES with glycosaminoglycans activates a p44/p42 mitogen-activated protein kinase-dependent signaling pathway and enhances human immunodeficiency virus type 1 infectivity. *J Virol* **76**, 2245–2254.

Chen, Z., Zhou, P., Ho, D. D., et al. (1997). Genetically divergent strains of simian immunodeficiency virus use CCR5 as a coreceptor for entry. *J. Virol.* **71**, 2705–2714.

Childs, E. A., Lyles, R. H., Selnes, O. A., et al. (1999). Plasma viral load and CD4 lymphocytes predict HIV-associated dementia and sensory neuropathy. *Neurology* **52**, 607–613.

Choe, H., Farzan, M., Konkel, M., et al. (1998). The orphan seven-transmembrane receptor APJ supports the entry of primary T-cell-line-tropic and dualtropic human immunodeficiency virus type 1. *J. Virol.* **72**, 6113–6118.

Cocchi, F., DeVico, A. L., Garzino-Demo, A., et al. (1995). Identification of RANTES, MIP-1 alpha, and MIP-1 beta as the major HIV-suppressive factors produced by CD8+ T cells. *Science* **270**, 1811–1815.

Collaborative Group on AIDS Incubation and HIV Survival Including the CASCADE EU Concerted Action. (2000). Time from HIV-1 seroconversion to AIDS and death before widespread use of highly-active antiretroviral therapy: a collaborative re-analysis. Concerted Action on SeroConversion to AIDS and Death in Europe. *Lancet* **355**, 1131–1137.

Combadiere, C., Ahuja, S. K. and Murphy, P. M. (1995). Cloning, chromosomal localization, and RNA expression of a human beta chemokine receptor-like gene. *DNA Cell Biol.* **14**, 673–680.

Connor, R. I., Sheridan, K. E., Ceradini, D., et al. (1997). Change in coreceptor use correlates with disease progression in HIV-1-infected individuals. *J. Exp. Med.* **185**, 621–628.

Cunningham, A. L., Li, S., Juarez, J., et al. (2000). The level of HIV infection of macrophages is determined by interaction of viral and host cell genotypes. *J. Leukoc. Biol.* **68**, 311–317.

Daniel, S., Brusic, V., Caillat-Zucman, S., et al. (1998). Relationship between peptide selectivities of human transporters associated with antigen processing and HLA class I molecules. *J. Immunol.* **161**, 617–624.

D'Aquila, R. T., Sutton, L., Savara, A., et al. (1998). CCR5/Δ(ccr5) heterozygosity: a selective pressure for the syncytium-inducing human immunodeficiency virus type 1 phenotype. *J. Infect. Dis.* **177**, 1549–1553.

Daugherty, B. L. and Springer, M. S. (1997). The beta-chemokine receptor genes CCR1 (CMKBR1), CCR2 (CMKBR2), and CCR3 (CMKBR3) cluster within 285 kb on human chromosome 3p21. *Genomics* **41**, 294–295.

De Clercq, E. (2000). Novel compounds in preclinical/early clinical development for the treatment of HIV infections. *Rev. Med. Virol.* **10**, 255–277.

Dean, M., Carrington, M., Winkler, C., et al. (1996). Genetic restriction of HIV-1 infection and progression to AIDS by a common deletion allele of the chemokine receptor 5 structural gene. *Science* **273**, 1856–1862.

den Dunnen, J. T. and Antonarakis, E. (2001). Nomenclature for the description of human sequence variations. *Hum. Genet.* **109**, 121–124.

den Dunnen, J. T. and Antonarakis, S. E. (2000). Mutation nomenclature extensions and suggestions to describe complex mutations: A discussion. *Hum. Mutat.* **15**, 7–12.

Deng, H., Liu, R., Ellmeir, W., et al. (1996). Identification of a major co-receptor for primary isolates of HIV-1. *Nature* **381**, 661–666.

Derdeyn, C. A., Decker, J. M., Sfakianos, J. N., et al. (2000). Sensitivity of human immunodeficiency virus type 1 to the fusion inhibitor T-20 is modulated by coreceptor specificity defined by the V3 loop of gp120. *J. Virol.* **74**, 8358–8367.

Doranz, B. J., Lu, Z. H., Rucker, J., et al. (1997). Two distinct CCR5 domains can mediate coreceptor usage by human immunodeficiency virus type 1. *J. Virol.* **71**, 6305–6314.

Doranz, B. J., Rucker, J., Yanjie, Y., et al. (1996). A dual-tropic primary HIV-1 isolate that uses fusin and the b-chemokine receptors CKR-5, CKR-2b as fusion cofactors. *Cell* **85**, 1149–1158.

Dyer, W. B., Geczy, A. F., Kent, S. J., et al. (1997). Lymphoproliferative immune function in the Sydney Blood Bank Cohort, infected with natural nef/long terminal repeat mutants, and in other long-term survivors of transfusion-acquired HIV-1 infection. *AIDS* **11**, 1565–1574.

Easterbrook, P. J., Rostron, T., Ives, N., et al. (1999). Chemokine receptor polymorphisms and human immunodeficiency virus disease progression. *J. Infect. Dis.* **180**, 1096–1105.

Edinger, A. L., Amedee, A., Miller, K., et al. (1997). Differential utilization of CCR5 by macrophage and T cell tropic simian immunodeficiency virus strains. *Proc. Natl. Acad. Sci. U S A* **94**, 4005–4010.

Efremov, R. G., Legret, F., Vergoten, G., et al. (1998). Molecular modeling of HIV-1 coreceptor CCR5 and exploring of conformational space of its extracellular domain in molecular dynamics simulation. *J. Biomol. Struct. Dynam.* **16**, 77–90.

Endres, M. J., Garlisi, C. G., Xiao, H., et al. (1999). The Kaposi's sarcoma-related herpesvirus (KSHV)-encoded chemokine vMIP-I is a specific agonist for the CC chemokine receptor (CCR)8. *J. Exp. Med.* **189**, 1993–1998.

Erdel, M., Theurl, M., Meyer, M., et al. (2001). High-resolution mapping of the human 4q21 and the mouse 5E3 SCYB chemokine cluster by fiber-fluorescence *in situ* hybridization. *Immunogenetics* **53**, 611–615.

Fan, P., Kyaw, H., Su, K., et al. (1998). Cloning and characterization of a novel human chemokine receptor. *Biochem. Biophys. Res. Commun.* **243**, 264–268.

Farber, J. M. and Berger, E. A. (2002). HIV's response to a CCR5 inhibitor: I'd rather tighten than switch! *Proc. Natl. Acad. Sci. USA* **99**, 1749–1751.

Faure, S., Meyer, L., Costagliola, D., et al. (2000). Rapid progression to AIDS in HIV+ individuals with a structural variant of the chemokine receptor CX3CR1. *Science* **287**, 2274–2277.

Feng, Y., Broder, C. C., Kennedy, P. E., et al. (1996). HIV-1 entry cofactor: functional cDNA cloning of a seven-transmembrane, G protein-coupled receptor. *Science* **272**, 872–877.

Fideli, U. S., Allen, S., Musunda, R., et al. (2001). Virologic and immunologic determinants of heterosexual transmission of human immunodeficiency virus type 1 (HIV-1) in Africa. *AIDS Res. Hum. Retroviruses* **17**, 901–910.

Finzer, P., Soto, U., Delius, H., et al. (2000). Differential transcriptional regulation of the monocyte-chemoattractant protein-1 (MCP-1) gene in tumorigenic and non-tumorigenic HPV 18 positive cells: the role of the chromatin structure and AP-1 composition. *Oncogene* **19**, 3235–3244.

Flores-Villanueva, P. O., Yunis, E. J., Delgado, J. C., et al. (2001). Control of HIV-1 viremia and protection from AIDS are associated with HLA-Bw4 homozygosity. *Proc. Natl. Acad. Sci. USA* **98**, 5140–5145.

Fukunaga, K., Asano, K., Mao, X. Q., et al. (2001). Genetic polymorphisms of CC chemokine receptor 3 in Japanese and British asthmatics. *Eur. Respir. J.* **17**, 59–63.

Gao, X., Nelson, G. W., Karacki, P., et al. (2001). Effect of a single amino acid change in MHC class I molecules on the rate of progression to AIDS. *N. Engl. J. Med.* **344**, 1668–1675.

Garcia, P. M., Kalish, L. A., Pitt, J., et al. (1999). Maternal levels of plasma human immunodeficiency virus type 1 RNA and the risk of perinatal transmission. Women and Infants Transmission Study Group. *N. Engl. J. Med.* **341**, 394–402.

Garzino-Demo, A., Moss, R. B., Margolick, J. B., et al. (1999). Spontaneous and antigen-induced production of HIV-inhibitory beta-chemokines are associated with AIDS-free status. *Proc. Natl. Acad. Sci. USA* **96**, 11986–11991.

Geijtenbeek, T. B., Kwon, D. S., Torensma, R., et al. (2000). DC-SIGN, a dendritic cell-specific HIV-1-binding protein that enhances trans-infection of T cells. *Cell* **100**, 587–597.

Glushakova, S., Baibakov, B., Zimmerberg, J., et al. (1997). Experimental HIV infection of human lymphoid tissue: correlation of CD4+ T cell depletion and virus syncytium-inducing/non-syncytium- inducing phenotype in histocultures inoculated with laboratory strains and patient isolates of HIV type 1. *AIDS Res. Hum. Retroviruses* **13**, 461–471.

Glushakova, S., Grivel, J. C., Fitzgerald, W., et al. (1998). Evidence for the HIV-1 phenotype switch as a causal factor in acquired immunodeficiency. *Nat. Med.* **4**, 346–349.

Goila, R. and Banerjea, A. C. (1998). Sequence specific cleavage of the HIV-1 coreceptor CCR5 gene by a hammer-head ribozyme and a DNA-enzyme-inhibition of the coreceptor function by DNA-enzyme. *FEBS Lett.* **436**, 233–238.

Gompels, U. A., Nicholas, J., Lawrence, G., et al. (1995). The DNA sequence of human herpesvirus-6: structure, coding content, and genome evolution. *Virology* **209**, 29–51.

Gonzalez, E., Bamshad, M., Sato, N., et al. (1999). Race-specific HIV-1 disease modifying effects associated with CCR5 haplotypes. *Proc. Natl. Acad. Sci. USA* **96**, 12004–12009.

Gonzalez, E., Dhanda, R., Bamshad, M., et al. (2001). Global survey of genetic variation in CCR5, RANTES, and MIP-1alpha: impact on the epidemiology of the HIV-1 pandemic. *Proc. Natl. Acad. Sci. USA* **98**, 5199–5204.

Gonzalez, M. A., Serrano, F., Llorente, M., et al. (1998). A hammerhead ribozyme targeted to the human chemokine receptor CCR5. *Biochem. Biophys. Res. Commun.* **251**, 592–596.

Graziosi, C. and Pantaleo, G. (1997). The multi-faceted personality of HIV. *Nat. Med.* **3**, 1318–1320.

Grivel, J. C. and Margolis, L. B. (1999). CCR5- and CXCR4-tropic HIV-1 are equally cytopathic for their T-cell targets in human lymphoid tissue. *Nat. Med.* **5**, 344–346.

Guerin, S., Meyer, L., Theodorou, I., et al. (2000). CCR5 delta32 deletion and response to highly active antiretroviral therapy in HIV-1-infected patients. *AIDS* **14**, 2788–27890.

Guignard, F., Combadiere, C., Tiffany, H. L., et al. (1998). Gene organization and promoter function for CC chemokine receptor 5 (CCR5). *J. Immunol.* **160**, 985–992.

Guo, R. F., Ward, P. A., Hu, S. M., et al. (1999). Molecular cloning and characterization of a novel human CC chemokine, SCYA26. *Genomics* **58**, 313–317.

Hendel, H., Caillat-Zucman, S., Lebuanec, H., et al. (1999). New class I and II HLA alleles strongly associated with opposite patterns of progression to AIDS. *J. Immunol.* **162**, 6942–6946.

Hill, C. M., Kwon, D., Jones, M., et al. (1998). The amino terminus of human CCR5 is required for its function as a receptor for diverse human and simian immunodeficiency virus envelope glycoproteins. *Virology* **248**, 357–371.

Hornung, F., Scala, G. and Lenardo, M. J. (2000). TNF-alpha-induced secretion of C-C chemokines modulates C-C chemokine receptor 5 expression on peripheral blood lymphocytes. *J. Immunol.* **164**, 6180–6187.

Houle, M., Thivierge, M., Le Gouill, C., et al. (1999). IL-10 up-regulates CCR5 gene expression in human monocytes. *Inflammation* **23**, 241–251.

Howard, O. M., Oppenheim, J. J., Hollingshead, M. G., et al. (1998a). Inhibition of in vitro and in vivo HIV replication by a distamycin analogue that interferes with chemokine receptor function: a candidate for chemotherapeutic and microbicidal application. *J. Med. Chem.* **41**, 2184–2193.

Howard, O. M., Shirakawa, A. K., Turpin, J. A., et al. (1999). Naturally occurring CCR5 extracellular and transmembrane domain variants affect HIV-1 co-receptor and ligand binding function. *J. Biol. Chem.* **274**, 16228–16234.

Howard, O. M. Z., Korte, T., Tarasova, N. I., et al. (1998b). Small molecule inhibitors of HIV-1 cell fusion blocks chemokine receptor-mediated function. *J. Leukocyte Biol.* **64**, 6–13.

Hsu Schmitz, S. (2000). Effects of treatment or/and vaccination on HIV transmission in homosexuals with genetic heterogeneity. *Math. Biosci.* **167**, 1–18.

Huang, Y., Paxton, W. A., Wolinsky, S. M., et al. (1996). The role of a mutant CCR5 allele in HIV-1 transmission and disease progression. *Nat. Med.* **2**, 1240–1243.

Hughes, A. L. and Yeager, M. (1999). Coevolution of the mammalian chemokines and their receptors. *Immunogenetics* **49**, 115–124.

Ioannidis, J. P., Rosenberg, P. S., Goedert, J. J., et al. (2001). Effects of CCR5-Δ32, CCR2-64I and SDF-1 3′A alleles on HIV disease progression: an international meta-analysis of individual-patient data. *Ann. Int. Med.* **135**, 782–795.

Ioannidis, J. P. A., Goedert, J. J., McQueen, P. G., et al. (1999). Comparison of viral load and human leukocyte antigen statistical and neural network predictive models for the rate of HIV-1 disease progression across two cohorts of homosexual men. *J. Acquir. Immune Defic. Syndr.* **20**, 129–136.

Kaslow, R. A., Carrington, M., Apple, R., et al. (1996). Influence of combinations of human major histocompatibility complex genes on the course of HIV-1 infection. *Nat. Med.* **2**, 405–411.

Kaslow, R. A., Dorak, M. T. and Tang, J. (2001). Reflection and reaction: Is protection in HIV infection due to Bw4 or not to Bw4? *Lancet Infect. Dis.* **1**, 221–222.

Kaslow, R. A. and McNicholl, J. M. (1999). Genetic determinants of HIV-1 infection and its manifestations. *Proc. Assoc. Am. Physicians* **111**, 299–307.

Kasten, S., Goldwich, A., Schmitt, M., et al. (2000). Positive influence of the Δ32CCR5 allele on response to highly active antiretroviral therapy (HAART) in HIV-1 infected patients. *Eur. J. Med. Res.* **5**, 323–328.

Kato, H., Tsuchiya, N. and Tokunaga, K. (2000). Single nucleotide polymorphisms in the coding regions of human CXC- chemokine receptors CXCR1, CXCR2 and CXCR3. *Genes Immun.* **1**, 330–337.

Katzenstein, T. L., Eugen-Olsen, J., Hofmann, B., et al. (1997). HIV-infected individuals with the CCR Δ32/CCR5 genotype have lower HIV RNA levels and higher CD4 cell counts in the early years of the infection than do patients with the wild type. Copenhagen AIDS Cohort Study Group. *J. Acquir. Immune Defic. Syndr. Hum. Retrovirol.* **16**, 10–14.

Keet, I. P., Tang, J., Klein, M. R., et al. (1999). Consistent associations of HLA class I and class II and transporter gene products with progression of human immunodeficiency virus-1 infection in homosexual men. *J. Infect. Dis.* **180**, 299–309.

Khoja, H., Wang, G., Ng, C. T., et al. (2000). Cloning of CCRL1, an orphan seven transmembrane receptor related to chemokine receptors, expressed abundantly in the heart. *Gene* **246**, 229–238.

Kilby, J. M., Hopkins, S., Venetta, T. M., et al. (1998). Potent suppression of HIV-1 replication in humans by T-20, a peptide inhibitor of gp41-mediated virus entry. *Nat. Med.* **4**, 1302–1307.

Klitz, W., Brautbar, C., Schito, A. M., et al. (2001). Evolution of the CCR5 Delta32 mutation based on haplotype variation in Jewish and Northern European population samples. *Hum. Immunol.* **62**, 530–538.

Kostrikis, L. G., Neumann, A. U., Thomson, B., et al. (1999). A polymorphism in the regulatory region of the CC-chemokine receptor 5 gene influences perinatal transmission of human immunodeficiency virus type 1 to African-American infants. *J. Virol.* **73**, 10264–10271.

Lee, B., Doranz, B. J., Rana, S., et al. (1998). Influence of the CCR2-V64I polymorphism on human immunodeficiency virus type 1 coreceptor activity and on chemokine receptor function of CCR2b, CCR3, CCR5, and CXCR4. *J. Virol.* **72**, 7450–7458.

Lee, B., Ratajczak, J., Doms, R. W., et al. (1999). Coreceptor/chemokine receptor expression on human hematopoietic cells: biological implications for human immunodeficiency virus-type 1 infection. *Blood* **93**, 1145–56.

Libert, F., Cochaux, P., Beckman, G., et al. (1998). The Δccr5 mutation conferring protection against HIV-1 in Caucasian populations has a single and recent origin in Northeastern Europe. *Hum. Mol. Genet.* **7**, 399–406.

Liu, H., Shioda, T., Nagai, Y., et al. (1999a). Distribution of HIV-1 disease modifying regulated on activation normal T cell expressed and secreted haplotypes in Asian, African and Caucasian individuals. French ALT and IMMUNOCO Study Group. *AIDS* **13**, 2602–2603.

Liu, H. L., Chao, D., Nakayama, E. E., et al. (1999b). Polymorphism in RANTES chemokine promoter affects HIV-1 disease progression. *Proc. Natl. Acad. Sci. USA* **96**, 4581–4585.

Liu, R., Paxton, W. A., Choe, S., et al. (1996). Homozygous defect in HIV-1 coreceptor accounts for resistance of some multiply-exposed individuals to HIV-1 infection. *Cell* **86**, 367–377.

Liu, R., Zhao, X. Q., Gurney, T. A., et al. (1998). Functional analysis of the proximal CCR5 promoter. *AIDS Res. Hum. Retroviruses* **14**, 1509–1519.

Lockett, S. F., Alonso, A., Wyld, R., et al. (1999). Effect of chemokine receptor mutations on heterosexual human immunodeficiency virus transmission. *J. Infect. Dis.* **180**, 614–620.

Loetscher, M., Loetscher, P., Brass, N., et al. (1998a). Lymphocyte-specific chemokine receptor CXCR3: regulation, chemokine binding and gene localization. *Eur. J. Immunol.* **28**, 3696–3705.

Loetscher, P. and Clark-Lewis, I. (2001). Agonistic and antagonistic activities of chemokines. *J. Leukoc. Biol.* **69**, 881–884.

Loetscher, P., Uguccioni, M., Bordoli, L., et al. (1998b). CCR5 is characteristic of Th1 lymphocytes. *Nature* **391**, 344–345.

Luther, S. A. and Cyster, J. G. (2001). Chemokines as regulators of T cell differentiation. *Nat. Immunol.* **2**, 102–107.

Maayan, S., Zhang, L., Shinar, E., et al. (2000). Evidence for recent selection of the CCR5-delta 32 deletion from differences in its frequency between Ashkenazi and Sephardi Jews. *Genes Immun.* **1**, 358–361.

MacDonald, K. S., Matukas, L., Embree, J. E., et al. (2001). Human leucocyte antigen supertypes and immune susceptibility to HIV-1, implications for vaccine design. *Immunol. Lett.* **79**, 151–157.

Magierowska, M., Theodorou, I., Debre, P., et al. (1999). Combined genotypes of CCR5, CCR2, SDF1, and HLA genes can predict the long-term nonprogressor status in human immunodeficiency virus-1-infected individuals. *Blood* **93**, 936–941.

Maho, A., Bensimon, A., Vassart, G., et al. (1999a). Mapping of the CCXCR1, CX3CR1, CCBP2 and CCR9 genes to the CCR cluster within the 3p21.3 region of the human genome. *Cytogenet. Cell Genet.* **87**, 265–268.

Maho, A., Carter, A., Bensimon, A., et al. (1999b). Physical mapping of the CC-chemokine gene cluster on the human 17q11. 2 region. *Genomics* **59**, 213–223.

Mangano, A., Gonzalez, E., Dhanda, R., et al. (2001). Concordance between the CC chemokine receptor 5 genetic determinants that alter risks of transmission and disease progression in children exposed perinatally to human immunodeficiency virus. *J. Infect. Dis.* **183**, 1574–1585.

Mangano, A., Prada, F., Roldan, A., et al. (1998). Distribution of CCR-5 Δ32 allele in Argentinian children at risk of HIV-1 infection: its role in vertical transmission. *AIDS* **12**, 109–110.

Mantovani, A., Locati, M., Vecchi, A., et al. (2001). Decoy receptors: A strategy to regulate inflammatory cytokines and chemokines. *Trends Immunol.* **22**, 328–336.

Margolis, L. B., Glushakova, S., Grivel, J. C., et al. (1998). Blockade of CC chemokine receptor 5 (CCR5)-tropic human immunodeficiency virus-1 replication in human lymphoid tissue by CC chemokines. *J. Clin. Invest.* **101**, 1876–1880.

Marmor, M., Sheppard, H. W., Donnell, D., et al. (2001). Homozygous and heterozygous CCR5-Δ32 genotypes are associated with resistance to HIV infection. *J. Acquir. Immune Defic. Syndr.* **27**, 472–481.

Martin, M. P., Dean, M., Smith, M. W., et al. (1998). Genetic acceleration of AIDS progression by a promoter variant of *CCR5*. *Science* **282**, 1907–1911.

Martinson, J. J., Chapman, N. H., Rees, D. C., et al. (1997). Global distribution of the CCR5 gene 32-basepair deletion. *Nat. Genet.* **16**, 100–103.

Martinson, J. J., Hong, L., Karanicolas, R., et al. (2000). Global distribution of the CCR2-64I/CCR5-59653T HIV-1 disease-protective haplotype. *AIDS* **14**, 483–489.

McDermott, D. H., Beecroft, M. J., Kleeberger, C. A., et al. (2000a). Chemokine RANTES promoter polymorphism affects risk of both HIV infection and disease progression in the Multicenter AIDS Cohort Study. *AIDS* **14**, 2671–2678.

McDermott, D. H., Colla, J. S., Kleeberger, C. A., et al. (2000b). Genetic polymorphism in CX3CR1 and risk of HIV disease. *Science* **290**, 2031.

McDermott, D. H., Zimmerman, P. A., Guignard, F., et al. (1998). CCR5 promoter polymorphism and HIV-1 disease progression. *Lancet* **352**, 866–870.

Mellado, M., Rodriguez-Frade, J. M., Manes, S., et al. (2001a). Chemokine signaling and functional responses: the role of receptor dimerization and TK pathway activation. *Annu. Rev. Immunol.* **19**, 397–421.

Mellado, M., Rodriguez-Frade, J. M., Vila-Coro, A. J., et al. (1999). Chemokine control of HIV-1 infection. *Nature* **400**, 723–724.

Mellado, M., Rodriguez-Frade, J. M., Vila-Coro, A. J., et al. (2001b). Chemokine receptor homo- or heterodimerization activates distinct signaling pathways. *EMBO J.* **20**, 2497–2507.

Mellors, J. W., Rinaldo, C. R., Jr., Gupta, P., et al. (1996). Prognosis in HIV-1 infection predicted by the quantity of virus in plasma. *Science* **272**, 1167–1170.

Meng, G., Wei, X.-P., Wu, X.-Y., et al. (2002). Primary intestinal epithelial cells selectively transfer R5 HIV-1 to CCR5^{+} cells. *Nat. Med.* **8**, 150–156.

Meyer, L., Magierowska, M., Hubert, J. B., et al. (1997). Early protective effect of CCR-5 Δ32 heterozygosity on HIV-1 disease progression: relationship with viral load. *AIDS* **11**, F73–78.

Michael, N. L., Chang, G., Louie, L. G., et al. (1997). The role of viral phenotype and CCR-5 gene defects in HIV-1 transmission and disease progression. *Nat. Med.* **3**, 338–340.

Michael, N. L., Nelson, J. A. E., Kewalramani, V. N., et al. (1998). Exclusive and persistent use of the entry coreceptor CXCR4 by human immunodeficiency virus type 1 from a subject homozygous for CCR5 Δ-32. *J. Virol.* **72**, 6040–6047.

Miyamasu, M., Sekiya, T., Ohta, K., et al. (2001). Variations in the human CC chemokine eotaxin gene. *Genes Immun.* **2**, 461–463.

Modi, W. S. and Chen, Z. Q. (1998). Localization of the human CXC chemokine subfamily on the long arm of chromosome 4 using radiation hybrids. *Genomics* **47**, 136–139.

Mummidi, S., Ahuja, S. S., Gonzalez, E., et al. (1998). Genealogy of the CCR5 locus and chemokine system gene variants associated with altered rates of HIV-1 disease progression. *Nat. Med.* **4**, 786–793.

Mummidi, S., Ahuja, S. S., McDaniel, B. L., et al. (1997). The human CC chemokine receptor 5 (CCR5) gene. Multiple transcripts with 5′-end heterogeneity, dual promoter usage, and evidence for polymorphisms within the regulatory regions and noncoding exons. *J. Biol. Chem.* **272**, 30662–30671.

Mummidi, S., Bamshad, M., Ahuja, S. S., et al. (2000). Evolution of human and non-human primate CC chemokine receptor 5 gene and mRNA. Potential roles for haplotype and mRNA diversity, differential haplotype-specific transcriptional activity, and altered transcription factor binding to polymorphic nucleotides in the pathogenesis of HIV-1 and simian immunodeficiency virus. *J. Biol. Chem.* **275**, 18946–18961.

Murphy, P. M., Baggiolini, M., Charo, I. F., et al. (2000). International union of pharmacology. XXII. Nomenclature for chemokine receptors. *Pharmacol. Rev.* **52**, 145–176.

Nakayama, E. E., Hoshino, Y., Xin, X., et al. (2000). Polymorphism in the interleukin-4 promoter affects acquisition of human immunodeficiency virus type 1 syncytium-inducing phenotype. *J. Virol.* **74**, 5452–5459.

Naruse, K., Ueno, M., Satoh, T., et al. (1996). A YAC contig of the human CC chemokine genes clustered on chromosome 17q11.2. *Genomics* **34**, 236–240.

Nelson, R. T., Boyd, J., Gladue, R. P., et al. (2001). Genomic organization of the CC chemokine mip-3alpha/CCL20/larc/exodus/SCYA20, showing gene structure, splice variants, and chromosome localization. *Genomics* **73**, 28–37.

Nibbs, R. J., Wylie, S. M., Yang, J., et al. (1997). Cloning and characterization of a novel promiscuous human beta-chemokine receptor D6. *J. Biol. Chem.* **272**, 32078–32083.

Nomiyama, H., Fukuda, S., Iio, M., et al. (1999). Organization of the chemokine gene cluster on human chromosome 17q11.2 containing the genes for CC chemokine MPIF-1, HCC-2, HCC-1, LEC, and RANTES. *J. Interferon Cytokine Res.* **19**, 227–234.

Nomiyama, H., Imai, T., Kusuda, J., et al. (1998). Human chemokines fractalkine (SCYD1), MDC (SCYA22) and TARC (SCYA17) are clustered on chromosome 16q13. *Cytogenet. Cell Genet.* **81**, 10–11.

Nomiyama, H., Mera, A., Ohneda, O., et al. (2001). Organization of the chemokine genes in the human and mouse major clusters of CC and CXC chemokines: diversification between the two species. *Genes Immun.* **2**, 110–113.

Oberlin, E., Amara, A., Bachelerie, F., et al. (1996). The CXC chemokine SDF-1 is the ligand for LESTR/fusin and prevents infection by T-cell-line-adapted HIV-1. *Nature* **382**, 833–835.

O'Brien, S. J., Gao, X. and Carrington, M. (2001). HLA and AIDS: a cautionary tale. *Trends Mol. Med.* **7**, 379–381.

O'Brien, S. J. and Moore, J. P. (2000). The effect of genetic variation in chemokines and their receptors on HIV transmission and progression to AIDS. *Immunol. Rev.* **177**, 99–111.

O'Brien, S. J., Nelson, G. W., Winkler, C. A., et al. (2000a). Polygenic and multifactorial disease gene association in man: lessons from AIDS. *Annu. Rev. Genet.* **34**, 563–591.

O'Brien, T. R., McDermott, D. H., Ioannidis, J. P., et al. (2000b). Effect of chemokine receptor gene polymorphisms on the response to potent antiretroviral therapy. *AIDS* **14**, 821–826.

Ogilvie, P., Bardi, G., Clark-Lewis, I., et al. (2001). Eotaxin is a natural antagonist for CCR2 and an agonist for CCR5. *Blood* **97**, 1920–1924.

Operskalski, E. A., Stram, D. O., Busch, M. P., et al. (1997). Role of viral load in heterosexual transmission of human immunodeficiency virus type 1 by blood transfusion recipients. *Am. J. Epidemiol.* **146**, 655–661.

O'Shea, S., Rostron, T., Hamblin, A. S., et al. (1991). Quantitation of HIV: correlation with clinical, virological, and immunological status. *J. Med. Virol.* **35**, 65–69.

Paxton, W. A., Kang, S., Liu, R., et al. (1999). HIV-1 infectability of CD4(+) lymphocytes with relation to beta-chemokines and the CCR5 coreceptor. *Immunol. Lett.* **66**, 71–75.

Paxton, W. A., Liu, R., Kang, S., et al. (1998). Reduced HIV-1 infectability of CD4+ lymphocytes from exposed-uninfected individuals: association with low expression of CCR5 and high production of beta-chemokines. *Virology* **244**, 66–73.

Paxton, W. A., Martin, S. R., Tse, D., et al. (1996). Relative resistance to HIV-1 infection of CD4 lymphocytes from persons who remain uninfected despite multiple high-risk sexual exposures. *Nat. Med.* **2**, 412–417.

Paxton, W. A., Neumann, A. U., Kang, S., et al. (2001). RANTES production from CD4+ lymphocytes correlates with host genotype and rates of human immunodeficiency virus type 1 disease progression. *J. Infect. Dis.* **183**, 1678–1681.

Pedraza, M. A., del Romero, J., Roldan, F., et al. (1999). Heterosexual transmission of HIV-1 is associated with high plasma viral load levels and a positive viral isolation in the infected partner. *J. Acquir. Immune Defic. Syndr. Hum. Retrovirol.* **21**, 120–125.

Penn, M. L., Grivel, J. C., Schramm, B., et al. 1999. CXCR4 utilization is sufficient to trigger CD4+ T cell depletion in HIV-1-infected human lymphoid tissue. *Proc. Natl. Acad. Sci. USA* **96**, 663–668.

Penton-Rol, G., Polentarutti, N., Luini, W., et al. (1998). Selective inhibition of expression of the chemokine receptor CCR2 in human monocytes by IFN-gamma. *J. Immunol.* **160**, 3869–3873.

Pogo, A. O. and Chaudhuri, A. (2000). The Duffy protein: a malarial and chemokine receptor. *Semin. Hematol.* **37**, 122–129.

Pohlmann, S., Baribaud, F. and Doms, R. W. (2001). DC-SIGN and DC-SIGNR: helping hands for HIV. *Trends Immunol.* **22**, 643–646.

Proost, P., Menten, P., Struyf, S., et al. (2000). Cleavage by CD26/dipeptidyl peptidase IV converts the chemokine LD78beta into a most efficient monocyte attractant and CCR1 agonist. *Blood* **96**, 1674–1680.

Quillent, C., Oberlin, E., Braun, J., et al. (1998). HIV-1 resistance phenotype conferred by combination of two separate inherited mutations of *CCR5* gene. *Lancet* **351**, 14–18.

Quinn, T. C., Wawer, M. J., Sewankambo, N., et al. (2000). Viral load and heterosexual transmission of human immunodeficiency virus type 1. Rakai Project Study Group. *N. Engl. J. Med.* **342**, 921–929.

Raport, C. J., Gosling, J., Schweickart, V. L., et al. (1996). Molecular cloning and functional characterization of a novel human CC chemokine receptor (CCR5) for RANTES, MIP-1beta, and MIP-1alpha. *J. Biol. Chem.* **271**, 17161–17166.

Rizzardi, G. P., Morawetz, R. A., Vicenzi, E., et al. (1998). CCR2 polymorphism and HIV disease. *Nat. Med.* **4**, 252–253.

Rizzuto, C. D., Wyatt, R., Hernandezramos, N., et al. (1998). A conserved HIV gp120 glycoprotein structure involved in chemokine receptor binding. *Science* **280**, 1949–1953.

Rodriguez-Frade, J. M., Mellado, M. and Martinez, A. C. (2001). Chemokine receptor dimerization: two are better than one. *Trends Immunol.* **22**, 612–617.

Rodriguez-Frade, J. M., Vila-Coro, A. J., de Ana, A. M., et al. (1999). The chemokine monocyte chemoattractant protein-1 induces functional responses through dimerization of its receptor CCR2. *Proc. Natl. Acad. Sci. USA* **96**, 3628–3633.

Rollins, B. J., Morton, C. C., Ledbetter, D. H., et al. (1991). Assignment of the human small inducible cytokine A2 gene, SCYA2 (encoding JE or MCP-1), to 17q11.2-12: Evolutionary relatedness of cytokines clustered at the same locus. *Genomics* **10**, 489–492.

Romagnani, P., Annunziato, F., Piccinni, M. P., et al. (2000). Cytokines and chemokines in T lymphopoiesis and T-cell effector function. *Immunol Today* **21**, 416–418.

Rucker, J., Samson, M., Doranz, B. J., et al. (1996). Regions in beta-chemokine receptors CCR5 and CCR2b that determine HIV-1 cofactor specificity. *Cell* **87**, 437–446.

Saah, A. J., Hoover, D. R., Weng, S., et al. (1998). Association of HLA profiles with early plasma viral load, CD4+ cell count and rate of progression to AIDS following acute HIV-1 infection. Multicenter AIDS Cohort Study. *AIDS* **12**, 2107–2113.

Sallusto, F., Lenig, D., Mackay, C. R., et al. (1998). Flexible programs of chemokine receptor expression on human polarized T helper 1 and 2 lymphocytes. *J. Exp. Med.* **187**, 875–883.

Sallusto, F., Mackay, C. R. and Lanzavecchia, A. (1997). Selective expression of the eotaxin receptor CCR3 by human T helper 2 cells. *Science* **277**, 2005–2007.

Salzwedel, K., Smith, E. D., Dey, B., et al. (2000). Sequential CD4-coreceptor interactions in human immunodeficiency virus type 1 Env function: soluble CD4 activates Env for coreceptor-dependent fusion and reveals blocking activities of antibodies against cryptic conserved epitopes on gp120. *J. Virol.* **74**, 326–333.

Samson, M., Soularue, P., Vassart, G., et al. (1996). The genes encoding the human CC-chemokine receptors CC-CKR1 to CC-CKR5 (CMKBR1-CMKBR5) are clustered in the p21.3-p24 region of chromosome 3. *Genomics* **36**, 522–526.

Scalatti, G., Tresoldi, E., Bjorndal, A., et al. (1997). *In vivo* evolution of HIV-1 co-receptor usage and sensitivity to chemokine-mediated suppression. *Nat. Med.* **3**, 1259–1265.

Schliekelman, P., Garner, C. and Slatkin, M. (2001). Natural selection and resistance to HIV. *Nature* **411**, 545–546.

Schols, D., Struyf, S., Van Damme, J., et al. (1997). Inhibition of T-tropic HIV strains by selective antagonization of the chemokine receptor CXCR4. *J. Exp. Med.* **186**, 1383–1388.

Scotet, E., Schroeder, S. and Lanzavecchia, A. (2001). Molecular regulation of CC-chemokine receptor 3 expression in human T helper 2 cells. *Blood* **98**, 2568–2570.

Semba, R. D., Kumwenda, N., Hoover, D. R., et al. (1999). Human immunodeficiency virus load in breast milk, mastitis, and mother-to-child transmission of human immunodeficiency virus type 1. *J. Infect. Dis.* **180**, 93–98.

Shan, L., Qiao, X., Oldham, E., et al. (2000). Identification of viral macrophage inflammatory protein (vMIP)-II as a ligand for GPR5/XCR1. *Biochem. Biophys. Res. Commun.* **268**, 938–941.

Shearer, W. T., Quinn, T. C., LaRussa, P., et al. (1997). Viral load and disease progression in infants infected with human immunodeficiency virus type 1. Women and Infants Transmission Study Group. *N. Engl. J. Med.* **336**, 1337–1342.

Shin, H., Winkler, C., Stevens, J. C., et al. (2000). Genetic restriction of HIV-1 pathogenesis to AIDS by promoter alleles of IL10. *Proc. Natl. Acad. Sci. USA* **97**, 14467–14472.

Smith, M. W., Dean, M., Carrington, M., et al. (1997). Contrasting genetic influence of *CCR2* and *CCR5* variants on HIV-1 infection and disease progression. *Science* **277**, 959–965.

Steinberger, P., Andris-Widhopf, J., Buhler, B., et al. (2000). Functional deletion of the CCR5 receptor by intracellular immunization produces cells that are refractory to CCR5-dependent HIV-1 infection and cell fusion. *Proc. Natl. Acad. Sci. USA* **97**, 805–810.

Stephens, J. C., Reich, D. E., Goldstein, D. B., et al. (1998). Dating the origin of the CCR5-Δ32 AIDS-resistance allele by the coalescence of haplotypes. *Am. J. Hum. Genet.* **62**, 1507–1515.

Strizki, J. M., Xu, S., Wagner, N. E., et al. 2001. SCH-C (SCH 351125), an orally bioavailable, small molecule antagonist of the chemokine receptor CCR5, is a potent inhibitor of HIV-1 infection *in vitro* and *in vivo*. *Proc. Natl. Acad. Sci. USA* **98**, 12718–12723.

Struyf, S., De Meester, I., Scharpe, S., et al. (1998). Natural truncation of RANTES abolishes signaling through the CC chemokine receptors CCR1 and CCR3, impairs its chemotactic potency and generates a CC chemokine inhibitor. *Eur. J. Immunol.* **28**, 1262–1271.

Struyf, S., Menten, P., Lenaerts, J. P., et al. (2001). Diverging binding capacities of natural LD78beta isoforms of macrophage inflammatory protein-1alpha to the CC chemokine receptors 1, 3 and 5 affect their anti-HIV-1 activity and chemotactic potencies for neutrophils and eosinophils. *Eur. J. Immunol.* **31**, 2170–2178.

Su, B., Jin, L., Hu, F., et al. (1999). Distribution of two HIV-1-resistant polymorphisms (SDF1-3′A and CCR2-64I) in East Asian and world populations and its implication in AIDS epidemiology. *Am. J. Hum. Genet.* **65**, 1047–1053.

Tang, J., Costello, C., Keet, I. P. M., et al. (1999a). HLA class I homozygosity accelerates disease progression in human immunodeficiency virus type 1 infection. *AIDS Res. Hum. Retroviruses* **15**, 317–324.
Tang, J., Rivers, C., Karita, E., et al. (1999b). Allelic variants of human beta-chemokine receptor 5 (CCR5) promoter: evolutionary relationships and predictable association with HIV-1 disease progression. *Genes Immun.* **1**, 20–27.
Tang, J., Shelton, B., Makhatadze, N. J., et al. (2002a). Distribution of chemokine receptor *CCR2*and *CCR5* genotypes and their relative contribution to human immunodeficiency virus type 1 (HIV-1) seroconversion, early HIV-1 RNA concentration in plasma, and later disease progression. *J. Virol.* **76**, 662–672.
Tang, J., Wilson, C. M., Schaen, M., et al. (2002b). *CCR2* and *CCR5* genotypes in HIV type 1-infected adolescents: limited contributions to variability in plasma HIV type 1 RNA concentration in the absence of antiretroviral therapy. *AIDS Res. Hum. Retroviruses* **18**, 403–412.
Tartakovsky, B., Turner, D., Vardinon, N., et al. (1999). Increased intracellular accumulation of macrophage inflammatory protein 1beta and its decreased secretion correlate with advanced HIV disease. *J. Acquir. Immune Defic. Syndr. Hum. Retrovirol.* **20**, 420–422.
Tasaki, Y., Fukuda, S., Lio, M., et al. (1999). Chemokine PARC gene (SCYA18) generated by fusion of two MIP-1alpha/LD78alpha-like genes. *Genomics* **55**, 353–357.
Taylor, J. M., Wang, Y., Ahdieh, L., et al. (2000). Causal pathways for CCR5 genotype and HIV progression. *J. Acquir. Immune Defic. Syndr.* **23**, 160–171.
Thelen, M. (2001). Dancing to the tune of chemokines. *Nat. Immunol.* **2**, 129–134.
Trkola, A., Kuhmann, S. E., Strizki, J. M., et al. (2002). HIV-1 escape from a small molecule, CCR5-specific entry inhibitor does not involve CXCR4 use. *Proc. Natl. Acad. Sci. USA* **99**, 395–400.
Trkola, A., Paxton, W. A., Monard, S. P., et al. (1998). Genetic subtype-independent inhibition of human immunodeficiency virus type 1 replication by CC and CXC chemokines. *J. Virol.* **72**, 396–404.
Unutmaz, D., Xiang, W., Sunshine, M. J., et al. (2000). The primate lentiviral receptor Bonzo/STRL33 is coordinately regulated with CCR5 and its expression pattern is conserved between human and mouse. *J. Immunol.* **165**, 3284–3292.
Valdez, H., Purvis, S. F., Lederman, M. M., et al. (1999). Association of the CCR5Δ32 mutation with improved response to antiretroviral therapy. *J. Am. Med. Assoc.* **282**, 734.
Van Coillie, E., Proost, P., Van Aelst, I., et al. (1998). Functional comparison of two human monocyte chemotactic protein-2 isoforms, role of the

amino-terminal pyroglutamic acid and processing by CD26/dipeptidyl peptidase IV. *Biochemistry* **37**, 12672–12680.

van Rij, R. P., Proersen, S., Goudsmit, J., et al. (1998). The role of a stromal cell-derived factor-1 chemokine gene variant in the clinical course of HIV-1 infection. *AIDS* **12**, F85–90.

Veazey, R. S., DeMaria, M., Chalifoux, L. V., et al. (1998). Gastrointestinal tract as a major site of CD4+ T cell depletion and viral replication in SIV infection. *Science* **280**, 427–431.

Veazey, R. S., Marx, P. A. and Lackner, A. A. (2001). The mucosal immune system: Primary target for HIV infection and AIDS. *Trends Immunol.* **22**, 626–633.

Wang, J., Guan, E., Roderiquez, G., et al. (1999). Inhibition of CCR5 expression by IL-12 through induction of beta-chemokines in human T lymphocytes. *J. Immunol.* **163**, 5763–5769.

Ward, S. G., Bacon, K. and Westwick, J. (1998). Chemokines and T lymphocytes: More than an attraction. *Immunity* **9**, 1–11.

Winkler, C., Modi, W., Smith, M. W., et al. (1998). Genetic restriction of AIDS pathogenesis by an SDF-1 chemokine gene variant. ALIVE Study, Hemophilia Growth and Development Study (HGDS), Multicenter AIDS Cohort Study (MACS), Multicenter Hemophilia Cohort Study (MHCS), San Francisco City Cohort (SFCC). *Science* **279**, 389–393.

Wu, L., Gerard, N. P., Wyatt, R., et al. (1996). CD4-induced interaction of primary HIV-1 gp120 glycoproteins with the chemokine receptor CCR-5. *Nature* **384**, 179–183.

Xin, X., Nakamura, K., Liu, H., et al. (2001). Novel polymorphisms in human macrophage inflammatory protein-1 alpha (MIP-1alpha) gene. *Genes Immun.* **2**, 156–158.

Yoshie, O., Imai, T. and Nomiyama, H. (1997). Novel lymphocyte-specific CC chemokines and their receptors. *J. Leukoc. Biol.* **62**, 634–644.

Youn, B. S., Zhang, S. M., Broxmeyer, H. E., et al. (1998). Characterization of CKbeta8 and CKbeta8-1: two alternatively spliced forms of human beta-chemokine, chemoattractants for neutrophils, monocytes, and lymphocytes, and potent agonists at CC chemokine receptor 1. *Blood* **91**, 3118–3126.

Zaitseva, M., Blauvelt, A., Lee, S., et al. (1997). Expression and function of CCR5 and CXCR4 on human Langerhans cells and macrophages: implications for HIV primary infection. *Nat. Med* **3**, 1369–1375.

Zella, D., Barabitskaja, O., Burns, J. M., et al. (1998). Interferon-gamma increases expression of chemokine receptors CCR1, CCR3, and CCR5, but not CXCR4 in monocytoid U937 cells. *Blood* **91**, 4444–4450.

Zhou, N., Luo, Z., Luo, J., et al. (2000). A novel peptide antagonist of CXCR4 derived from the N-terminus of viral chemokine vMIP-II. *Biochemistry* **39**, 3782–3787.

Zimmerman, P. A., Buckler-White, A., Alkhatib, G., et al. (1997). Inherited resistance of HIV-1 conferred by an inactivating mutation in CC chemokine receptor 5: Studies in populations with contrasting clinical phenotypes, defined racial background, and quantified risk. *Mol. Med.* **3**, 23–36.

Zimmermann, N., Bernstein, J. A. and Rothenberg, M. E. (1998). Polymorphisms in the human CC chemokine receptor-3 gene. *Biochim. Biophys. Acta Gen. Struct. Express* **1442**, 170–176.

Zlotnik, A., and Yoshie, O. (2000). Chemokines: a new classification system and their role in immunity. *Immunity* **12**, 121–127.

Zoeteweij, J. P., Golding, H., Mostowski, H., et al. (1998). Cutting edge cytokines regulate expression and function of the HIV coreceptor CXCR4 on human mature dendritic cells. *J. Immunol.* **161**, 3219–3223.

CHAPTER 8

NRAMP1 and resistance to intracellular pathogens

Philippe Gros

Department of Biochemistry; Centre for the Study of Host Resistance; McGill Cancer Centre

Erwin Schurr

Centre for the Study of Host Resistance; Departments of Medicine and Human Genetics, McGill University, Montreal, Quebec Canada

1. INTRODUCTION

The role of genetic factors in predisposition to infectious diseases has long been recognised in humans (reviewed by Cooke and Hill, 2001), and some infections such as tuberculosis and leprosy were long believed to be inheritable diseases. One of the clearest examples of the effect of the host genetic makeup on susceptibility to infection in humans is malaria (Kwiatkowski, 2000; Fortin et al., 2002), where the blood-borne parasite itself may have exerted a positive selective pressure for the retention of otherwise disease-associated alterations in certain erythrocyte proteins. Indeed, in sickle-cell anemia, heterozygosity for mutant hemoglobin alleles confers survival advantage over homozygosity for either mutant or wild-type alleles (Pasvol et al., 1978; Hill et al., 1991; Shear et al., 1993). On the other hand, functional polymorphisms affecting transcriptional control of key host response genes such as tumour necrosis factor-α (TNF-α) have been shown to drastically affect disease progression and outcome (McGuire et al., 1994). However, in most serious infectious diseases, the molecular identification of the genetic component of susceptibility has remained an extremely difficult task with few successes. Indeed, reduced penetrance, variable expressivity, a wide disease spectrum associated with variations in microbe-encoded virulence determinants, together with poor diagnostic criteria make it very difficult to decipher and map single gene effects, even if major, in human populations. Complex genetic traits can, however, be dissected in genetically well-defined inbred, recombinant inbred, and recombinant congenic strains of mice in which single gene effects may have either naturally segregated or have been experimentally isolated by breeding. These genes can then be localised in linkage studies and, in certain cases, can be identified by transcription mapping and positional cloning (Vidal et al., 1993; Poltorak et al., 2000; Qureshi et al., 1999; Lee et al., 2001; Brown et al., 2001). The mouse is the experimental model of choice for this type of analysis for the following reasons: (1) The virulence status of the infectious agent, as well as the dose and route of infection, can be tightly controlled, thereby reducing microbial-induced variability; (2) large numbers of wild-type isolates and mutant stocks of mice are available in an inbred status; (3) informative segregating animals can be generated in large numbers for linkage mapping and positional cloning; (4) the sequence of the mouse genome is soon to be completed, providing a compendium of candidate genes for a particular region; (5) null alleles at candidate genes can be readily obtained by gene targeting; and (6) mutant variants of the gene can be reintroduced on a null background to analyse genotype/phenotype correlations. Such candidates identified in the mouse can then be tested for relevance in human populations in association or linkage studies either in endemic areas of disease, or in the outbreak situation or in first contact epidemics, but ultimately by direct biochemical assays of the protein function. Finally, knowledge of the cellular pathway in which the gene and protein play a role may open opportunities for new prophylactic and therapeutic strategies in the corresponding disease. Here, we will review one instance where such an approach has led to the isolation and validation of a gene (*NRAMP1*) that plays a role in determining susceptibility to infections in both mouse and humans. In addition, functional studies of the Nramp proteins have yielded valuable information on a normal mechanism of macrophage defense against certain types of infection and may prove to be a valuable novel target for pharmacological intervention in the corresponding diseases.

2. STUDIES IN MICE

2.1 Discovery of the *Ity/Lsh/Bcg* Locus

The concept of a genetic basis for innate susceptibility to mouse typhoid was initially proposed by Webster in the early part of the twentieth century. In a series of seminal papers (Webster, 1923, 1924, 1933), he documented strain differences in susceptibility to infection and susbsequently derived by inbreeding substrains of mice that were either resistant (BR) or susceptibile (BS). Subsequent segregation analysis in F1, F2, and backcross mice indicated that the genetic component was simple, in which resistance to *Bacillus enteritidis* was autosomal and inherited in a dominant fashion. More recent strain surveys by Robson and Vas (1972) and Plant and Glynn (1976), using *Salmonella typhimurium* as an infectious agent, confirmed that modern inbred strains fell into two nonoverlapping groups with respect to susceptibility. Using liver bacterial load as a phenotypic marker, it was shown that the interstrain difference was caused by a single, dominant, autosomal locus which was named *Ity* (Plant and Glynn, 1976). *Ity* was initially linked to *ln* on mouse chromosome 1 (Plant and Glynn, 1979) and subsequently mapped to a 15cM interval in a five-point test cross (O'Brien et al., 1980). Independently, Bradley and his colleagues, working with the intracellular parasite *Leishmania donovani*, noted a striking difference in the early response and replication rates of the parasite in the liver of 25 different inbred strains of mice (Bradley, 1974, 1977). Segregation analyses in informative F1 and backcross mice bred between resistant and susceptible progenitors gave susceptibility ratios consistent with a single dominant autosomal gene which was designated *Lsh*. Using recombinant inbred mice as well as a standard three-point test cross, *Lsh* was mapped on the proximal portion of chromosome 1 (Bradley et al., 1979). Interestingly, a comparison of the strain distribution pattern of various recombinant inbred mouse strains for susceptibility to *S. typhimurium* (*Ity*) and *L. donovani* (*Lsh*) infections suggested that the two loci were either tightly linked or were in fact the same gene (Plant et al., 1982; O'Brien et al., 1980). In parallel, our group also noted differences in susceptibility of inbred strains to infection with small doses of *Mycobacterium bovis* (BCG). Although interstrain differences in susceptibility to infection with Mycobacteria (e.g., *M. tuberculosis*) had been previously documented (Pierce et al., 1947; Donovick et al., 1949; Lynch et al., 1965), studies of BCG replication in the spleen after intravenous infection first indicated that inbred strains fall into two nonoverlapping groups, being either resistant (no replication) or susceptible (rapid replication) in the early phase of infection (Forget et al., 1981). The interstrain difference was determined by a single, dominant (resistance) and autosomal gene, designated *Bcg* (Gros et al., 1981). Strikingly, studies in recombinant inbred strains, as well as progeny testing in informative crosses indicated that *Bcg* may be either tightly linked or most probably identical to *Ity* and *Lsh* (Skamene et al., 1982). These results indicated that a single locus on mouse chromosome 1 may control innate resistance and susceptibility to antigenically and phylogenetically unrelated pathogens.

The fact that all three pathogens infect and replicate inside host macrophages suggested that the genetic difference controlled by this locus may be phenotypically expressed by this cell type. Indeed, a number of studies *in vivo* or *in vitro* indicated that the genetic difference between R and S alleles at the *Ity/Lsh/Bcg* locus was detectable within a few hours after infection and was expressed by a bone marrow-derived, radiation-resistant cell which could be poisoned by silica (Gros et al., 1983). Furthermore, experiments *in vitro* with explanted macrophages showed that *Ity/Lsh/Bcg* modulates the capacity of these cells to control intracellular replication of *L. donovani* (Crocker et al., 1984); *S. typhimurium* (Lissner et al., 1983); and several Mycobacteria, including *M. bovis* (Stach et al., 1984), *M. smegmatis* (Denis et al., 1990), and *M. avium* (Stokes et al., 1986; De Chastellier et al., 1993). The spectrum of microbial agents for which replication may be influenced by *Ity/Lsh/Bcg* was later studied *in vivo* in inbred mouse strains harboring S or R alleles at the locus, in mouse strains congenic for the target portion of chromosome 1, in syngeneic animals bearing a null mutation at this locus (see below), or in explanted macrophage populations or in macrophage cell lines derived from such animals. These studies have shown that *Ity/Lsh/Bcg* affects replication not only of *S. typhimurium, L. donovani,* and several mycobacteria but also of a variety of ovine (Gautier et al., 1998) and avian (Hu et al., 1995) strains of *Salmomella,* as well as *Brucella abortus* (Barthel et al., 2001) and *Pasteurella pneumotropica* (Chapes et al., 2001). Interestingly, *Ity/Lsh/Bcg* alleles also seem to affect susceptibility to infection with *Toxoplasma gondii* (McLeod et al., 1989; Blackwell et al., 1994) and *Francisella tularensis* (Kovarova et al., 2000, 2002), although in these cases, the susceptibility allele at *Ity/Lsh/Bcg* is associated with increased resistance. Finally, *in vivo* or *in vitro* replication of other pathogens such as *Chlamydia, Legionella, Listeria, Pseudomonas, Bacillus subtilis,* and *Staphylococcus aureus* is not affected by *Ity/Lsh/Bcg* alleles (Yoshida et al., 1991; De Chastellier et al., 1993; Gruenheid et al., 1999; Pal et al., 2000; Chapes et al., 2001). The role of *Ity/Lsh/Bcg* in susceptibility to infection with virulent strains of human *Mycobacterium tuberculosis* (MTB) has been debated (North and Medina, 1998; Buschman and Skamene, 2001). Nonconcordance of susceptibility to MTB (bacterial replication and survival) with *Bcg* alleles in inbred strains have argued against a role for this locus (Medina and North, 1998); When studied, differential susceptibility in inbred strains has been observed to be complex (Mitsos et al., 2000; Kramnik et al., 1998). Studies in recombinant mice either congenic or mutated for *Ity/Lsh/Bcg* have also argued against a role for this locus (North et al., 1999). On the other hand, studies in humans (see below) have clearly established a role for this locus in susceptibility to tuberculosis and leprosy. Thus, the pleiotropic effect of *Ity/Lsh/Bcg* alleles on infections with different types of intracellular parasites suggests that the gene affects an important mechanism of macrophage defenses against such infectious agents.

2.2 Positional Cloning of the Mouse Nramp1 Gene

The *Ity/Lsh/Bcg* locus was characterised by positional cloning, using an experimental approach based on the production of high-resolution linkage maps of the region with a large number of informative meioses (Malo et al., 1993a), the creation of a corresponding physical map (Malo et al., 1993b), the identification of the transcription units in the interval by exon trapping, and the characterisation of the tissue and cell-specific patterns of expression of the candidate genes (Vidal et al., 1993). One of them was found to be expressed in spleen and macrophages and was given the appellation *Nramp1* (natural resistance associated macrophage protein 1, OMIM No. 600266; also classified as solute carrier family 11 member 1, *Slc11a1*). The full-length cDNA sequence predicts a 56-kDa protein which displays structural features characteristic of membrane proteins, including 12 highly hydrophobic membrane domains, one predicted glycosylated extracytoplasmic loop, and one sequence signature found in many membrane transporters (Vidal et al., 1993; Cellier et al., 1995). Studies with anti-Nramp1 antibodies indicate that the protein is indeed expressed in macrophages as a mature 90- to 110-kDa to protein extensively modified by glycosylation and phosphorylation (Vidal et al., 1996). Sequence analysis of the *Nramp1* gene in inbred strains reveal that the susceptibility allele of the *Ity/Lsh/Bcg* locus is linked to a single G169D substitution in predicted transmembrane domain 4 of the protein (Malo et al., 1994). This mutation was shown to affect protein stability or targeting, which results in the absence of a mature protein species in macrophages from susceptible $Nramp1^{Asp169}$ mice (Vidal et al., 1996). On the other hand, creation of a null allele at *Nramp1* by homologous recombination confers susceptibility to infection with *Salmonella*, *Leishmania*, and *Mycobacterium* spp. in otherwise resistant 129Sv mice (Vidal et al., 1995). In these experiments, animals bearing allelic combinations $Nramp1^{Asp169/Asp169}$, $Nramp1^{null/Asp169}$, or $Nramp1^{null/null}$ were found to be equally sensitive to infection, indicating that the Asp169 allele of *Nramp1* results in complete loss of function. Finally, introduction of a genomic DNA fragment containing the $Nramp1^{Gly169}$ allele in transgenic animals of $Nramp1^{Asp169}$ genetic background restores resistance to infections (Govoni et al., 1996).

2.3 Expression of Nramp1 in Professional Phagocytes

In the mouse, *Nramp1* mRNA is expressed primarily in the spleen and liver and is abundant in primary macrophages and in macrophage cell lines such as J774A and RAW264.7, but is also detected in cells of the granulocytic lineage (Govoni et al., 1995, 1997). Expression studies in RAW macrophages show that the gene can be strongly induced by exposure of these cells to lipolysaccharide and interferon-γ, but also by inflammatory stimuli. This inducible expression is concomitant to the presence of NF-IL6 and IFN response elements in the proximal promotor (−500-bp) region of *Nramp1* (Govoni et al., 1995, 1997). In humans, *NRAMP1* mRNA is also expressed in spleen, but is very abundant in lungs and is most abundant in peripheral blood leukocytes, where it is primarily expressed in monocytes and polymorphonuclear leukocytes (Cellier et al., 1994, 1997). Migration of immature macrophages to tissues or maturation *in vitro* is associated with increased expression of *NRAMP1*. Parallel studies in the promyelocytic leukemia cell line, HL-60, experimentally induced to differentiate into either granulocytes or monocytes demonstrate strong induction of *NRAMP1* mRNA expression in both lineages (Cellier et al., 1997). Studies on the cellular and subcellular localisation of the protein in macrophages have been carried out either (1) in primary macrophages using polyclonal antisera directed against the amino terminus of the protein or (2) in RAW macrophages stably transfected and overexpressing a recombinant Nramp1 protein tagged with a c-Myc antigenic epitope at its amino terminus (Govoni et al., 1999; Gruenheid et al., 1997; Searle et al., 1998). Immunofluorescence experiments indicate that Nramp1 is not expressed at the plasma membrane in macrophages, but rather is present in a punctate subcellular compartment positive for the lysosomal protein LAMP-1, suggesting that Nramp1 is expressed in the late endosome/lysosome compartment of macrophages (Gruenheid et al., 1997; Searle et al., 1998). Additional experiments by confocal microscopy of latex beads phagocytosed by primary macrophages, as well as immunoblotting with purified latex phagosomes, indicate that soon after phagocytosis, Nramp1 is recruited to the membrane of maturing phagosomes with kinetics similar to that of Rab5 but clearly distinct of Rab7 (Gruenheid et al., 1997). Similar recruitment of Nramp1 protein to the phagosomal membrane was demonstrated in macrophages which have ingested *Salmonella typhimurium*, *Leishmania*, *Mycobacterium avium*, and *Yersinia enterocolitica* (Searle et al., 1998; Govoni et al., 1999; Cuellar-Mata et al., 2002). These results demonstrated that the Nramp1 protein is expressed at the phagosomal membrane, and its transport activity may modulate the intraphagosomal milieu to affect microbial replication at that site, either through a bacteriostatic or bacteriocidal mechanism. Thus, the nature of the substrate(s) transported by Nramp1 was of great interest.

2.4 Divalent Cation Transport by Nramp Proteins

Nramp1 is not unique but is a member of a very large gene family which has been remarkably conserved throughout evolution (for reviews, see Cellier et al., 1996; Forbes and Gros, 2001; Cellier et al., 2001). In mammals, a second Nramp protein (Nramp2; OMIM No. 600523, now designated *SLC11A2* and also known as *DCT1* and *DMT1*) has been identified and shares 65% sequence identity (74% overall similarity) with Nramp1 (Gruenheid et al., 1995; Gunshin et al., 1997). As opposed to its macrophage-/monocyte-specific Nramp1 counterpart, Nramp2 is ubiquitously expressed, and this protein plays a pivotal role in iron acquisition and metabolism (see below). Sequencing projects in different model organisms, including insects, plants, yeast, and bacteria have revealed structurally and functionally conserved Nramp homologs (Belouchi et al., 1997; Agranoff et al., 1999; Rodrigues et. al., 1995, Makui et al., 2000; Curie et al., 2000; Dorschner et al., 1999; Feng et al., 1996; Girard-Santosuosso et al., 1997; Mathews et al., 1998). The functional characterisation of these homologs, including complementation studies in heterologous hosts and model organisms, has provided important clues on the mechanism of action of these proteins as transporters for divalent cations. Structural features in the Nramp superfamily include 12 transmembrane domains, several of which contain highly conserved, but thermodynamically disfavored, charged residues, including a pair of invariant histidines in TM6. A predicted loop delineated by TMs 7 and 8 is poorly conserved; it contains predicted asparagine-linked glycosylation sites (N-X-S/T) and for Nramp2 was recently shown to be extracellular (Picard et al., 2000). Finally, the TM8-TM9 intracellular loop contains a transport motif previously recognised on the cytoplasmic face of the membrane subunits of many bacterial periplasmic permeases (Kerppola and Ames, 1992). Although the role of this transport signature in Nramp proteins remains unknown, mutations at certain of its conserved positions abrogate Nramp2 function (Pinner et al., 1997).

Mutations in the *Drosophila melanogaster* homologue *Malvolio* (57% identity, 65% similarity) cause a sensory neuron defect in taste discrimination (Rodrigues et al., 1995) which can corrected by addition of supplementary divalent metals in the diet (Orgad et al., 1998) or by expression of the human NRAMP1 gene in transgenic flies (D'Sousa et al., 1999). In plants, *Oryza sativa* has three *OsNramp* homologues, whereas *Arabidopsis thaliana* has six such *AtNramp* sequences that share between 47 and 61% similarity with their human counterparts (Belouchi et al., 1997; Curie et al., 2000; Thomine et al., 2000). In *A. thaliana*, *AtNramp3* and *AtNramp4* mRNAs are induced by iron starvation, and *AtNramp3* overexpression causes Cd^{2+} hypersensitivity and Fe^{2+} overaccumulation in the roots (Thomine et al., 2000). The yeast *Saccharomyces cerevisiae* has three *Nramp* homologues, *SMF1*, *-2*, and *-3*. Overexpression of the *SMF1* gene permits growth on medium containing the chelator EGTA, suggestive of divalent metal transport (Supek et al., 1996). Studies in yeast and in *Xenopus laevis* oocytes suggest a pH-dependent, divalent metal transport mechanism with a broad substrate specificity (Mn^{2+}, Cd^{2+}, Cu^{2+}) for SMF1 and SMF2, whereas SMF3 seems to be more specific for Fe^{2+} (Chen et al., 1999). Finally, expression of mammalian Nramp2 (but not Nramp1) in a *smf1/smf2* mutant can complement the defect and restore growth on chelators and on alkaline medium (Pinner et al., 1997). Likewise, *smf1* mutants can be complemented by expression of plant *AtNramp1*, *AtNramp3*, or *AtNramp4* genes (Curie et al., 2000; Thomine et al., 2000). Together, these results indicate that structurally similar but evolutionarily distant Nramp proteins share a common pH-dependent divalent metals transport mechanism.

The mammalian Nramp2 protein has been the most studied and is the best understood member of the Nramp family. *Nramp2* produces two alternatively spliced transcripts generated by alternative use of two 3′ exons encoding distinct C-termini of the protein as well as distinct 3′ untranslated regions (Lee et al., 1998). One *Nramp2* mRNA contains an iron responsive element (IRE) in its 3′-UTR. IREs are RNA secondary structures present in the 5′- or the 3′-UTR of mRNAs encoding proteins involved in iron metabolism and that either enhance the stability or inhibit translation of the tagged RNAs in response to different iron conditions (Thiel, 1998). The second *Nramp2* splice isoform (non-IRE) encodes a protein (isoform II) where the C-terminal 18 amino acids of the IRE form (isoform I) are replaced by a novel 25-amino-acid segment and codes for a distinct 3′-UTR lacking the IRE. *Nramp2* mRNA and proteins (IRE) are expressed at the highest levels in the intestine and in the kidney (Gruenheid et al., 1995; Gunshin et al., 1997), and iron deprivation results in dramatic induction of the protein at the brush border of the duodenum (Canonne-Hergaux et al., 1999, 2000). Likewise, Nramp2 (non-IRE) is expressed in erythropopietic precursors, and in reticulocytes and phenylhydrazine- or erythropoietin-induced reticulocytosis results in strong increased expression of Nramp2 in peripheral blood (Canonne-Hergaux et al., 2001). Identical mutations (G185R) have been described in mouse (*mk*) and rat (*belgrade*) models of microcytic anemia and are associated with severe impairment of iron uptake at the intestinal brush border and in peripheral tissues (Andrews, 2000). This mutation affects both the transport properties of the protein but also the normal maturation and/or targeting of the protein in enterocytes, reticulocytes, and kidney proximal tubule epithelial cells (Canonne-Hergaux et al., 1999, 2000, 2001). These data have established that Nramp2/DMT1 functions as the major transferrin-independent iron acquisition system in the intestine. The subcellular localisation of Nramp2/DMT1 to the early, tranferrin receptor positive recycling endosome compartment (Gruenheid et al., 1999), together with studies in reticulocytes from *mk/mk* mice (Canonne-Hergaux et al., 2001) have also suggested that Nramp2 transports transferrin iron from acidified endosomes to the cytoplasm (please see Andrews, 2000, for a comprehensive review). Finally, transport studies in *Xenopus* oocytes (Gunshin et al., 1997; Tandy et al., 2000) and more recently in cultured mammalian cells (Picard et al., 2000) have shown that Nramp2/DMT1 can transport a broad range of divalent metals such as Fe^{2+}, Zn^{2+}, Cd^{2+}, Mn^{2+}, Cu^{2+}, and Co^{2+}. Transport is saturable and pH dependent and is accompanied by an inward proton movement (electrogenic), suggesting a proton cotransport mechanism. These results have established Nramp2/DMT1 as a key regulator of iron homeostasis in the body, but have in turn suggested that Nramp1 could also function as a pH-dependent divalent cation transporter at the membrane of acified phagosomes to affect microbial survival or replication.

2.5 Nramp1 Transport at the Phagosomal Membrane

The mechanism of transport and substrate of Nramp1 at the phagosomal membrane has been studied by several groups and has remained a matter of considerable debate. The effect of divalent cations on replication of mycobacteria *in vivo* or *in vitro* has been studied. In one study, dietary iron loading of Bcg^r mice was found to increase replication of *M. avium in vivo* (Gomes and Appleberg, 1998), suggesting that excess iron may overwhelm the Nramp1 advantage of these mice. Conversely, Zwilling et al. (1999) reported that low iron concentrations (0.005–0.05 μM) have either an inhibitory or a stimulatory effect on the growth of *M. avium* in Bcg^r and Bcg^s macrophages, respectively. Iron concentrations above 0.05 μM stimulated bacterial growth in both macrophages. Additional studies by the same group using either primary macrophages or RAW macrophages, transfected or not with a wild-type *Nramp1*^{Gly169} gene, identified a stimulatory effect of Nramp1 on association (binding/transport) of radioactive iron with phagosomes containing either latex beads or *M. avium* (Khun et al., 1999, 2001). The authors proposed a model in which Nramp1 would transport iron into the phagosome to increase hydroxyl radical formation by the Fenton/Haber-Weiss reaction and thus reduce microbial viability in the phagosomal space. Although plausible and attractive, this model is difficult to reconcile with known structural and functional characteristics of the Nramp family. Indeed, direction of transport would be opposite to that demonstrated for Nramp2/DMT1, with respect to the topological orientation of the two proteins and with respect to the proton gradient across the phagosomal membrane. Also, because iron has been shown essential to mycobacterial growth *in vitro* and promotes the development of active tuberculosis and the growth of other pathogens in humans (reviewed by Ratledge and Dover, 2000), a mechanism based in increased supply of iron to the bacterium seems somehow counterintuitive. Independently, our group has used the fluorescent probe fura-FF6 to monitor in real time the flux of divalent metals across the phagosomal membrane of wild-type and *Nramp1*$^{-/-}$ macrophages (Jabado et al., 2000). Fura-FF6 fluorescence is a pH-resistant and divalent cation-sensitive dye and it can be covalently attached to zymosan particles for phagocytosis by macrophages. Using microfluorescence imaging, it was observed that when exposed to Mn^{2+}, phagosomes formed in wild-type cells accumulated less Mn^{2+} than phagosomes formed in *Nramp1*$^{-/-}$ macrophages. Likewise, wild-type macrophages could release quenching of Mn^{2+}-preincubated FF6-labeled zymosan more efficiently than *Nramp1*$^{-/-}$ cells. The *Nramp1* specific effect was pH dependent and could be abrogated by inhibition of the vacuolar H^{+}-ATPase, suggesting that Nramp1 functions as a pH-dependent divalent cation efflux pump at the phagosomal membrane in a manner very similar to that previously demonstrated for Nramp2 at the plasma membrane (Gunshin et al., 1997; Picard et al., 2000). Also, Atkinson and Barton (1998, 1999) working with radioisotopic iron and with a calcein quenching assay in Nramp1-transfected cells reported decreased intracellular accumulation and increased efflux of iron in these cells. They concluded a dual effect of Nramp1, whereby Nramp1 would efflux iron out of the phagosome and into the cytoplasm, where it could be effluxed out of the cells by other means, possibly including Nramp2. Finally, Goswami et al. (2001) obtained evidence that in *Xenopus* oocytes Nramp1 functions as a divalent cation antiporter that can flux divalent cations in either direction against a proton gradient.

2.6 Consequences of Nramp1-Dependent Divalent Cation Transport on Microbial Virulence

Intracellular parasites have evolved a number of strategies to successfully evade the bacteriostatic and bactericidal defense mechanims of macrophages. In many cases, these strategies involve modulating the fusogenic properties of the phagosome in which they are enclosed, including maturation into the phagolysosome (reviewed by Knodler et al., 2001). Different pathogens can alter or arrest phagosome maturation in an individually distinct fashion, possibly creating intraphagosomal conditions most favorable for their respective survival and replication. Although the precise molecular basis for this modulation is poorly understood, recent studies have clearly indicated that synthesis of pathogen-encoded soluble factors is required for this process (Knodler et al., 2001). In the case of *Mycobacterium* spp., the phagosome is blocked in its maturation, but retains the ability to interact with the endosomal recycling pathway. This phagosome is positive for Cathepsin D and for the transferrin receptor, but remains negative for Rab7 and for the vacuolar H^+-ATPase. As a result the phagosome does not acidify fully and is less bactericidal possibly due to the lack of activation of lysosomal proteases (reviewed by Russell, 2001). The effect of Nramp1 recruitment to the phagosomal membrane on the ability of *Mycobacterium* spp. to inhibit phagosome maturation was studied for *Mycobacterium bovis* (Hackam et al., 1998) and for *Mycobacterium avium* (De Chastellier et al., 1993; Frehel, Gros and De Chastellier, unpublished observations). Microfluorescence ratio imaging of 129sv (wild type) and *Nramp*$^{-/-}$ primary macrophages was used to measure pH of individual phagosomes containing either live or dead *M. bovis* labeled with a combination of two pH-sensitive fluorophores. The pH of phagosomes containing live *M. bovis* was significantly more acidic in wild-type macrophages (pH 5.5 ± 0.06) than in *Nramp*$^{-/-}$ cells (pH 6.6 ± 0.05; $P < 0.005$). The enhanced acidification could not be accounted for by differences in proton consumption during dismutation of superoxide, phagosomal buffering power, counterion conductance, or in the rate of proton "leak". Rather, following ingestion of live *M. bovis*, 129sv cells exhibited increased concanamycin-sensitive H^+-pumping across the phagosomal membrane, associated with enhanced recruitment of V-ATPase-positive endosomes and/or lysosomes. The Nramp1 effect on pH was only seen with live *M. bovis* and not in phagosomes containing dead *M. bovis* or latex beads (Hackam et al., 1998). Independently, electron microscopy was used to monitor the effect of Nramp1 on the maturation and characteristics of *M. avium*-containing phagosomes (Frehel, Gros and De Chastellier, unpublished observations). In these experiments, fusion of phagosomes to endosomes and lysosomes was monitored by acquisition of specific markers or after labeling with fluid-phase markers (BSA-Au). Increased bacteriostatic activity of wild-type macrophages over *Nramp1*$^{-/-}$ cells was associated with increased intraphagosomal damage of *M. avium*. This was concomitant to increased fusion to lysosomes, and increased acidification, as opposed to *Nramp1*$^{-/-}$ phagosomes which fused preferentially with early endosomal vesicles. Together, these studies indicate that recruitment of Nramp1 affects the fusogenic properties and degree of acidification of mycobacterial phagosomes. This can be most easily interpreted as Nramp1 antagonising the ability of *Mycobacterium* spp. to block phagosome maturation.

In the case of *Salmonella*, the bacterial phagosome formed in macrophages is distinct from that formed by *Mycobacterium*: it is positive for LAMP, Rab7, and for vacuolar H^+-ATPase but remains negative for the mannose

6-phosphate receptor. As opposed to the mycobacterial phagosome, its *Salmonella* counterpart can fully acidify but becomes inaccessible to early endosomal markers (reviewed by Knodler et al., 2001). We investigated (Cuellar-Mata et al., 2002) whether *Nramp1* can also affect the maturation and acidification of *Salmonella*-containing vacuoles (SCV). In these studies, RAW macrophages (*Nramp1^{D169}*, devoid of functional Nramp1 and susceptible to *Salmonella* infection) were compared with isogenic clones stably expressing a transfected *Nramp1* RAW/Nramp1^{+}, resistant to infection (Govoni et al., 1999). Intravacuolar pH, measured *in situ*, was similar in Nramp1-expressing and -deficient cells. SCV acquired LAMP1 and fused with preloaded fluid-phase markers in both cell types, suggesting similar fusion to late endosomes/lysosomes. In contrast, whereas few vacuoles in RAW cells acquired M6PR, many more contained M6PR in RAW/Nramp1^{+} cells. Shortly after formation, SCV in RAW became inaccessible to extracellular markers, suggesting inability to fuse with newly formed endosomes. Expression of *Nramp1* markedly increased the access to extracellularly added markers (Cuellar-Mata et al., 2002). These results suggest that *Nramp1* counteracts the ability of *Salmonella* to become secluded in a compartment that limits access of bactericidal agents, allowing the normal degradative pathway of the macrophage to proceed.

2.7 Conclusions and Perspectives from Studies in Mice

Studies in the mouse model have not only allowed the positional cloning of *Nramp1* but also provided a genetic and biochemical framework to understand the normal physiological role of *Nramp1* and *Nramp2* in transport of key divalent cations such as Fe^{2+} and Mn^{2+} (Forbes and Gros, 2001). They have identified these two proteins as playing a key role in the ability of macrophages to resist infection with intracellular parasites and in the acquisition of iron at the intestine and in peripheral tissues, respectively.

One of the central, yet unresolved issues is how does Nramp1-mediated modulation of divalent cation content of the phagosome affect replication of unrelated pathogens at that site. Several nonmutually exclusive hypotheses can be put forward to account for the effect. In the near future, the experimental testing and/or validation of these models *in vitro* and *in vivo* will constitute a major challenge but necessary step to the possible development of novel therapeutic strategies based on the Nramp1 effect. One possibility is that divalent cations such as Fe^{2+}, Mn^{2+}, and Zn^{2+} are essential nutrients as cofactors in many enzymatic reactions. Therefore, depletion of such ions from the phagosomal space may simply reduce the availability of rate-limiting nutrient and thus may have a simple but pleitropic bacteriostatic effect. In another model, one can envision that removal of divalent metals from the phagosomal lumen may actively increase the bactericidal activity of macrophage enzymes by inhibiting a competing and bacterially encoded detoxifying mechanism. Indeed, Fe^{2+}, Cu^{2+}, Mn^{2+}, and Zn^{2+} are essential cofactors for the activity of superoxide dismutases produced by *Mycobacterium* spp., *Salmonella*, and *Leishmania* (Dey and Datta, 1994; Tsolis et al., 1995; Zhang et al., 1991) which detoxify superoxide ions and hydroxyl radicals produced by the macrophage NADPH oxidase. Finally, and as discussed above, intracellular pathogens such as *Mycobacterium* spp., *Salmonella*, and *Leishmania* can actively modulate the fusogenic properties of the phagosome in which they reside. In the case of *Mycobacterium* where the block on maturation has been well characterised, it requires the bacteria to be alive and metabolically active. Likewise, successful intracellular survival of *Salmonella* in macrophages requires the activation, synthesis, and secretion of specific protein mediators (reviewed by Russell, 2001; Knodler et al., 2001). Thus, the expression, processing, and/or targeting of these pathogenicity determinants may itself be dependent on divalent cations, and an Nramp1-mediated removal would affect the ability of each of these pathogens to modulate the maturation of the phagosome. Results discussed above are in agreement with such a model and suggest that Nramp1 action counteracts the specific strategy developed by *Mycobacterium* and *Salmonella* to block phagosome maturation.

It is interesting to note that a large body of published data support a key role of divalent cations for microbial virulence *in vivo* (Ratledge and Dover, 2000), in particular in the case of intracellular parasites. For example, *Salmonella* displays at least five high-affinity acquisition systems for iron (Ratledge and Dover, 2000; Tsolis et al., 1996). These include siderophore-dependent acquisition systems for Fe^{3+} (TonB-dependent), Fe^{2+} transport systems such as *feoAB*, as well as other transporters of the ABC type such as SitC encoded by a specific pathogenicity island. Interestingly, we and others have recently described *Nramp* homologues in *E. coli*, *Salmonella*, and *Mycobacterium* but also in many other bacterial species that function as transporters for Fe^{2+} and Mn^{2+} and that import these cations in a pH-dependent fashion (Cellier et al., 2001; Kehres et al., 2000; Agranoff et al., 1999; Makui et al., 2000). Therefore, it is likely that all these transport systems play a role in the scavenging of divalent metals in the limiting intracellular environment of the phagosome. Thus, it is tempting to speculate that mammalian Nramp1 could compete with these bacterially encoded transporters for acquisition of divalent metals from the phagosomal space, at the interface of host–parasite interactions.

Finally, one consequence of the work in mice reviewed above is that it has provided a testable candidate for parallel studies in humans aimed at evaluating the role of the human *NRAMP1* homologues in livestock and human diseases.

3. STUDIES IN SPECIES OTHER THAN MICE

3.1 Livestock Species

A major incentive for the cloning of *Nramp1* was that orthologues of *Nramp1* may be important genetic modulators of resistance/susceptibility to pathogenic intracellular parasites in livestock. Consequently, *Nramp1* orthologues have been cloned for a number of livestock species, including chicken (Hu et al., 1996), sheep (Bussmann et al., 1998), horse (Horin and Matiasovic, 2000), pig (Zhang et al., 2000), cow, and bison (Feng et al., 1996). Gene expression across all mammalian homologues is similar with highest expression levels in macrophages and neutrophils. In a notable difference from mammalian orthologues, chicken *Nramp1* is also highly expressed in the thymus (Hu et al., 1996). To date, only cow and chicken *Nramp1* orthologues have been employed for systematic genetic studies of susceptibility to infectious disease. In cattle, a study of *bovNRAMP1* expression levels in *M. bovis*-infected animals showed highly increased expression of *bovNRAMP1* in peripheral blood leukocytes, in tissue immediately adjacent to caseous necrotic lesions, and in lymph node granulomas (Estrada-Chavez et al., 2001). Employing cattle from herds with naturally occurring tuberculosis, the association of a polymorphism in the 3′-UTR of *bovNRAMP1* was investigated in a small case-control study involving 9 healthy control animals and 24 tuberculosis affected cattle (Barthel et al., 2000). No evidence for association between a 3′-UTR *bovNRAMP1* polymorphism and tuberculosis susceptibility was observed (Barthel et al., 2000). This suggests that the genetic control of naturally occurring *M. bovis* disease is not under exclusive control of the *bovNRAMP1* gene. The result does not, however, rule out the contribution of *bovNRAMP1* to tuberculosis risk as part of a multigenic genetic control system. Likewise, it is possible that unknown *bovNRAMP1* polymorphisms could have a major effect on tuberculosis susceptibility in cattle. Given the high costs and logistical difficulties associated with performing genetic crosses between tuberculosis-resistant and -susceptible cattle, it seems unlikely that more decisive conclusions can be accomplished in the near future.

Nontyphoidal *Salmonella* species originating from contaminated poultry products are a common cause of food-borne disease in humans. The ability of chickens to resist infection with *Salmonella* is known to be under genetic control (Bumstead and Barrow, 1988). Considering that resistance to infection with *Salmonella typhimurium* is part of the *Nramp1* phenotype, the chicken orthologue appeared to be an excellent candidate for a chicken *Salmonella* susceptibility locus. By performing comparative gene sequence analysis, Hu et al. (1997) identified a nonconservative R223G amino acid substitution between *Salmonella*-resistant "W1" and *Salmonella*-susceptible 'C' inbred chicken. The R223G substitution is located at the predicted extracellular border of Nramp1 TM5, a peptide segment that is highly conserved across species suggesting that replacement of a bulky, basic amino acid residue by a single hydrogen atom will impact on protein function. Cosegregation of the R223G polymorphism with susceptibility to *Salmonella* was assessed in a panel of 425 (W1 × C)F1 × C backcross animals segregating the phenotype. If susceptible C chicken were infected with *S. typhimurium* a majority of infected birds died during the initial 7 days. Birds surviving past this time showed a greatly reduced rate of mortality presumably due to the emergence of effective acquired immune responses. Hence, overall mortality rates displayed a biphasic appearance, a rapid initial phase during which most of the chicken died, and a second phase after 7 days during which the mortality rate was much slower. By selecting the initial phase of rapid infection-induced death as the phenotype, the R223G *Nramp1* allele was shown to be significantly linked to death (Hu et al., 1997). By contrast, there was little evidence for an impact of *Nramp1* alleles on survival of birds after 7 days. In perfect analogy to the mouse model, these results suggest that chicken *Nramp1* alleles are involved in early, innate immune responses while a different set of gene(s) is involved in determining late phase resistance/susceptibility.

3.2 Human NRAMP1

Following the molecular isolation of mouse *Nramp1*, efforts were initiated to obtain the human orthologue. Employing a mouse cDNA *Nramp1* probe, a human spleen library was screened under conditions of high stringency and a set of overlapping clones was obtained from which the full-length human *NRAMP1* cDNA could be assembled (Cellier et al., 1994). The predicted amino acid sequence of *NRAMP1* showed that 88% of amino acid residues were identical to those found in mouse *Nramp1*. Hence, the human gene, in perfect homology to the mouse gene, encodes a predicted polytopic membrane protein that carries a conserved sequence motif characteristic for a class of ubiquitous transport proteins (Cellier et al., 1995). The regions of sequence divergence were found mainly at the amino- and carboxy-terminal ends as well as in predicted extracellular loops in the amino-terminal half of the polypeptides (Cellier et al., 1994). These sequence studies indicated that essential functional activities of the Nramp1 proteins and their participation in physiological processes should be conserved between mouse and human.

Human *NRAMP1* has been localised by radiation hybrid analysis and YAC mapping to chromosome region 2q35 in close proximity to the *VIL* gene (Cellier et al., 1994; Liu et al., 1995). The exon/intron organisation of the gene has been fully elucidated and a total of 15 exons have been identified that conform to the correlation between the organisation of exons and functional protein domains (Cellier et al., 1994; Marquet et al., 2000). The translation initiation site was mapped to exon 1, the trancription initiation site was localised to 148 bp (Blackwell et al., 1995) and 176 bp (Kishi et al., 1996) 5′ of the initiation codon, respectively, and a putative polyadenylation site was localised in exon 15 (Marquet et al., 2000). Analysis of RNA expression indicated the presence of a transcript carrying a 74-bp insertion fragment (Cellier et al., 1994) that was due to an alternative splicing event within intron 5 that inserted 74 bp of an alternatively spliced exon into the mRNA. The alternatively spliced exon 4a is derived entirely from an Alu element that had inserted into intron 5 of the gene. Alternative splicing is predicted to result in the introduction of a premature translational stop site in exon 5 of *NRAMP1*. As shown by RT–PCR analysis, both transcriptional forms of *NRAMP1* are expressed in human cell lines (Cellier et al., 1994). The functional consequence of NRAMP1 truncation by the alternative splice event is unknown but would be expected to produce a nonfunctional protein. It is presently unknown if specific physiological trigger mechanisms exist that can shift the balance of *NRAMP1* transcript expression towards the alternatively spliced form.

The *NRAMP1* promoter region carries a number of regulatory sequence motifs, including a binding motif for the PU.1 myeloid specific transcription factor and several interferon-γ response elements (Blackwell et al., 1995; Kishi et al., 1996; Cellier et al., 1997). The presence of a binding site for a myeloid-specific transcription factor was confirmed by analysis of *NRAMP1* expression in a panel of cell lines induced to differentiate along the myeloid differentiation pathway (Cellier et al., 1997). In such cells, as well as in mature macrophages and PMN, a strong expression of *NRAMP1* was detected, whereas *NRAMP1* expression was absent from erythroid or lymphoid type cells. These results suggest that *NRAMP1* expression is specific for myeloid cells (Cellier et al., 1997).

The *NRAMP1* gene is located in a chromosomal region that is unusually enriched for repeat elements. Of approximately 20 kb of chromosomal DNA in the immediate vicinity of *NRAMP*, close to 60% of all nucleotides belong to repeat elements (Marquet et al., 2000) and the vast majority of *NRAMP1* intronic sequence is made up of Alu repeat elements. During successive waves of Alu transpositions, younger Alu family members evidently had integrated into more ancestral Alu repeats, giving rise to extended genome stretches of pure repeat element sequence. Specifically, the chromosomal region immediately upstream of the *NRAMP1* promoter is composed of several kilobases of Alu and Mer sequences (Roger et al., 1998; Kishi et al., 1996). The possible role of this repeat region as mutation generator and origin of chromosomal instabilities has been investigated in a comparative sequence analysis between humans and great apes (Roger et al., 1998). Despite extensive genomic sequence comparisons there was no evidence for increased genomic instability or enrichment of sequence polymorphisms. Hence, the biological function, if any, of the high density of Alu and other repeat elements in the *NRAMP1* chromosomal segment remains unknown.

3.3 Genetic Studies Employing *NRAMP1* as Candidate Gene

The identification of *NRAMP1* polymorphisms was a prerequisite to conduct genetic studies employing *NRAMP1* as candidate gene. Overall, the *NRAMP1* gene has limited sequence diversity suggesting that changes in the primary amino acid sequence of the gene are evolutionarily not well tolerated. Only one amino acid polymorphism, an aspartate-to-asparagine change at codon 543 in exon 15 of *NRAMP1*, has been described that is present in allele frequencies above 1% (Liu et al., 1995). There are two known additional low frequency variants, an alanine-to-valine change at codon 318 in exon 9 and a rare three-amino-acid insertion-deletion variant in exon 2 (Liu et al., 1995; White et al., 1994). Not surprisingly, a homologue of the Nramp1 G169D variant has never been observed in human populations. In addition, there are a number of silent nucleotide substitutions in exons as well as intronic sequences (Liu et al., 1995). Finally, attention has been focused on putative regulatory polymorphisms in the 5′ and 3′ untranslated *NRAMP1* regions. In the 3′-UTR, two 4-bp insertion-deletion polymorphisms have been described with unknown functional relevance (Liu et al., 1995; Buu et al., 1995). In the 5′-UTR, a single nucleotide polymorphism (SNP) was identified at position −236, and a promoter repeat polymorphism of the general structure $t(gt)_5ac(gt)_4ac(gt)_nggcaga(g)_6$ ($n = 9,10$) was identified at approximately 300 bp upstream of the transcriptional start site (Liu et al., 1995; Blackwell et al., 1995). In most populations, allele $n = 9$ is found with approximately 75–80% frequency and allele $n = 10$ with approximately 20–25%

frequency. In addition to these two predominant repeat alleles, additional configurations exist that may give rise to PCR-amplified DNA fragments of indistinguishable size (Graham et al., 2000). Employing transient transfections of luciferase promoter constructs into U937 cells, it was shown that the $n = 9$ allele drives significantly higher levels of *NRAMP1* expression than the $n = 10$ allele (Searle and Blackwell, 1999). The same allelic imbalance in luciferase expression was also found in transfected cells that had been stimulated either with INFγ alone or a combination of INFγ and LPS. Interestingly, addition of LPS enhanced activity of the $n = 9$ allele but decreased $n = 10$ allele expression (Searle and Blackwell, 1999). To further address the biological relevance of these findings it will be necessary to show that allelic imbalances exist in *NRAMP1* expressing tissues and to compare the magnitude of allelic imbalance with interindividual differences in *NRAMP1* expression levels. Nevertheless, these results are so far the only experimental data connecting *NRAMP1* sequence polymorphisms with a possible change in functional activity of *NRAMP1*.

4. *NRAMP1* AND TUBERCULOSIS

Tuberculosis and leprosy are the two major mycobacterial diseases that have plagued humankind for centuries. Despite effective chemotherapeutic treatments both diseases remain severe global health problems. In 2000, an estimated 8.4 million new tuberculosis cases were recorded resulting in approximately 2 million deaths (WHO, 2001). In the same year, close to 700,000 new leprosy cases were reported worldwide (WHO, 2000). These staggering numbers suggest that present control mechanisms fail to effectively intercept the spread of the causative mycobacteria through exposed populations. A reasonable hope is that a better understanding of the host–pathogen interplay and the host genetic factors that govern susceptibility to these diseases will be critical in devising more effective control strategies.

Results obtained in the mouse model of BCG infection imply that the human *NRAMP1* gene also impacts on tuberculosis and leprosy susceptibility. Consequently, a number of studies have investigated the possible contribution of *NRAMP1* to mycobacterial disease susceptibility. Employing a population-based case-control design, a study in Gambia investigated the contribution of *NRAMP1* alleles to tuberculosis risk among 410 adult pulmonary tuberculosis patients (Bellamy et al., 1998). These patients were members of one of six well-defined ethnic groups. A control group of 417 individuals was matched by ethnicity to the patients, but contained a large excess of males. Four polymorphisms within *NRAMP1* were analysed for association with tuberculosis: the promoter $(GT)_n$ repeat, the *NRAMP1* 469 + 14 G/C SNP in intron 4, the D543N amino acid variant in exon 15, and the 4-bp insertion/deletion polymorphism *NRAMP1* 1729 + 55del4 immediately 3′ of the *NRAMP1* stop codon (Bellamy et al., 1998). All four polymorphisms were found associated with tuberculosis when patient stratification according to ethnic group was taken into account. Quite consistently, an odds ratio (OR) of 1.8–1.9 was found for the rare alleles, i.e., the rare alleles were found to be significant risk factors for development of tuberculosis. Interestingly, the two 5′ polymorphisms were associated with tuberculosis independently of the two 3′ polymorphisms. Heterozygotes for the intron 4 SNP and the 3′ UTR insertion/polymorphism were at significantly increased risk of tuberculosis (OR = 4.07; 95%CI 1.86–9.12). These results provided evidence for a role of *NRAMP1* in tuberculosis susceptibility; however, the excess number of tuberculosis cases associated with the *NRAMP1* polymorphisms are still modest.

The role of *NRAMP1* alleles as risk factors for clinically defined tuberculosis was analysed in several additional studies in Japan, Korea, and Guinea-Conakry. In the Japanese study, 267 smear-positive, HIV-negative pulmonary tuberculosis patients and 202 healthy controls were enrolled (Gao et al., 2000). This study detected independent association of the *NRAMP1* D543N polymorphism and the promoter (GT)*n* repeat polymorphism with tuberculosis. In agreement with the study in Gambia (Bellamy et al., 1998), the common promoter repeat allele was found to be a significant independent protective factor for tuberculosis ($P = 0.039$) (Gao et al., 2000). In a similar population-based case control study involving 192 smear-positive tuberculosis cases and 192 healthy controls in Korea, the *NRAMP1* 1729 + 55del4 polymorphism was found associated with tuberculosis ($P = 0.02$; Ryu et al., 2000). Finally, in a small family-based control study the transmission of *NRAMP1* alleles to affected children was assessed in 44 families from Guinea-Conakry (Cervino et al., 2000). Of the three polymorphisms analysed, promoter (GT)*n*, *NRAMP1* 469 + 14G/C, and *NRAMP1* 1729 + 55del4, only the 469 + 14G/C polymorphism in intron 4 showed significant evidence of association with tuberculosis ($P = 0.036$). None of the case-control studies can exclude the possibility that a polymorphism in an unknown gene in close linkage disequilibrium with *NRAMP1* is responsible for the observed association of *NRAMP1* alleles with tuberculosis. However, since *NRAMP1* alleles have been found associated with tuberculosis in four distinct ethnic populations this would require that linkage disequilibrium of *NRAMP1* alleles with the unknown gene would have to be preserved in all four populations. Hence, the association of *NRAMP1* with tuberculosis across ethnicities supports the suggestion that it is indeed *NRAMP1* itself that is modulating tuberculosis risk.

Although close to two-thirds of the world's population is infected with *Mycobacterium tuberculosis*, only an estimated 10% of those infected develop tuberculosis over the course of their lifetime. An interesting question is whether *NRAMP1* alleles are involved in susceptibility to infection with *M. tuberculosis* or with progression from infection to clinically overt disease. Bellamy et al. (1998) favoured an impact of *NRAMP1* alleles on progression due to the presumed high rate of infected individuals in the control group (Bellamy et al., 1998). However, as pointed out by Borgdorff (1998) only an estimated 60% of control individuals in Gambia are expected to be infected with *M. tuberculosis*, making it formally possible that *NRAMP1* alleles exert their major effect on infection susceptibility. More recently, the view that *NRAMP1* alleles impact on progression from infection to clinical disease has obtained additional support from the linkage analysis of *NRAMP1* during an outbreak of tuberculosis in Canada (see below; Greenwood et al., 2000).

The potential role of *NRAMP1* in tuberculosis susceptibility was analysed in two linkage studies. In northern Brazil, 98 multiplex tuberculosis families comprising over 700 individuals were enrolled in a genetic study. Complex segregation analyses provided strong evidence for the presence of major tuberculosis susceptibility gene(s) in this family panel, yet no significant evidence to support linkage of *NRAMP1* with tuberculosis was detected (Shaw et al., 1997). Instead, a trend ($P = 0.025$) in favour of linkage of tuberculosis with *D2S1471*, a microsatellite repeat in close proximity to the *NRAMP1* gene, was found. The implications of this finding are difficult to assess. Certainly, the data appear to argue against a major role of *NRAMP1* alleles in susceptibility to tuberculosis and suggest the presence of other unknown susceptibility genes in the studied family panel. Alternatively, it is possible that major genetic effects can be missed if gene–environment interactions are not taken into account. This was shown by the genetic analysis of an outbreak of tuberculosis in an Aboriginal Canadian family (Greenwood et al., 2000). In this familial outbreak situation, it was possible to establish that nearly all family members had become infected with *M. tuberculosis* and that genetic analysis, indeed, was focused on progression from infection to disease. By modeling the exposure histories of all family members, four liability classes were defined that were associated with variable penetrance of a putative susceptibility locus. Assuming a relative risk of 10 for the susceptibility allele, a major tuberculosis susceptibility locus was mapped to the immediate vicinity of *NRAMP1* with high significance ($P < 10^{-5}$). Interestingly, if family members were not assigned to liability classes no significant evidence for linkage was obtained (Greenwood et al., 2000). Although it is possible that the Greenwood et al. study detected the genetic effect of a susceptibility gene different from *NRAMP1*, this seems improbable. A scan of the draft human genome sequence does not identify any obvious susceptibility gene candidates, and the odds of having an unknown tuberculosis susceptibility gene located in the *NRAMP1* vicinity appear slim.

Why would a major effect be associated with *NRAMP1* in the Greenwood et al. (2000) study and not others? One possible explanation is given by the need to model gene–environment interaction to detect the major gene effect (see above). Alternatively, certain factors such as the specific population history of the studied family, the absence of environmental mycobacteria, and/or the absence of evolutionary selection against major susceptibility genes due to the lack of extensive historical exposure to *M. tuberculosis* influence the study outcome (Abel and Casanova, 2000). Therefore, dependent on these factors, tuberculosis susceptibility would manifest itself in a spectrum of genetic control ranging from the clearly defined Mendelian defects in hypersusceptible families to polygenic control in endemic populations (Abel and Casanova, 2000). Finally, an aspect that has received little attention is the fact that an outbreak of tuberculosis was studied by Greenwood et al. and that all cases can be classified as primary tuberculosis patients (in contrast to reactivational tuberculosis cases). It is possible that different mechanisms of genetic control are acting on different forms of tuberculosis and that *NRAMP1* acts as a major susceptibility gene for primary tuberculosis disease but not reactivational tuberculosis.

5. *NRAMP1* AND LEPROSY

The role of *NRAMP1* has also been analysed in the context of leprosy susceptibility. In a small sample of families from French Polynesia, no evidence for linkage between *NRAMP1* and leprosy was detected (Levee et al.,1994; Roger et al., 1997). Likewise, in a population-based case control design, Roy et al. (1999) failed to detect association between *NRAMP1* alleles and clinical subtypes of leprosy. By contrast, in a sample of 20 multiplex leprosy families from Southern Vietnam, linkage of an *NRAMP1* haplotype with leprosy *per se*, i.e., leprosy irrespective of its specific clinical manifestations, was observed (Abel et al., 1998). Evidence for linkage among 16 ethnic Vietnamese families was strong ($P < 0.005$) but absent among the 4 ethnic Chinese families ($P > 0.6$). These results support a prior complex segregation analysis of leprosy susceptibility that detected evidence for a major leprosy susceptibility gene among ethnic Vietnamese but not among ethnic Chinese (Abel et al., 1995). Finally, a significantly different distribution of the *NRAMP1* 1729 + 55del4 alleles was observed among 92 paucibacillary and 181 multibacillary leprosy cases from Mali ($P = 0.012$; Meisner et al., 2001). No effect of *NRAMP1* alleles on leprosy *per se* susceptibility was noted in this association study. These results suggest the presence of genetic heterogeneity in the control of leprosy susceptibility, i.e., *NRAMP1* alleles function as risk factors in some but not other populations. Alternatively, it is possible that, for example, due to small sample size the power in the negative studies was too low to detect the effect of *NRAMP1* alleles.

The "Mitsuda reaction" skin test measures the human immune response to intradermally injected heat-killed *M. leprae* bacilli 4 weeks postinjection. In a Brazilian population, complex segregation analysis suggested the presence of major genetic effects in the control of Mitsuda reactivity (Feitosa et al., 1996). In an unexpected twist, strong linkage of the *NRAMP1* genome region was detected with the "Mitsuda reaction" (Alcais et al., 2000) in the same Vietnamese families that showed linkage of leprosy *per se* susceptibility with *NRAMP1*. Whether considering the extent of Mitsuda reactivity as a quantitative trait, as a binary trait or by focusing only on children with extreme values (<3 mm and >10 mm) very strong evidence for linkage with *NRAMP1* was obtained ($P = 0.001$; Alcais et al., 2000). Strikingly, the proportion of alleles shared among sibs with Mitsuda reaction <3 mm and Mitsuda reaction >10 mm was 65%. This finding suggested that, in contrast to the *Nramp1* mouse model, human *NRAMP1* alleles have a strong impact on the antimycobacterial immune response. A follow-up study in Brazil failed to detect linkage between *NRAMP1* polymorphisms and Mitsuda reactivity (Hatagima et al., 2001). However, the latter study suffered from a small sample size and a lack of clearly defined differences of identity-by-state vs identity-by-descent alleles; both factors potentially having a strong detrimental impact on the power of linkage detection. Further studies are needed to establish possible genetic heterogeneity in the control of Mitsuda reactivity.

6. *NRAMP1* AND INFECTIONS OTHER THAN TUBERCULOSIS AND LEPROSY

A possible involvement of *NRAMP1* in a variety of infectious diseases has been tested. For example, two studies have investigated a possible role of *NRAMP1* in susceptibility to *Mycobacterium avium-intracellulare* complex (MAC) disease in HIV-free individuals. In one small study, *NRAMP1* alleles in eight sporadic cases of MAC were compared with allele frequencies in 22 healthy controls (Huang et al., 1998). Not surprisingly, no significant difference was noted. In a second study, the *NRAMP1* mRNA was reverse transcribed and sequenced in four cases of familial MAC belonging to two

Japanese families (Tanaka et al., 2000). The only abnormality was one copy of a nonconservative amino acid substitution, R419Q, in one patient, making it unlikely to be the underlying cause of MAC in this family. The possible role of *NRAMP1* in cutaneous leishmaniasis was investigated in a small case-control study from an endemic region of the Ethiopian highlands but no evidence in favour of association was obtained (Maasho et al., 1998). Similarly, no evidence for linkage of the *NRAMP1* region with visceral leishmaniasis was detected in Brazilian families (Blackwell et al., 1997). In a large case-control study in Vietnam no evidence for association of *NRAMP1* alleles with typhoid fever was obtained (Dunstan et al., 2001). Likewise, no evidence for association between *NRAMP1* alleles and Chagas disease was found in a small Peruvian case control study (Calzada et al., 2001). Finally, significant association was observed between HIV infection and *NRAMP1* alleles in a case-control study of a Colombian population (Marquet et al., 1999). The biological significance of the latter finding is unknown but may be related to the increased state of macrophage activation that can be mediated by increased levels of NRAMP1 protein expression.

7. *NRAMP1* AND AUTOIMMUNE DISORDERS

Why is it that *NRAMP1* alleles predisposing to tuberculosis have not been selected out of exposed populations? One possible answer is that other common diseases, such as autoimmune disorders, provide a counterbalancing force for allele maintenance. In this view, *NRAMP1* alleles that are risk factors for tuberculosis would act as protective factors for autoimmune disorders (Blackwell and Searle, 1999). These considerations have prompted experiments aimed at detecting *NRAMP1* allele distortions in patients with a variety of autoimmune diseases. For example, in two association studies in Korean rheumatoid arthritis (RA) patients, significant associations were observed between the 3′ *NRAMP1* D543N and 1729 + 55del4 polymorphisms and RA (Yang et al., 2000; Singal et al., 2000). In a case-control study of 119 Latvian patients with juvenile rheumatoid arthritis (JRA) and 111 controls, significant associations both on the allelic and genotypic level were observed with the *NRAMP1* $(GT)_n$ promoter polymorphism (Sanjeevi et al., 2000). Homozygosity of the common $n = 9$ allele (increased promoter activity) was associated with significantly increased risk of JRA (OR = 2.32), whereas protection was associated with homozygosity in the less common $n = 10$ allele (reduced promoter activity; OR = 0.33). These results have been interpreted as support for the counterbalancing hypothesis and as evidence for a functional role of the promoter polymorphism (Searle and Blackwell, 1999). Sarcoidosis is a hypersensitivity to unknown antigens. A population-based case-control study of 157 sarcoidosis patients and 111 healthy controls detected a significant depletion of the non-$n = 9$ alleles among patients (Maliarik et al., 2000). In inflammatory bowel disease, a first report detected an association between *D2S434* and *D2S1323* alleles and Crohn's disease and concluded from these findings an association of *NRAMP1* with Crohn's disease (Hofmeister et al., 1997). However, a later study using two intragenic *NRAMP1* polymorphisms, silent substitutions 274 C/T in exon 3 and 823 C/T in exon 8, failed to detect a significant association with inflammatory bowel disease (Stokkers et al., 1999). The involvement of *NRAMP1* alleles as risk factors for Crohn's disease and related inflammatory bowel disorder needs additional study. The two microsatellites reported to be associated with Crohn's disease are at significant distance from *NRAMP1* (approximately 1 and 3 Mbp, respectively) making it unlikely that *NRAMP1* associations have been detected in the 1997 study. On the other hand, the two *NRAMP1* polymorphisms used in the 1999 study are only in weak linkage disequilibrium with the 3′ *NRAMP1* polymorphisms making it possible that an association of the D543N and 1729 + 55del4 polymorphisms with inflammatory bowel disease may have gone unnoticed. Finally, stimulated by the suggestion that *NRAMP1* may be involved in iron transport, *NRAMP1* promoter alleles have been studied as risk factors for multiple sclerosis and a significant enrichment of the common $n = 9$ allele was found in 104 patients from South Africa (Kotze et al., 2001). Taken together, the studies in autoimmune disorders suggest that *NRAMP1* functions as a general immunoregulatory molecule affecting both infectious and noninfectious immune disorders.

8. CONCLUSIONS FROM STUDIES IN HUMAN POPULATIONS

A possible contribution of *NRAMP1* to disease susceptibility has been tested in many human diseases. In several of the studies that detected linkage or association with a disease phenotype the contribution of *NRAMP1* alleles to disease risk was of modest scale. Conversely, some studies that failed to detect an *NRAMP1* effect frequently were of modest sample size. Taken together, this makes it likely that present studies detected both false-positive and false-negative results. Probably the best control for false-positive results is replication of significant association and linkage results in independent studies. Such replication has been achieved most extensively in the case of tuberculosis and it has been firmly established that *NRAMP1* alleles are risk factors for clinical tuberculosis. To follow-up on this very encouraging finding two avenues of research need to be pursued. First, it is necessary to develop assays of *NRAMP1* function that allow correlative studies of biological activity with *NRAMP1* alleles to provide more direct support for a causal role of *NRAMP1* in disease susceptibility. Second, it is necessary to refine the analysed disease phenotypes. Pulmonary tuberculosis is a complex mix of clinical pictures and underlying mechanisms of pathogenesis. To enhance the penetrance of genetic factors and the strength of genetic effects it is of pivotal importance to employ refined clinical definitions and/or to focus on specific pathways of pathogenesis by employing quantitative trait genetics to human populations. This will eventually allow us to pinpoint the genetic lesion caused by *NRAMP1* alleles and open the way for possible pharmacological repair of the defect.

REFERENCES

Abel, L. and Casanova, J. L. (2000). Genetic predisposition to clinical tuberculosis: bridging the gap between simple and complex inheritance. *Am. J. Hum. Genet.* **67**, 274–277.

Abel, L., Sanchez, F. O., Oberti, J., et al. (1998). Susceptibility to leprosy is linked to the human *NRAMP1* gene. *J. Infect. Dis.* **177**, 133–145.

Abel, L., Vu, D. L., Oberti, J., et al. (1995). Complex segregation analysis of leprosy in southern Vietnam. *Genet. Epidemiol.* **12**, 63–82.

Agranoff, D., Monahan, I. M., Mangan, J. A., et al. (1999). *Mycobacterium tuberculosis* expresses a novel pH-dependent divalent cation transporter belonging to the *Nramp* family. *J. Exp. Med.* **190**, 717–724.

Alcais, A., Sanchez, F. O., Thuc, N. V., et al. (2000). Granulomatous reaction to intradermal injection of lepromin (Mitsuda reaction) is linked to the human NRAMP1 gene in Vietnamese leprosy sibships. *J. Infect. Dis.* **181**, 302–308.

Andrews, N. C. (2000). Iron homeostasis: insights from genetics and animal models. *Nat. Rev. Genet.* **1**, 208–217.

Atkinson, P. G. and Barton, C. H. (1998). Ectopic expression of Nramp1 in COS-1 cells modulates iron accumulation. *FEBS Lett.* **425**, 239–242.

Atkinson, P. G. and Barton, C. H. (1999). High level expression of Nramp1^{G169} in RAW264.7 cell transfectants: analysis of intracellular iron transport. *Immunology* **96**, 656–662.

Barthel, R., Feng, J., Piedrahita, J. A., et al. (2001). Stable transfection of the bovine *NRAMP1* gene into murine RAW264.7 cells: effect on *Brucella abortus* survival. *Infect. Immun.* **69**, 3110–3119.

Barthel, R., Piedrahita, J. A., McMurray, D. N., et al. (2000). Pathologic findings and association of *Mycobacterium bovis* infection with the bovine NRAMP1 gene in cattle from herds with naturally occurring tuberculosis. *Am. J. Vet. Res.* **61**, 1140–1144.

Bellamy, R., Ruwende, C., Corrah, T., et al. (1998). Variations in the *NRAMP1* gene and susceptibility to tuberculosis in West Africans. *N. Engl. J. Med.* **338**, 640–644.

Belouchi, A. M., Cellier, M., Kwan, T., et al. (1997). Cloning and characterization of the *OsNramp* family from *Oryza sativa*, a new family of membrane proteins possibly implicated in the transport of metal ions. *Plant Mol. Biol.* **33**, 1085–1092.

Beutler, B. (1998). Defective LPS signaling in C3H/HeJ and C57BL/10ScCr mice: mutations in Tlr4 gene. *Science* **282**, 2085–2088.

Blackwell, J. M. and Searle, S. (1999). Genetic regulation of macrophage activation: understanding the function of Nramp1 (=Ity/Lsh/Bcg). *Immunol. Lett.* **65**, 73–80.

Blackwell, J. M., Barton, C. H., White, J. K., et al. (1995). Genomic organization and sequence of the human NRAMP gene: identification and mapping of a promoter region polymorphism. *Mol. Med.* **1**, 194–200.

Blackwell, J. M., Black, G. F., Peacock, C. S., et al. (1997). Immunogenetics of leishmanial and mycobacterial infections: the Belem Family Study. *Philos. Trans. R. Soc. Lond. B Biol. Sci.* **352**, 1331–1345.

Blackwell, J. M., Roberts, C. W., Roach, T. I. A., et al. (1994). Influence of macrophage resistance gene *Lsh/Ity/Bcg* (candidate Nramp) on *Toxoplasma gondii* infection in mice. *Clin. Exp. Immunol.* **97**, 107–112.

Borgdorff, M. W. (1998). The NRAMP1 gene and susceptibility to tuberculosis. *N. Engl. J. Med.* **339**, 199–200.

Bradley, D. J. (1974). Genetic control of natural resistance to *Leishmania donovani*. *Nature* **250**, 353–354.

Bradley, D. J. (1977). Regulation of *Leishmania* populations within the host. II. genetic control of acute susceptibility of mice to *Leishmania donovani* infection. *Clin. Exp. Immunol.* **30**, 130–140.

Bradley, D. J., Taylor, B. A., Blackwell, J., et al. (1979). Regulation of *Leishmania* populations within the host. III. Mapping of the locus controlling susceptibility to visceral leishmaniasis in the mouse. *Clin. Exp. Immunol.* **37**, 7–14.

Brown, M. G., Dokun, A. O., Heusel, J. W., et al. (2001). Vital involvement of a natural killer cell activation receptor in resistance to viral infection. *Science* **292**, 934–937.

Bumstead, N. and Barrow, P. A. (1988). Genetics of resistance to *Salmonella typhimurium* in newly hatched chicks. *Br. Poult. Sci.* **29**, 521–529.

Buschman, E. and Skamene, E. (2001). From *Bcg/Lsh/Ity* to *Nramp1*: Three decades of search and research. *Drug Metabol. Dispos.* **29**, 471–473.

Bussmann, V., Lantier, I., Pitel, F., et al. (1998). cDNA cloning, structural organization, and expression of the sheep NRAMP1 gene. *Mamm. Genome* **9**, 1027–1031.

Buu, N. T., Cellier, M., Gros, P., et al. (1995). Identification of a highly polymorphic length variant in the 3′UTR of *NRAMP1*. *Immunogenetics* **42**, 428–429.

Calzada, J. E., Nieto, A., Lopez-Nevot, M. A., et al. (2001). Lack of association between NRAMP1 gene polymorphisms and *Trypanosoma cruzi* infection. *Tissue Antigens* **57**, 353–357.

Canonne-Hergaux, F., Fleming, M., Levy, J., et al. (2000). Increased expression and inappropriate targeting of the DMT1/Nramp2 iron transporter at the intestinal brush border of microcytic anemia *mk* mice. *Blood* **96**, 3964–3970.

Canonne-Hergaux, F., Gruenheid, S., Ponka, P., et al. (1999). Cellular and subcellular localization of the Nramp2 iron transporter in the intestinal brush border and regulation by iron. *Blood* **93**, 4406–4417.

Canonne-Hergaux, F., Zhang, A-S., Ponka, P., et al. (2001). Characterization of the iron transporter DMT1 (Nramp2/DCT1) in red blood cells of normal and anemic *mk/mk* mice. *Blood* **98**, 3823–3830.

Cellier, M. F., Bergevin, I., Boyer, E., et al. (2001). Polyphyletic origins of bacterial Nramp transporters. *Trends Genet.* **17**, 365–370.

Cellier, M., Belouchi, A. and Gros, P. (1996). Resistance to intracellular infections: comparative genome of NRAMP. *Trends Genet.* **12**, 201–204.

Cellier, M., Govoni, G., Vidal, S., et al. (1994). The human *NRAMP* gene: cDNA cloning, chromosomal mapping, genomic organization and tissue specific expression. *J. Exp. Med.* **180**, 1741–1752.

Cellier, M., Govoni, G., Vidal, S., et al. (1994). Human natural resistance-associated macrophage protein: cDNA cloning, chromosomal mapping, genomic organization, and tissue-specific expression. *J. Exp. Med.* **180**, 1741–1752.

Cellier, M., Prive, G., Belouchi, A., et al. (1995). Nramp defines a family of membrane proteins. *Proc. Natl. Acad. Sci. USA* **92**, 10089–10093.

Cellier, M., Prive, G., Belouchi, A., et al. (1995). The natural resistance associated macrophage protein (Nramp) defines a new family of membrane proteins conserved throughout evolution. *Proc. Natl. Acad. Sci. USA* **92**, 10089–10094.

Cellier, M., Shustik, C., Dalton, W., et al. (1997). Expression of the human NRAMP1 gene in professional primary phagocytes: Studies in blood cells and in HL-60 promyelocytic leukemia. *J. Leuk. Biol.* **61**, 96–105.

Cellier, M., Shustik, C., Dalton, W., et al. (1997). The human *NRAMP1* gene as a marker of professional primary phagocytes: Studies in blood cells and in induced HL-60 promyelocytic leukemia. *J. Leuk. Biol.* **61**, 96–105.

Cervino, A. C., Lakiss, S., Sow, O., et al. (2000). Allelic association between the NRAMP1 gene and susceptibility to tuberculosis in Guinea-Conakry. *Ann. Hum. Genet.* **64**, 507–512.

Chapes, S. K., Mosier, D. A., Wright, A. D., et al. (2001). MHCII, *Tlr4* and *Nramp1* gene control host pulmonary resistance against the opportunistic bacterium *Pasteurella pneumotropica*. *J. Leuk. Biol.* **69**, 381–386.

Chen, X-Z, Peng, J-B., Cohen, A., et al. (1999). Yeast SMF1 mediates H(+)-coupled iron uptake with concomitant uncoupled cation currents. *J. Biol. Chem.* **274**, 35089–35094.

Cooke, G. S. and Hill, A. V. S. (2001). Genetics of susceptibility to human infectious disease. *Nature Genet. Rev.* **2**, 967–977.

Crocker, P. R., Blackwell, J. M. and Bradley, D. J. (1984). Expression of the natural resistance gene *Lsh* in resident liver macrophages. *Infect. Immun.* **43**, 1033–1040.

Cuellar-Mata, P., Jabado, N., Liu, J., et al. (2002). Nramp1 modifies the fusion of *Salmonella typhimurium* containing vacuoles with cellular endomembranes in macrophages. *J. Biol. Chem.* **277**, 2258–2265.

Curie, C., Alonso, J. M., Le Jean, M., et al. (2000). Involvement of *NRAMP1* from *Arabidopsis thaliana* in iron transport. *Biochem. J.* **347**, 749–755.

D'Sousa, J., Cheah, P. W., Gros, P., et al. (1999). Functional complementation of the *malvolio* mutation in the taste pathway of *Drosophila* by the Human Natural Resistance-Associated Macrophage Protein 1 (NRAMP-1). *J. Exp. Biol.* **202**, 1909–1915.

De Chastellier, C., Frehel, C., Offreso, C., et al. (1993). Implication of phagosome–lysosome fusion in restriction of *Mycobacterium avium* growth in bone marrow macrophages from genetically resistant mice. *Infect. Immun.* **61**, 3775–3784.

Denis, M., Forget, A., Pelletier, M., et al. (1990). Killing of *Mycobacterium smegmatis* by macrophages from genetically susceptible and resistant mice. *J. Leuk. Biol.* **47**, 25–30.

Dey, R. and Datta, S. C. (1994). Leishmanial glycosomes contain superoxide dismutase. *Biochem. J.* **301**, 317–319.

Donovick, R., McKee, C. M., Jambor, W. P., et al. (1949). The use of the mouse in a standardized test for antituberculous activity of compounds of natural or synthetic origin. II. Choice of mouse strain. *Am. Rev. Tub.* **60**, 109–120.

Dorschner, M. O. and Phillips, R. B. (1999). Comparative analysis of two *Nramp* loci from rainbow trout. *DNA Cell Biol.* **18**, 573–583.

Dunstan, S. J., Ho, V. A., Duc, C. M., et al. (2001). Typhoid fever and genetic polymorphisms at the natural resistance-associated macrophage protein 1. *J. Infect. Dis.* **183**, 1156–1160.

Estrada-Chavez, C., Pereira-Suarez, A. L., Meraz, M. A., et al. (2001). High-level expression of NRAMP1 in peripheral blood cells and tuberculous granulomas from *Mycobacterium bovis*-infected bovines. *Infect. Immun.* **69**, 7165–7168.

Feitosa, M., Krieger, H., Borecki, I., et al. (1996). Genetic epidemiology of the Mitsuda reaction in leprosy. *Hum. Hered.* **46**, 32–53.

Feng, J., Li, Y., Hashad, M., et al. (1996). Bovine natural resistance associated macrophage protein 1 (Nramp1) gene. *Genome Res.* **6**, 956–964.

Forbes, J. R. and Gros, P. (2001). Divalent metal transport by Nramp proteins at the interface of host:parasite interactions. *Trends Microbiol.* **9**, 397–403.

Forget, A., Skamene, E., Gros, P., et al. (1981). Differences in response among inbred strains of mice to infection with small doses of *Mycobacterium bovis* (BCG). *Infect. Immun.* **32**, 42–47.

Fortin, A., Stevenson, M. M. and Gros, P. (2002). Complex genetic control of susceptibility to malaria in mice. *Genes Immun.* **3**, 177–186.

Gao, P. S., Fujishima, S., Mao, X. Q., et al. (2000). Genetic variants of NRAMP1 and active tuberculosis in Japanese populations. *Clin. Genet.* **58**, 74–76.

Gautier, A. V., Lantier, I. and Lantier, F. (1998). Mouse susceptibility to infection with the *Salmonella abortusovis* vaccine strain Rv6 is controlled by the *Ity/Nramp1* gene and influences the antibody but not the complement responses. *Microbial. Pathogen.* **24**, 47–55.

Girard-Santosuosso, O., Bumstead, N., Lantier, I., et al. (1997). Partial conservation of the mammalian *NRAMP* syntenic group on chicken chromosome 7. *Mamm. Genome* **8**, 614–616.

Gomes, M. S. and Appelberg, R. (1998). Evidence for a link between iron metabolism and *Nramp1* gene function in innate resistance against *Mycobacterium avium*. *Immunology* **95**, 165–168.

Goswami, T., Bhattacharjee, A., Babal, P., et al. (2001). Natural-resistance associated macrophage protein 1 is a H+/bivalent cation antiporter. *Biochem. J.* **354**, 511–519.

Govoni, G., Canonne-Hergaux, F., Pfeifer, C. G., et al. (1999). Functional expression of Nramp1 *in vitro* after transfection into the murine macrophage line RAW264.7. *Infect. Immun.* **67**, 2225–2232.

Govoni, G., Gauthier, S., Iscove, N. N., et al. (1997). Cell specific and inducible *Nramp1* gene expression in macrophages *in vitro* and *in vivo*. *J. Leuk. Biol.* **62**, 277–286.

Govoni, G., Vidal, S., Gauthier, S., et al. (1996). The *Bcg/Ity/Lsh* locus: Genetic transfer of resistance to infections in C57BL/6J mice transgenic for the *Nramp1Gly169* allele. *Infect. Immun.* **64**, 2923–2929.

Govoni, G., Vidal, S., Lepage, P., et al. (1995). Genomic Organization and cell specific expression of the mouse *Nramp 1* gene. *Genomics* **27**, 9–19.

Graham, A. M., Dollinger, M. M., Howie, S. E., et al. (2000). Identification of novel alleles at a polymorphic microsatellite repeat region in the human NRAMP1 gene promoter: analysis of allele frequencies in primary biliary cirrhosis. *J. Med. Genet.* **37**, 150–152.

Greenwood, C. M., Fujiwara, T. M., Boothroyd, L. J., et al. (2000). Linkage of tuberculosis to chromosome 2q35 loci, including NRAMP1, in a large aboriginal Canadian family. *Am. J. Hum. Genet.* **67**, 405–416.

Gros, P., Skamene E., and Forget, A. (1983). Cellular mechanisms of genetically controlled host resistance to *Mycobacterium bovis* (BCG). *J. Immunol.* **131**, 1966–1972.

Gros, P., Skamene, E., and Forget, A. (1981). Genetic control of natural resistance to *Mycobacterium bovis* (BCG) in mice. *J. Immunol.* **127**, 2417–2421.

Gruenheid, S., Canonne-Hergaux, F., Gauthier, S., et al. (1999). The iron transport protein Nramp2 is an integral membrane protein that co-localizes with transferrin in recycling endosomes. *J. Exp. Med.* **189**, 831–841.

Gruenheid, S., Cellier, M., Vidal, S., et al. (1995). Identification and characterization of a second mouse *Nramp* gene. *Genomics* **25**, 514–525.

Gruenheid, S., Pinner, E., Desjardins, M., et al. (1997). Natural resistance to infection with intracellular parasites: The Nramp1 protein is recruited to the membrane of the phagosome. *J. Exp. Med.* **185**, 717–730.

Gruenheid, S., Skamene, E. and Gros, P. (1999). Nramp1: A novel macrophage protein with a key function in resistance to intracellular pathogens. In: *Advances in Cell and Molecular Biology of Membranes and Organelles: Phagocytosis: The Host.* Gordon, S. (ed.). New York: Jai Press, Vol. **5**, pp. 345–362.

Gunshin, H., B. Mackenzie, U. V. Berger, Y., et al. (1997). Cloning and characterization of a mammalian proton-coupled metal-ion transporter. *Nature* **388**, 482–488.

Hackam, D. J., Rotstein, O. D., Zhang, W-J., et al. (1998). Host resistance to intracellular infections: mutation at *Nramp1* impair phagosomal acidification. *J. Exp. Med* **188**, 351–364.

Hatagima, A., Opromolla, D. V., Ura, S., et al. (2001). No evidence of linkage between Mitsuda reaction and the NRAMP1 locus. *Int. J. Lepr. Other Mycobact. Dis.* **69**, 99–103.

Hill, AV., Allsopp, CE., Kwiatkowski, D., et al. (1991). Common west African HLA antigens are associated with protection from severe malaria. *Nature* **352**, 595–600.

Hofmeister, A., Neibergs, H. L., Pokorny, R. M., et al. (1997). The natural resistance-associated macrophage protein gene is associated with Crohn's disease. *Surgery* **122**, 173–178.

Horin, P. and Matiasovic, J. (2000). Two polymorphic markers for the horse SLC11A1 (NRAMP1) gene. *Anim. Genet.* **31**, 152.

Hu J., Bumstead N., Burke D., et al. (1995). Genetic and physical mapping of the natural resistance associated macrophage protein 1 (NRAMP1) in chicken. *Mamm. Genome* **6**, 809–815.

Hu, J., Bumstead, N., Barrow, P., et al. (1997). Resistance to salmonellosis in the chicken is linked to NRAMP1 and TNC. *Genome Res.* **7**, 693–704.

Hu, J., Bumstead, N., Skamene, E., et al. (1996). Structural organization, sequence, and expression of the chicken NRAMP1 gene encoding the natural resistance-associated macrophage protein 1. *DNA Cell Biol.* **15**, 113–123.

Huang, J. H., Oefner, P. J., Adi, V., et al. (1998). Analyses of the NRAMP1 and IFN-gammaR1 genes in women with *Mycobacterium avium*-intracellulare pulmonary disease. *Am. J. Resp. Crit. Care Med.* **157**, 377–381.

Jabado, N., Jankowsky, A., Dougaparsad, S., et al. (2000). Natural resistance to intracellular infections: Nramp1 functions as a pH-dependent Manganese transporter at the phagosomal membrane. *J. Exp. Med.* **192**, 1237–1248.

Jothy, S. and Gros. P. (1995). The *Ity/Lsh/Bcg* locus: Natural resistance to infection with intracellular parasites is abrogated by disruption of the *Nramp 1* gene. *J. Exp. Med.* **182**, 655–666.

Kehres, D. G., Zaharik, M. L., Finlay, B.B., et al. (2000). The NRAMP proteins of *Salmonella typhimurium* and *Escherichia coli* are selective manganese transporters involved in the response to reactive oxygen. *Mol. Microbiol.* **36**, 1085–1100.

Kerppola, R. E. and Ames, G. F. (1992). Topology of the hydrophobic membrane-bound components of the histidine periplasmic permease: Comparison with other members of the family. *J. Biol. Chem.* **267**, 2329–2336.

Kishi, F., Tanizawa, Y. and Nobumoto, M. (1996). Structural analysis of human natural resistance-associated macrophage protein-1 promoter. *Mol. Immunol.* **33**, 265–268.

Knodler, L. A., Celli, J. and Finlay, B. B. (2001). Pathogenicity trickery: Deception of host cell processes. *Nat. Rev. Mol. Cell Biol.* **2**, 578–588.

Kotze, M. J., de Villiers, J. N., Rooney, R. N., et al. (2001). Analysis of the NRAMP1 gene implicated in iron transport: association with multiple sclerosis and age effects. *Blood Cells Mol. Dis.* **27**, 44–53.

Kovarova, H., Halada, P., Man, P., et al. (2002). Proteome study of *Francisella tularensis* live vaccine strain-containing phagosomes in *Bcg/Nramp1* congenic macrophages: resistant allele contributes to permissive environment and susceptibility to infection. *Proteomics* **2**, 85–93.

Kovarova, H., Hernychova, L., Hajduch, M., et al. (2000). Influence of the *Bcg* locus on natural resistance to primary infection with the facultative intracellular bacterium *Francisella tularensis* in mice. *Infect. Immun.* **68**, 1480–1484.

Kramnik, I., Demant, P. and Bloom, B.B. (1998). Susceptibility to tuberculosis as a complex genetic trait: analysis using recombinant congenic strains of mice. *Novartis Foundation Symp.* **217**, 120–131.

Kuhn, D. E., Baker, B.D., Lafuse, W. P., et al. (1999). Differential iron transport into phagosomes isolated from the RAW264. macrophage cell lines transfected with $Nramp1^{Gly169}$ or $Nramp1^{Asp169}$. *J. Leuk. Biol.* **66**, 113–119.

Kuhn, D. E., Lafuse, W. P. and Zwilling, B. S. (2001). Iron transport into *Mycobacterium avium*-containing phagosomes from an *Nramp1*(Gly169)-transfected RAW264.7 macrophage cell line. *J. Leuk. Biol.* **69**, 43–49.

Kwiatkowski, D. (2000). Genetic susceptibility to malaria getting complex. *Curr. Opin. Genet. Dev.* **10**, 314–319.

Lee, P. L., Gelbart, T., West, C., et al. (1998). The human Nramp2 gene: Characterization of the gene structure, alternative splicing, promoter region and polymorphisms. *Blood Cells Mol. Dis.* **24**, 199–207.

Lee, S-H., Girard, S., Macina, D., et al. (2001). Susceptibility to cytomegalovirus infection is associated with deletion of the natural killer cell lectin-like receptor *ly49h* gene. *Nat. Genet.* **28**, 42–45.

Levee, G., Liu, J., Gicquel, B., et al. (1994). Genetic control of susceptibility to leprosy in French Polynesia: No evidence for linkage with markers on telomeric human chromosome 2. *Int. J. Lepr. Other Mycobact. Dis.* **62**, 499–511.

Lissner, C. R., Swanson, R. N. and O'Brien, A. D. (1983). Genetic control of the innate resistance of mice to *Salmonella typhimurium*: Expression of the *Ity* gene in peritoneal and splenic macrophages isolated *in vitro*. *J. Immunol.* **131**, 3006–3013.

Liu, J., Fujiwara, T. M., Buu, N. T., et al. (1995). Identification of polymorphisms and sequence variants in the human homologue of the mouse natural resistance-associated macrophage protein gene. *Am. J. Hum. Genet.* **56**, 845–853.

Lynch, C. J., Pierce-Chase, C. H. and Dubos, R. (1965). A genetic study of susceptibility to experimental tuberculosis in mice infected with mammalian tubercle bacilli. *J. Exp. Med.* **121**, 1051–1070.

Maasho, K., Sanchez, F., Schurr, E., et al. (1998). Indications of the protective role of natural killer cells in human cutaneous leishmaniasis in an area of endemicity. *Infect. Immun.* **66**, 2698–2704.

Makui, H., Roig, E., Cole, S. T., et al. (2000). Identification of *Escherichia coli* K-12 Nramp (MntH) as a selective divalent metal ion transporter. *Mol. Microbiol.* **35**, 1065–1078.

Maliarik, M. J., Chen, K. M., Sheffer, R. G., et al. (2000). The natural resistance-associated macrophage protein gene in African Americans with sarcoidosis. *Am. J. Resp. Cell Mol. Biol.* **22**, 672–675.

Malo, D., Vidal, S., Lieman, J., et al. (1993b). Physical delineation of the minimal chromosomal segment encompassing the host resistance locus *Bcg*. *Genomics* **17**, 667–675.

Malo, D., Vidal, S. M., Hu, J., et al. (1993a). High-resolution linkage map in the vicinity of the host resistance locus Bcg. *Genomics* **16**, 655–663.

Malo, D., Vogan, K., Vidal, S., et al. (1994). Haplotype mapping and sequence analysis of the mouse *Nramp* gene predict susceptibility to infection with intracellular parasites. *Genomics* **23**, 51–61.

Marquet, S., Lepage, P., Hudson, T. J., et al. (2000). Complete nucleotide sequence and genomic structure of the human NRAMP1 gene region on chromosome region 2q35. *Mamm. Genome* **11**, 755–762.

Marquet, S., Sanchez, F. O., Arias, M., et al. (1999). Variants of the human NRAMP1 gene and altered human immunodeficiency virus infection susceptibility. *J. Infect. Dis.* **180**, 1521–1525.

Matthews, G. D. and Crawford, A. M. (1998). Cloning, sequencing and linkage mapping of the *NRAMP1* gene of sheep and deer. *Anim. Genet.* **29**, 1–6.

McGuire, W., Hill, A. V., Allsopp, C. E., et al. (1994). Variation in the TNF-alpha promoter region associated with cerebral malaria. *Nature* **371**, 508–510.

McLeod, R., Skamene, E., Brown, C. R., et al. (1989). Genetic regulation of early survival and cyst number after peroral *Toxoplasma gondii* infection in A × B/B × A recombinant inbred and B10 congenic mice. *J. Immunol.* **143**, 3031–3034.

Medina, E. and North, R. J. (1998). Resistance ranking of some common inbred mouse strains to *Mycobacterium tuberculosis* and relationship to the major histocompatibility haplotype and *Nramp1* genotype. *Immunology* **93**, 270–274.

Meisner, S. J., Mucklow, S., Warner, G., et al. (2001). Association of NRAMP1 polymorphism with leprosy type but not susceptibility to leprosy per se in west Africans. *Am. J. Trop. Med. Hyg.* **65**, 733–735.

Mitsos, L. M., Cardon, L., Fortin, A., et al. (2000). Genetic control of susceptibility to infection with mycobacterium tuberculosis in mice. *Genes. Immun.* **1**, 467–477.

North, R. J., LaCourse, R., Ryan, L., et al. (1999). Consequence of Nramp1 deletion on *Mycobacterium tuberculosis* infection in mice. *Infect. Immun.* **67**, 5811–5814.

North, R. J. and Medina, E. (1998). How important is *Nramp1* in tuberculosis. *Trends Microbiol.* **6**, 441–443.

O'Brien, A. D., Rosenstreich, D. L. and Taylor, B. A. (1980). Control of natural resistance to *Salmonella typhimurium* and *Leishmania donovani* in mice by closely linked but distinct genetic loci. *Nature* **287**, 440–442.

Orgad, S., Nelson, H., Segal, D., et al. (1998). Metal ions suppress the abnormal taste behavior of the *Drosophila* mutant *malvolio*. *J. Exp. Biol.* **201**, 115–120.

Pal, S., Peterson, E. M. and De La Maza, L. M. (2000). Role of *Nramp1* in *Chlamydia* infection in mice. *Infect. Immun.* **68**, 4831–4833.

Pasvol, G., Weatherall, D. J. and Wilson, R. J. (1978). Cellular mechanism for the protective effect of haemoglobin S against *P. falciparum* malaria. *Nature* **274**, 701–703.

Picard, V., Govoni, G., Jabado, N., et al. (2000). Functional analysis of mammalian Nramp2 in intact cells by a calcein quenching assay. *J. Biol. Chem.* **275**, 35738–35745.

Pierce, C., Dubos, R. J. and Middlebrook, G. (1947). Infection of mice with mammalian tubercle bacilli grown in Tween-albumin liquid medium. *J. Exp. Med.* **86**, 159–174.

Pinner, E., Gruenheid, S., Raymond, M., et al. (1997). Functional complementation of the yeast divalent cation transporter family SMF by a member of the mammalian Natural Resistance Associated Macrophage Family, Nramp2. *J. Biol. Chem.* **272**, 28933–28938.

Plant, J. and Glynn, A. A. (1976). Genetics of resistance to infection with *Salmonella typhimurium* in mice. *J. Infect. Dis.* **133**, 72–78.

Plant, J. and Glynn, A. A. (1979). Locating *Salmonella* resistance gene on mouse chromosome 1. *Clin. Exp. Immunol.* **37**, 1–6.

Plant, J. and Glynn, A. A. (1974). Natural resistance to *Salmonella* infection, delayed hypersensitivity and Ir genes in different strains of mice. *Nature* **248**, 345–347.

Plant, J. E., Blackwell, J. M., O'Brien, A. D., et al. (1982). Are the *Lsh* and *Ity* disease resistance genes at one locus on mouse chromosome 1? *Nature* **297**, 510–511.

Poltorak, A., He, X., Smirnova, I., et al. (2000). *Saccharomyces cerevisiae* expresses three functionally distinct homologues of the *Nramp* family of metal transporters. *Mol. Cell. Biol.* **20**, 7893–7902.

Qureshi, S. T., Lariviere, L., Leveque, G., et al. (1999). Endotoxin-tolerant mice have mutations in Toll-like receptor 4 (*Tlr4*). *J. Exp. Med.* **189**, 615–625.

Ratledge, C. and Dover, L. G. (2000). Iron metabolism in pathogenic bacteria. *Ann. Rev. Microbiol.* **54**, 881–941.

Robson, H. G. and Vas, S. I. (1972). Resistance of inbred mice to *Salmonella typhimurium*. *J. Infect. Dis.* **126**, 378–86.

Rodrigues, V., Cheah, P. Y., Ray, K., et al. (1995). *Malvolio,* the *Drosophila* homologue of mouse NRAMP-1 (*Bcg*), is expressed in macrophages and in the nervous system and is required for normal taste behavior. *EMBO J.* **14**, 3007–3020.

Roger, M., Levee, G., Chanteau, S., et al. (1997). No evidence for linkage between leprosy susceptibility and the human natural resistance-associated macrophage protein 1 (NRAMP1) gene in French Polynesia. *Int. J. Lepr. Other Mycobact. Dis.* **65**, 197–202.

Roger, M., Sanchez, F. O., and Schurr, E. (1998). Comparative study of the genomic organization of DNA repeats within the 5′-flanking region of the natural resistance-associated macrophage protein gene (NRAMP1) between humans and great apes. *Mamm. Genome* **9**, 435–439.

Roy, S., Frodsham, A., Saha, B., et al. (1999) Association of vitamin D receptor genotype with leprosy type. *J. Infect. Dis.* **179**, 187–191.

Russell, D. G. (2001). *Mycobacterium tuberculosis:* Here today, and there tomorrow. *Nature Rev. Cell Mol. Biol.* **2**, 569–586.

Ryu, S., Park, Y. K., Bai, G. H., et al. (2000). 3′UTR polymorphisms in the NRAMP1 gene are associated with susceptibility to tuberculosis in Koreans. *Int. J. Tuberc. Lung Dis.* **4**, 577–580.

Sanjeevi, C. B., Miller, E. N., Dabadghao, P., et al. (2000). Polymorphism at NRAMP1 and D2S1471 loci associated with juvenile rheumatoid arthritis. *Arthritis Rheum.* **43**, 1397–1404.

Searle, S. and Blackwell, J. M. (1999). Evidence for a functional repeat polymorphism in the promoter of the human NRAMP1 gene that correlates with autoimmune versus infectious disease susceptibility. *J. Med. Genet.* **36**, 295–299.

Searle, S., Bright, N. A., Roach, T. I., et al. (1998). Localisation of Nramp1 in macrophages: Modulation with activation and infection. *Journal of Cell Science,* **111**(19), 2855–2866.

Shaw, M. A., Collins, A., Peacock, C. S., et al. (1997). Evidence that genetic susceptibility to *Mycobacterium tuberculosis* in a Brazilian population is under oligogenic control: Linkage study of the candidate genes NRAMP1 and TNFA. *Tuberc. Lung Dis.* **78**, 35–45.

Shear, H. L., Roth, E. F. Fabry, M. E., et al. (1993). Transgenic mice expressing human sickle hemoglobin are partially resistant to rodent malaria. *Blood* **81**, 222–226.

Singal, D. P., Li, J., Zhu, Y., et al. (2000). NRAMP1 gene polymorphisms in patients with rheumatoid arthritis. *Tissue Antigens* **55**, 44–47.

Skamene, E., Gros, P., Forget, A., et al. (1982). Genetic regulation of resistance to intracellular pathogens. *Nature* **297**, 506–509.

Stach, J. L., Gros, P., Forget, A., et al. (1984). Phenotypic expression of genetically controlled natural resistance to *Mycobacterium bovis* (BCG). *J. Immunol.* **132**, 888–892.

Stokes, R. W., Orme, I. and Collins, F. M. (1986). Role of mononuclear phagocytes in expression of resistance and susceptibility to *Mycobacterium avium* infections in mice. *Infect. Immun.* **54**, 811–819.

Stokkers, P. C. F., de Heer, K., Leegwater, A. C., et al. (1999). Inflammatory bowel disease and the genes for the natural resistance-associated macrophage protein-1 and the interferon-gamma receptor 1. *Int. J. Colorectal Dis.* **14**, 13–17.

Supek, F., Supekova, L., Nelson, H., et al. (1996). A yeast manganese transporter related to the macrophage protein involved in conferring resistance to mycobacteria. *Proc. Natl. Acad. Sci. USA* **93**, 5105–5110.

Tanaka, E., Kimoto, T., Matsumoto, H., et al. (2000). Familial pulmonary *Mycobacterium avium* complex disease. *Am. J. Respir. Crit. Care Med.* **161**, 1643–1647.

Tandy, S., Williams, M., Leggett, A., et al. (2000). Nramp2 expression is associated with pH-dependent iron uptake across the apical membrane of human intestinal Caco-2 cells. *J. Biol. Chem.* **275**, 1023–1029.

Theil, E. C. (1998). The iron responsive element (IRE) family of mRNA regulators. Regulation of iron transport and uptake compared in animals, plants, and microorganisms. *Met. Ions Biol. Syst.* **35**, 403–434.

Thomine, S., Wang, R., Ward, J. M., et al. (2000). Cadmium and iron transport by members of a plant metal transporter family in *Arabidopsis* with homology to *Nramp* genes. *Proc. Natl. Acad. Sci. U S A.* **97**, 4991–4996.

Tsolis, R. M., Adams, L. G., Hantman, M. J., et al. (1995). Role of *Salmonella typhimurium* Mn-superoxide dismutase (SodA) in protection against early killing by J774 macrophages. *Infect. Immun.* **63**, 1739–1744.

Tsolis, R. M., Baumler, A. J., Heffron, F., et al. (1996). Contribution of TonB- and Feo-mediated iron uptake to growth of *Salmonella typhimurium* in the mouse. *Infect. Immun.* **64**, 4549–4556.

Vidal, S. M., Malo, D., Vogan, K., et al. (1993). Natural resistance to infection with intracellular parasites: isolation of a candidate for Bcg. *Cell* **73**, 469–485.

Vidal, S., Tremblay, M. L., Govoni, G., et al. (1995). The *Ity/Lsh/Bcg* locus: natural resistance to infection with intracellular parasites is abrogated by disruption of the *Nramp1* gene. *J. Exp. Med.* **182**, 655–666.

Vidal, S. M., Pinner, E., Lepage, P., et al. (1996). Natural resistance to intracellular infections: *Nramp1* encodes a membrane phosphoglycoprotein absent in macrophages from susceptible ($Nramp1^{D169}$) mouse strains. *J. Immunol.* **157**, 3559–3568.

Webster, L. T. (1933). Inherited and acquired factors in resistance to infection. I. Development of resistant and susceptible lines of mice through selective inbreeding. *J. Exp. Med.* **57**, 793–817.
Webster, L. T. (1923). Microbic virulence and host susceptibility in mouse typhoid. *J. Exp. Med.* **37**, 231–267.
Webster, L. T. (1924). Microbic virulence and host susceptibility in paratyphoid enteritidis infection of white mice: Effect of selective breeding on host resistance. *J. Exp. Med.* **39**, 879–886.
White, J. K., Shaw, M.- A., Barton, C. H., et al. (1994). Genetic and physical mapping of 2q35 in the region of the NRAMP and IL8R genes: Identification of a polymorphic repeat in exon 2 of NRAMP. *Genomics* **24**, 295–302.
WHO. (2000). Leprosy – Global situation. *Wkly. Epidemiol. Rec.* **75**, 226–231.
WHO. (2001). *Global Tuberculosis Control.* Geneva, Switzerland: WHO Report. WHO/CDS/TB/2001.28.
Yang, Y. S., Kim, S. J., Kim, J. W., et al. (2000). NRAMP1 gene polymorphisms in patients with rheumatoid arthritis in Koreans. *J. Korean Med. Sci.* **15**, 83–87.
Yoshida, S.- I., Goto, Y., Mizuguchi, Y., et al. (1991). Genetic control of natural resistance in mouse macrophages regulating intracellular *Legionella pneumophila* replication *in vitro*. *Infect. Immun.* **59**, 428–435.
Zhang, G., Wu, H., Ross, C. R., et al. (2000). Cloning of porcine NRAMP1 and its induction by lipopolysaccharide, tumor necrosis factor alpha, and interleukin-1beta: role of CD14 and mitogen-activated protein kinases. *Infect. Immun.* **68**; 1086–1093.
Zhang, Y., Lathigra, R., Garbe, T., et al. (1991). Genetic analysis of superoxide dismutase, the 23 kilodalton antigen of *Mycobacterium tuberculosis*. *Mol. Microbiol.* **5**, 381–91.
Zwilling, B. S., Kuhn, D. E., Wikoff, L., et al. (1999). Role of iron in Nramp1-mediated inhibition of mycobacterial growth. *Infect. Immun* **67**, 1386–1392.

CHAPTER 9

The interleukin-12/interferon-γ loop is required for protective immunity to experimental and natural infections by *Mycobacterium*

Marion Bonnet, Claire Soudais, and Jean-Laurent Casanova
Laboratory of Human Genetics of Infectious Diseases, Université René Descartes, Necker Medical School, Paris

Mendelian susceptibility to poorly pathogenic mycobacteria, such as bacillus Calmette-Guérin (BCG) and environmental nontuberculous mycobacteria (EM), is a rare human syndrome. Some patients present with mutations in the genes encoding IL-12p40 or IL12Rβ1, associated with impaired production of IFNγ. Others carry mutations in the genes encoding IFNγR1, IFNγR2, or STAT1, associated with impaired response to IFNγ. Knockout mice for IL-12, IFNγ, or their receptors are also vulnerable to experimental infection with nonvirulent mycobacteria. Studies with knockout mice also implicate other molecules involved in the induction of, or response to, IFNγ, such as IL-18, IL-1, TNFα, IRF-1, and NOS2, in the control of mycobacterial infection. It is now clear that the IL-12-IFNγ loop is crucial for protective immunity to experimental and natural mycobacterial infection in both mice and men.

1. INTRODUCTION

Bacillus Calmette-Guérin (BCG) vaccines and environmental mycobacteria (EM) are poorly pathogenic in humans. However, they cause disseminated disease in severely immunodeficient individuals, but such patients are also susceptible to a broad range of pathogens (Reichenbach et al., 2001; Casanova and Abel, 2002). BCG and EM may also lead to severe disease in otherwise healthy individuals without any overt immunodeficiency. These patients present a selective vulnerability to mycobacterial infections (Casanova et al., 1995; 1996; Levin et al., 1995; Frucht and Holland, 1996), without any other associated infections, apart from salmonellosis, which affects less than half of the cases (reviewed in Dorman and Holland, 2000, and Casanova and Abel, 2002). Parental consanguinity and familial forms are frequently observed, suggesting a Mendelian disorder. Consequently the condition has been called

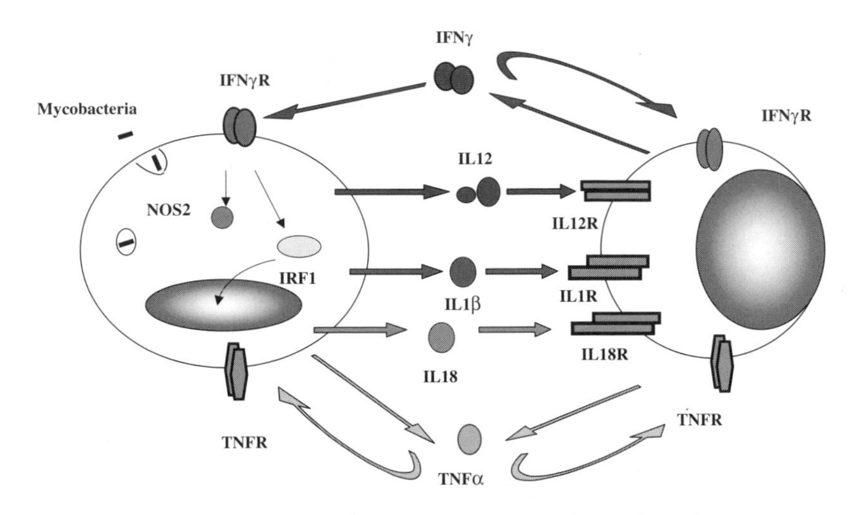

Figure 9.1. Cytokine interactions between macrophage/dendritic cells and NK/T lymphocytes. (See color plate.)

Mendelian susceptibility to mycobacterial infection (MIM 209950) (McKusick, 1998). In most cases inheritance is autosomal and recessive, but autosomal dominant inheritance has been reported in some families (Jouanguy et al., 1999) and X-linked recessive inheritance in another (Frucht and Holland, 1996). The clinical outcome correlates with the type of granulomatous lesions (Emile et al., 1997), i.e., patients with lepromatous-like granulomas (poorly delimited, multibacillary, with no epithelioid or giant cells) die from overwhelming infection, whereas tuberculoid granulomas (well-delimited, paucibacillary, with epithelioid and giant cells) are associated with a favourable outcome.

Patients with Mendelian susceptibility to mycobacteria have abnormal production of, or response to interferon-γ (IFNγ). IFNγ is a pleiotropic cytokine secreted by natural killer (NK) and T cells (Fig. 9.1). The condition is clinically heterogeneous because of its genetic heterogeneity. Mutations of five different genes have been identified as being causative, defining 10 disorders involving impairment of IFNγ-mediated immunity. The genes involved are *IFNGR1* and *IFNGR2*, encoding the two chains of the IFNγ receptor; *STAT1* encoding the transcription factor signal transducer and activator of transcription 1 (STAT1) activated in response to IFNγ; *IL12B*, encoding the p40 subunit of interleukin-12 (IL-12), a potent IFNγ-inducing cytokine secreted by macrophages and dendritic cells; and *IL12RB1* encoding the β1-chain of the IL-12 receptor, expressed on NK and T cells. There is also allelic heterogeneity: dominant and recessive mutations have been found in *IFNGR1* and null or hypomorphic mutations in *IFNGR1* and *IFNGR2*. In this chapter, we will review the molecular genetic basis of vulnerability to mycobacterial disease in natural conditions of infection – a hallmark of the human disease (Casanova and Abel, 2002). We will then review the vulnerability to experimental mycobacterial infection in mice with targeted genetic defects in the IL12-IFNγ loop.

2. MOLECULAR BASIS OF THE HUMAN SYNDROME OF *MENDELIAN SUSCEPTIBILITY TO MYCOBACTERIAL INFECTION*

2.1 *IFNGR1* Mutations

2.1.1 Complete IFNγR1 Deficiency

Complete recessive IFNγ receptor ligand-binding chain (IFNγR1) deficiency was the first causative genetic defect identified. It was found to be associated with vulnerability to BCG vaccination in two kindreds (Jouanguy et al., 1996; Newport et al., 1996). More kindreds (23 patients) have since been identified (Pierre-Audigire et al., 1997; Altare et al., 1998; Holland et al., 1998; Roesler et al., 1999; Jouanguy et al., 2000; Cunningham et al., 2000; Allende et al., 2001). The patients are of various ethnic origins. There are two forms of complete IFNγR1 deficiency. Null mutations are the most common. They preclude cell surface expression of the receptor due to a premature stop codon upstream from the segment encoding the transmembrane segment. In three other families, children have complete IFNγR1 deficiency despite normal amounts of IFNγR1 on the cell surface (Jouanguy et al., 2000). In these cases, mutations in the segment encoding the extracellular ligand-binding domain prevent the surface receptors binding to their natural ligand, IFNγ. All these mutations are associated with a complete lack of cellular response to exogenous recombinant IFNγ. Clinically, complete IFNγR1 deficiency results in a selective susceptibility to mycobacterial infection such that it is early onset and severe. Pathogens associated with the disease are either slow-growing mycobacteria, including *Mycobacterium avium* and *Mycobacterium kansasii* or *Mycobacterium szulgai*, or fast-growing mycobacteria, including *Mycobacterium fortuitum, Mycobacterium chelonae, Mycobacterium smegmatis,* and *Mycobacterium peregrinum*. The last two are among the least virulent mycobacteria and had never previously been reported to cause disseminated disease in humans. The clinical phenotype is characterised by lepromatous-like lesions, and tuberculoid granulomas almost certainly rule out complete IFNγR1 deficiency. These patients also have high levels of IFNγ in serum, allowing rapid diagnosis of IFNγR deficiency (Fieschi et al., 2001). Most patients die before they are 12 years old (Table 9.1).

2.1.2 Partial IFNγR1 Deficiency

Partial IFNγR1 deficiency may be caused by recessive or dominant mutated *IFNGR1* alleles. Cells from patients with partial, as opposed to complete, IFNγR1 deficiency have impaired but not abolished response to IFNγ *in vitro*. These patients are also vulnerable to both slow growing mycobacteria, including *M. avium and M. kansasii*. Susceptibility to the fast-growing mycobacteria *M. chelonae* is, surprisingly, shown in only one patient. The clinical phenotype of children with partial IFNγR1 deficiency is milder than that of children with complete IFNγR1 deficiency: they suffer well-circumscribed and differentiated tuberculoid granulomas (Lamhamedi et al., 1998). Two siblings with recessive IFNγR1 deficiency have been reported (Jouanguy et al., 1997). They carry a homozygous recessive missense mutation resulting in an amino-acid substitution in the extracellular domain of the receptor. The receptor is produced normally and translocated to the cell surface but has an abnormally low affinity for its ligand, IFNγ, although IFNγ binding is not totally abolished. A dominant form of partial IFNγR1 deficiency has been identified in 49 patients from 29 unrelated kindreds (Jouanguy et al., 1999; Villella et al., 2001; Dorman and Holland, 2000; Aksu et al., 2001; Arend et al., 2001). These patients have a small heterozygous frameshift deletion in *IFNGR1* exon 6, downstream from the segment encoding the transmembrane domain. The mutant alleles encode a truncated receptor with only five intracellular amino acids; however, it is present at the cell surface and binds IFNγ correctly. These truncated receptors dimerise normally and form tetramers with two IFNγR2 molecules. Nevertheless, this receptor fails to transduce the IFNγ-triggered signal, due to the lack of intracellular binding domains for the signalling cascade molecules JAK-1 and STAT-1. The receptors accumulate on the cell surface of cells from these patients due to the absence of an intracellular recycling site. The combination of normal IFNγ binding, absence of signalling, and accumulation at the cell surface is responsible for their dominant-negative effect. Even though most IFNγR1 dimers are not functional in heterozygous cells, the few wild-type IFNγR1 dimers account for a partial rather than a complete defect. Interestingly, this disorder led to the first identification of a small deletion hotspot in the human genome, at position 818 of *IFNGR1* (Jouanguy et al., 1999). Overlapping small deletions (818del4

Table 9.1. *Susceptibility to mycobacterial species among humans and mice with IL12-IFNγ gene defects*

Mycobacterium strains	IL-12-IFNg loop	
	Human	Mouse
BCG	S	S
M. tuberculosis	S	S
Slow-growing strains		
M. avium	S	S
M. kansassi	S	
M. szulgai	S	
M. genavensae		S (IFNG-KO)
M. asiaticum	S	
M. gordonae	S	
M. leprae		MS (IFNG-KO)
M. celatum		S
M. branderi		S
M. xenopi		S
M. malmoense		S
M. bohemicum		S
M. conspicuum		S
M. interiectum		S
M. intermedium		S
clinical isolate 223/96		S
M. heildelbergense		MS
M. lentiflavum		R
clinical isolate 11867/96		R
Fast-growing strains		
M. fortuitum	S	
M. chelonae	S	S*
M. smegmatis	S	MS (TNF-KO)
M. peregrinum	S	
M. abscessus	S	S*
M. confluentis		R

Abbreviations: R, resistent; S, susceptible; MS, moderately susceptible; TNFKO, TNFgene disrupted mouse; IFNGKO, INFγ gene disrupted mouse.

* Unpublished data.

in 11 kindreds and 818delT in one) were found to have occurred independently in 12 unrelated families. A model of slipped mispairing events and subsequent repair during replication was proposed, based on the presence of two direct repeats and small deletion consensus motifs in the vicinity of nucleotide 818. Another small deletion hotspot has been recently identified in *IFNGR1*, in exon 5 at position 561 (561del4). This deletion has been found in three unrelated heterozygous carriers and two patients (Rosenzweig et al., 2002).

2.2 *IFNGR2* Mutations

2.2.1 Complete IFNγR2 Deficiency

Two patients with complete IFNγR2 deficiency have been reported (Dorman and Holland, 1998, 2000). They carry a homozygous recessive frameshift deletion in the coding region of *IFNGR2*, resulting in a premature stop codon upstream from the segment encoding the transmembrane domain. The clinical phenotype is analogous to that of patients with complete IFNγR1 deficiency, with early-onset and severe mycobacterial infections, no mature granulomas and high levels of IFNγ in serum (Fieschi et al., 2001). Deficient patients are also vulnerable to fast growing mycobacteria such as *M. abscessus* and *M. fortuitum*.

2.2.2 Partial IFNγR2 Deficiency

A 20-year-old patient with a history of BCG and *M. abscessus* infection was found to have a partial IFNγR2 deficiency. It was due to a homozygous missense mutation in the extracellular domain, resulting in single amino-acid substitution (Doffinger et al., 2000). The encoded receptor was normally localised at the cell surface, but was associated with an abnormally low cellular response to IFNγ. Clinically, the patient's phenotype was similar to that of partial IFNγR1 deficiency.

2.3 *STAT1* Mutations

STAT1 is an essential element of IFN-induced signal transduction, either as a STAT1 homodimer, also designated the gamma-activated factor (GAF), or as a heterotrimeric transcription complex, the interferon stimulated gene factor 3 (ISGF3), which also includes STAT2 and IRF9 (previously known as p48). Two unrelated patients presented an identical heterozygous missense mutation leading to partial dominant STAT-1 deficiency (Dupuis et al., 2001). This mutation is null for both activation of GAF and ISGF3, but in heterozygous patients' cells; the mutation is dominant for GAF activation and recessive for ISGF3. Despite having received BCG vaccinations, patients were vulnerable to *M. avium*. No severe viral infection was detected in these patients, suggesting that IFN type I-mediated viral immunity is not impaired. The clinical and cellular phenotypes of these patients, with respect to mycobacterial infection and GAF activation, were similar to those of patients with partial recessive IFNγR deficiency. This observation also implies that IFNγ mediated antimycobacterial immunity is strictly STAT-1 dependent. New STAT-1 mutations have been identified in patients with BCG and HSV infection. These mutations cause a complete defect in STAT-1. Patients' cells are vulnerable to viral infection in vitro, and this implicates IFN type I-mediated immunity (Dupuis et al., 2003).

2.4 *IL-12B* Mutations

IL-12 is a heterodimeric cytokine, consisting of the subunits p40 and p35. It is produced mainly by macrophages and dendritic cells. A kindred with a loss-of-function recessive mutation in the *IL12B* gene, encoding the p40 subunit, has been identified (Altare et al., 1998). The patient presents a homozygous frameshift deletion of 4.4 kb encompassing two coding exons. Subsequently, five other kindreds (12 patients) with IL-12p40 deficiency have been reported (Picard et al., 2002; Elloumi-Zghal et al., 2002). The patients suffered BCG infection associated (in only one case) with *M. chelonae* and in the other cases with *M. avium*. One patient was also infected by *Salmonella enterica*. No IL-12 secretion could be detected in these patients, and their lymphocytes produced less IFNγ than those from normal individuals. This defect can be complemented by treatment with exogenous recombinant IL-12. Thus, IFNγ deficiency appears as a consequence of inherited IL-12 deficiency. The clinical phenotype of IL12-deficient patients is, however, milder than that of complete IFNγR patients, probably due to residual IL-12-independent IFNγ production.

2.5 *IL-12RB1* Mutations

Recessive mutations in the *IL12RB1* gene, encoding the β1 subunit of IL-12 receptor have been reported in 9 kindreds (Altare et al., 1998; de Jong et al., 1998; Verhagen et al., 2000; Aksu et al., 2001; Altare et al., 2001; Sakai et al., 2001; Elloumi-Zghal et al., 2002). Patients were found to be infected mostly with *M. avium*, and one child was infected with *M. chelonae*. Half of the patients had *Salmonella enterica* infections, but no other infections were reported. All patients were homozygous for the mutations, which preclude expression of IL-12Rβ1 at the cell surface. The loss of IL-12 signalling through its receptor leads to impairment of IFNγ production *in vitro* by otherwise functional NK or T cells. Another patient was found to respond poorly to IL-12 despite normal IL-12Rβ1 (Gollob et al., 2000). The clinical phenotype of IL-12Rβ1 patients is similar to that of IL-12 p40 children. Thus, IL-12-mediated mycobacterial immunity occurs mainly through IL-12Rβ1. Deficiencies in the IL-12 signalling pathway seem to lead to a greater phenotypic heterogeneity than deficiencies of IFNγR, for which there is a strict correlation between genotype and phenotype (Dupuis et al., 2000). Interestingly, three patients with IL-12p40 or IL-12Rβ1 deficiency were resistant to BCG (Picard et al., 2002; Aksu et al., 2001; Altare et al., 2001). One of them was vaccinated three times with live BCG without any adverse effect, whereas his brother had disseminated BCG-osis (Altare et al., 2001). Two of these patients did not even develop atypical mycobacteriosis. Alternative pathways may compensate for the loss of IL-12 signalling, and in such individuals IL-12-independent IFNγ production may be sufficient to control mycobacterial infection. The efficiency of any such compensatory pathways may differ between individuals, leading to different cellular phenotypes in the patients. Our recent identification of 37 patients from 26 kindreds indicates that IL12Rβ1 deficiency is associated with a broad resistance, low penetrance, and favourable outcome (Fieschi et al., 2003).

3. SUSCEPTIBILITY TO EXPERIMENTAL MYCOBACTERIAL INFECTION IN MICE WITH IMPAIRED IL-12/IFNγ LOOP

3.1 IFNγ o/o Mice and IFNγR o/o Mice

IFNγ o/o (GKO) mice are highly susceptible to BCG infection. GKO have 10 to 100 times more CFU than controls in lung, liver, and spleen after injection with BCG intravenously (Dalton et al., 1993; Murray et al., 1998). IFNγR o/o (GRKO) infected with BCG die between 7 to 9 weeks after inoculation and suffer massive granulomatous pulmonary lesions (Fig. 9.2) (Kamijo et al., 1993). GKO, like GRKO, animals are unable to control *M. tuberculosis* infection and there is substantial dissemination of intralesional acid-fast bacteria throughout spleen, liver, lungs and kidneys (Figs. 9.3 and 9.4) (Cooper et al., 1993; Flynn et al., 1993; Erb et al., 1998). GKO mice are also more susceptible than wild type to poorly virulent mycobacteria, including

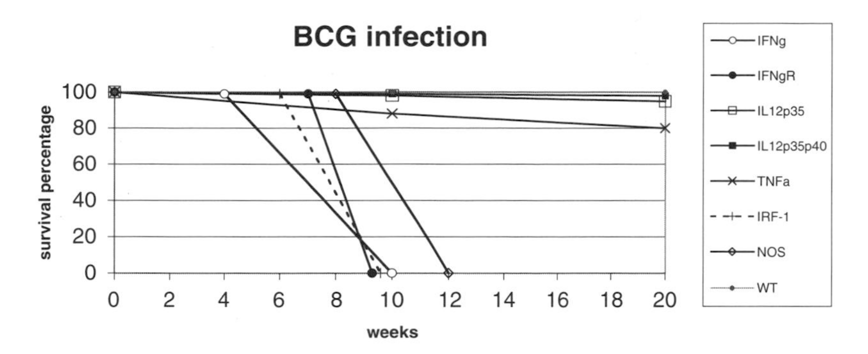

Figure 9.2. Survival following inoculation with BCG among mice with IL12-IFNγ, TNFα, IRF-1 and NOS gene defects. WT = wild-type.

M. avium. However, this phenomenon is apparent only at later stages of the infection (Doherty and Sher, 1997). GKO infected with *M. genavense* also show a chronic inflammatory response with diffuse granulomatous lesions in liver (Ehlers and Richter, 2001). Some newly identified nontuberculous mycobacteria (most are slow growing but one is fast growing) have also been tested on GKO mice. The mice were susceptible to most of the strains tested (Ehlers and Richter, 2001). It thus appears that IFNγ is a major determinant of the outcome of infection with several mycobacterial species.

3.2 IL12p40 o/o, IL12p35 o/o, and IL12p35 o/o p40 o/o Mice

Bioactive heterodimeric IL-12p70 (p40 + p35) is a key factor for induction of IFNγ during mycobacterial infection: IL-12 neutralisation experiments

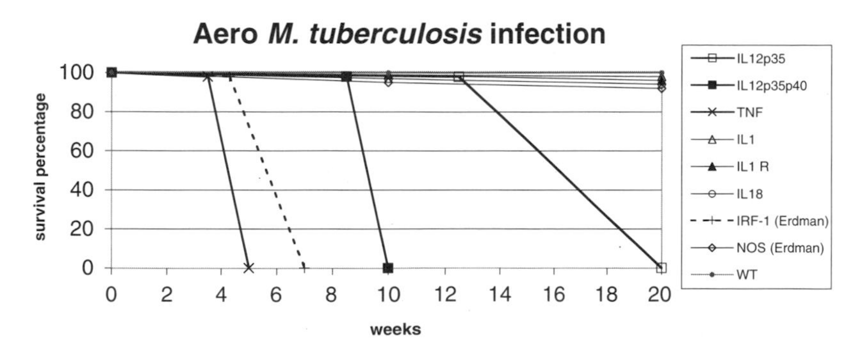

Figure 9.3. Survival following aerosol *M. tuberculosis* infection among mice with IL12-IFNγ, TNFα, IRF-1 and NOS gene defects.

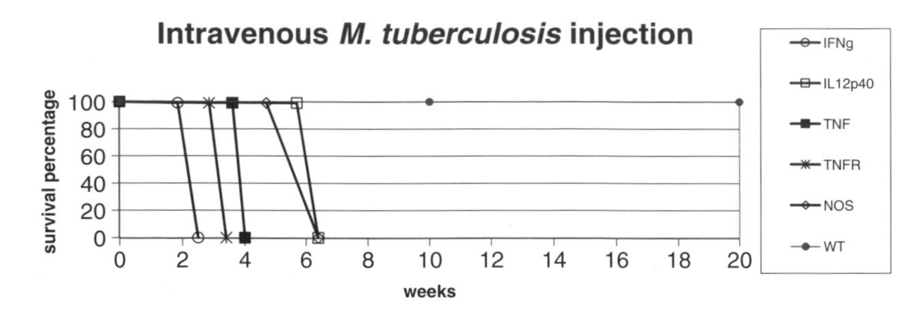

Figure 9.4. Survival following intravenous *M. tuberculosis* injection among mice with IL12-IFNγ, TNFα, IRF-1 and NOS gene defects.

demonstrate the importance of IL-12p70 in the host defense against mycobacteria. Infection of IL-12 p40 o/o mice with *M. tuberculosis* results in uncontrolled growth of the bacteria in all target organs. Mice deficient for the p40 subunit are more susceptible than the p35 o/o mice to bacterial growth (Figs. 9.3 and 9.4) (Cooper et al., 2002). Increased susceptibility of IL-12 p40 o/o animals correlates with very low, but not abolished, IFNγ mRNA abundance. This results in a marked decrease in the formation of granulomas containing large numbers of bacteria. The expression of TNFα, another protective cytokine is also delayed in IL-12 p40 o/o mice. IL-12Rβ2 expression is not up-regulated in IL-12 p40 o/o animals but it is in wild-type animals during infection (Cooper et al., 1997). A similar phenotype is observed after pulmonary infection of IL-12 p40 o/o mice with BCG (Wakeham et al., 1998). Thus, IL-12 appears to be an essential initiator of the development of the granulomatous response and subsequent inhibition of mycobacterial growth and spread. Infection with *M. tuberculosis* by inhalation leads to mortality in IL-12p35 o/o p40 o/o mice and in IL-12p35 o/o mice (Fig. 9.3). Double knockout mice survive BCG infection but are chronically infected (Fig. 9.2) (Holscher et al., 2001). The chronic disease after BCG infection contrasts with the phenotype of IFNγ o/o mice, which show a high mortality rate. However, it agrees with the milder phenotype of IL-12p40- or IL12Rβ1-deficient patients than IFNγR-deficient patients do. However, only mutations of the p40 subunit have been described in humans (Casanova and Abel, 2002). Another possible resistance mechanism of IL-12p35 o/o mice to mycobacterial infection may involve the newly described heterodimeric IL-23, composed of IL-12p40 and a p19 subunit, that has an activity similar to, but distinct from IL-12. The p19 subunit mRNA is similarly induced by *M. tuberculosis* infection in WT, p35 o/o, and p40 o/o mice, such that IL-23 is available in IL-12 p35 o/o animals.

3.3 IL-18 and IL-1α/β and IL-1R1 o/o Mice: Parallel IFNγ Inducing Pathways

IL-18 was first described as IFNγ-inducing factor (IGIF) (Okamura et al., 1995). It has similarities with IL-1β in terms of structure, secretion, and signalling. IL-18, in synergy with IL-12, promotes IFNγ production (Micallef et al., 1996; Ahn et al., 1997). IL-18 o/o mice are slightly more susceptible than WT mice to *M. tuberculosis* but not to BCG (Figs. 9.2–9.4). Splenic IFNγ production, but not that of IL-12 or TNFα, is slightly low, implicating IL-18 in IFNγ production. Thus, reduced IFNγ production is not secondary to reduced IL-12 production. These animals also show very low IL-1β production (Sugawara et al., 1999). IL-1 is a proinflammatory cytokine mainly secreted by macrophages and dendritic cells and which synergizes with IL-12 to induce IFNγ production (for review see Dinarello, 1996). IL-1 α/β o/o mice survive and develop larger granulomatous lesions after infection with *M. tuberculosis* and are able to control the infection (Fig. 9.3). Neither defect in IFNγ nor IL-12 production has been detected in these animals, but they display an abnormally high production of TNFα, which may compensate for loss of expression of IL-1. Injection of IL-1 α/β o/o mice with IL1 α/β failed to cure the granulomatous lesions completely (Yamada et al., 2000). IL-1R1 o/o mice, infected with the less virulent H37Rv strain of *M. tuberculosis* survived with discrete developing granulomas (Fig. 9.3) (Juffermans et al., 2000; Sugawara et al., 2001). When the more virulent Kurono strain was used, mice succumbed after between 6 and 7 weeks and showed a much higher density than WT mice of bacilli in their lung. IL-1R1 o/o mice produce less IFNγ than WT associated with a less mRNA for IL-1 and iNOS (Sugawara et al., 2001).

3.4 TNFα, TNFR1 o/o Mice: IFNγ-Inducible Genes

Tumour necrosis factor-α (TNFα) is an IFNγ-inducible proinflammatory cytokine that synergizes with IFNγ to activate the bactericidal activity of macrophages (Rook et al., 1987; Flynn et al., 1995). TNFα o/o mice exhibit a slightly increased susceptibility to administration of BCG: in terms of mortality and the number of bacilli in lungs and spleen which are higher than in WT, but there is no granulomas formation (Fig. 9.2) (Kaneko et al., 1999). TNFα o/o mice also have more bacilli in lung and spleen following infection with *M. tuberculosis* and are killed by disseminated infection (Figs. 9.3 and 9.4) (Bean et al., 1999; Kaneko et al., 1999). Infected animals show poorly formed and multibacillary granulomas containing few differentiated epithelioid cells, which is consistent with the report that TNF is central to granuloma formation (Kindler et al., 1989). IL-12 and IFNγ production is impaired from TNFα o/o mice splenocytes after BCG stimulation, whereas IL-1β secretion is higher than WT and this may compensate for the absence of TNFα. NO production by peritoneal macrophages is not affected by stimulation with BCG, but is impaired by stimulation with *M. tuberculosis* (Kaneko et al., 1999). Clearance of bacteria, using *M. smegmatis* as the model infectious agent, is delayed in TNF o/o mice. This could be due to the delay in cell recruitment and in chemokine secretion (Roach et al., 2002). Thus, TNFα appears to have a protective role against *M. tuberculosis* infection due to its involvement in the generation of structurally effective granulomas. Olleros et al. (2002) showed that transgenic mice, constitutively expressing the transmembrane form of TNF, are protected against BCG and *M. tuberculosis* infection.

As observed for TNF o/o, TNFR1 o/o mice infected with BCG display few and small granulomas (Senaldi et al., 1996). TNFR1 o/o mice, infected with *M. tuberculosis*, die quickly (Fig. 9.4). TNFα expression is not affected by the absence of its receptor but macrophages from TNFR1 o/o mice produce only very small amounts of reactive nitrogen intermediates (Flynn et al., 1995). In contrast, bacterial loads in lung and spleen of WT and TNFR1 o/o mice are almost identical following infection with *M. avium*, but the TNFR1 o/o mice die following such infection with a hyperinflammatory response. Granuloma formation is delayed in TNFR1 o/o mice and cellular integrity is severely impaired (Ehlers et al., 1999; Benini et al., 1999). One week before death, TNFR1 o/o mice show increased levels of IFNγ and IL12p40, at both the mRNA and at the protein levels, and higher plasma levels of TNFα. iNOS expression and activity are unchanged. This hyperinflammatory response is associated with increased numbers of cells of both CD4+ and CD8+ T subsets. Depletion of the T-cell subset reverses the lethal phenotype of infection in TNFR1 o/o mice, allowing normal granuloma formation and epithelioid cell differentiation. Neither TNFα nor IFNγ neutralisation reverse the lethal inflammation in TNFR1 o/o mice, whereas neutralisation of IL-12p40 allows survival of the animals and normal granuloma formation with few infiltrating T cells. Thus, TNFR1-mediated signalling seems to be involved in a feedback loop, which controls T-cell recruitment and activation at the site of infection, and their positive effect on macrophage activation, which maintains the inflammatory process.

4. IRF-1 AND NOS2: OTHER IFNγ-INDUCIBLE GENES

IRF-1 is a transcription factor strongly induced by IFNs α/β, IFNγ, TNFα, and IL-1β (Fujita et al., 1989). IRF-1 is thought to be responsible for induction of mycobactericidal effectors by IFNγ. IRF-1 o/o mice appear moribund after infection with BCG. Many acid-fast bacteria are found in abundant granulomatous lesions in liver, lungs, and spleen (Kamijo et al., 1994). All IRF-1 o/o mice die of uncontrolled bacterial expansion after *M. tuberculosis* infection (Fig. 9.3) (Cooper et al., 2000). Thus IRF-1 is important in IFNγ-induced antimycobaterial immunity.

NOS-2 is an IFNγ-induced enzyme, involved in production of nitrogen reactive species (NO). NOs are believed to play an important role in macrophage mycobactericidal activity (Chan et al., 1992; Fang et al., 1997). NOS2 o/o mice, infected with BCG, are unable to control bacterial growth: the mice died with large granulomas with extensive lesions in spleen (Fig. 9.2). The serum IFNγ concentration is lower than in WT (Garcia et al., 2000). NOS2 o/o mice are highly susceptible to intravenous but not to aerosol *M. tuberculosis* infection (Figs. 9.3 and 9.4) (MacMicking et al., 1997; Cooper et al., 2000): the control of infection by the liver is mostly dependent on nitric oxide, whereas in the lung, control is mainly dependent on IFNγ and TNFα. NOS2 o/o mice are a remarkably permissive environment for *M. tuberculosis* growth (MacMicking et al., 1997). Several new clinical strains of *M. tuberculosis* have been tested on NOS2 o/o mice. The observations demonstrate the crucial role of NOS2 in the control of infection regardless of the route of infection (Scanga et al., 2001). The granulomas induced in NOS2 o/o mice by *M. avium* are larger than those in WT, but contain similar numbers of bacteria (Ehlers et al., 1999; Gomes et al., 1999). Even with various strains of *M. avium*, NOS2 o/o mice are not abnormally susceptible to infection. Moreover NOS2 o/o mice appear to clear mycobacteria more efficiently at later than earlier time points. The increased clearance is associated with higher levels of IFNγ in the serum (Gomes et al., 1999).

5. CONCLUSIONS

Although many proteins have been implicated in granuloma formation and elimination of mycobacteria, IL-12 and IFN-γ seem to be the key partners. The identification of patients deficient in the IL-12-IFNγ loop underlines the critical role of these cytokines in immunity to mycobacteria in natural conditions of infection – a hallmark of the human model. These patients are from diverse ethnic groups and were infected by various species of environmental mycobacteria. Inbred mice genetically deficient for genes in the IL12-IFNγ loop are similarly vulnerable to experimental infection with mycobacteria. Animal and human studies converge to highlight the central role of the IL-12-IFNγ loop in antimycobacterial immunity. However, many BCG and EM infections in humans remain unexplained both genetically and immunologically. Further susceptibility genes will undoubtedly be identified and available knockout mice are likely to indicate new candidates genes. Cytokines, such as IL-1β or IL-18, which cooperate, with IL-12 to induce IFNγ production, appear to be responsible for Mendelian susceptibility to mycobacteria. Nevertheless, in other patients, susceptibility may be due to downstream effectors (for example TNFα, IRF-1, or NOS2) of IFNγ which promote granuloma formation and mycobactericidal activity.

REFERENCES

Ahn, H. J., Maruo, S., Tormura, M., et al. (1997). A mechanism underlying synergy between IL-12 and IFN-gamma-inducing factor in enhanced production of IFN-gamma. *J. Immunol.* **159**, 2125–2131.

Aksu, G., Tirpan, C., Cavusoglu, C., et al. (2001). *Mycobacterium fortuitum-chelonae* complex infection in a child with complete interleukin-12 receptor beta 1 deficiency. *Pediatr. Infect. Dis. J.* **20**, 551–553.

Allende, L. M., Lopez-Goyanes, A., Paz-Artal, E., et al. (2001). A point mutation in a domain of gamma interferon receptor 1 provokes severe immunodeficiency. *Clin. Diagn. Lab. Immunol.* **8**, 133–137.

Altare, F., Jouanguy, E., Lamhamedi, S., et al. (1998). A causative relationship between mutant IFNgR1 alleles and impaired cellular response to IFNgamma in a compound heterozygous child. *Am. J. Hum. Genet.* **62**, 723–726.

Altare, F., Durandy, A, Lammas, D., et al. (1998). Impairment of mycobacterial immunity in human interleukin-12 receptor deficiency. *Science* **280**, 1432–1435.

Altare, F., Lammas, D., Revy, R., et al. (1998). Inherited interleukin 12 deficiency in a child with bacille Calmette–Guerin and *Salmonella enteritidis* disseminated infection. *J. Clin. Invest.* **102**, 2035–2040.

Altare, F., Ensser, A., Breiman, A., et al. (2001). Interleukin-12 receptor beta1 deficiency in a patient with abdominal tuberculosis. *J. Infect. Dis.* **184**, 231–236.

Arend, S. M., Janssen, R., Gosen, J. J., et al. (2001). Multifocal osteomyelitis caused by nontuberculous mycobacteria in patients with a genetic defect of the interferon-gamma receptor. *Neth. J. Med.* **59**, 140–151.

Bean, A. G., Roach, D. R., Briscoe, H., et al. (1999). Structural deficiencies in granuloma formation in TNF gene-targeted mice underlie the heightened susceptibility to aerosol *Mycobacterium tuberculosis* infection, which is not compensated for by lymphotoxin. *J. Immunol.* **162**, 3504–3511.

Benini, J., Ehlers, E. M. and Ehlers, S. (1999). Different types of pulmonary granuloma necrosis in immunocompetent vs. TNFRp55-gene-deficient mice aerogenically infected with highly virulent *Mycobacterium avium*. *J. Pathol.* **189**, 127–137.

Casanova, J. L. and Abel, L. (2002). Genetic dissection of immunity to mycobacteria: The human model. *Annu. Rev. Immunol.* **20**, 581–620.

Casanova, J. L., Jouanguy, E., Lamhamedi, S., et al. (1995). Immunological conditions of children with BCG disseminated infection. *Lancet* **346**, 581.

Casanova, J. L., Blanche, S., Emile, J. F., et al. (1996). Idiopathic disseminated bacillus Calmette-Guerin infection: a French national retrospective study. *Pediatrics* **98**, 774–778.

Chan, J., Xing, Y., Magliozzo, R. S., et al. (1992). Killing of virulent *Mycobacterium tuberculosis* by reactive nitrogen intermediates produced by activated murine macrophages. *J. Exp. Med.* **175**, 1111–1122.

Cooper, A. M., Dalton, D. K., Stewart, T. A., et al. (1993). Disseminated tuberculosis in interferon gamma gene-disrupted mice. *J. Exp. Med.* **178**, 2243–2247.

Cooper, A. M., Magram, J., Ferranti, J., et al. (1997). Interleukin 12 (IL-12) is crucial to the development of protective immunity in mice intravenously infected with *Mycobacterium tuberculosis*. *J. Exp. Med.* **186**, 39–45.

Cooper, A. M., Pearl, J. E., Brooks, J. V., et al. (2000). Expression of the nitric oxide synthase 2 gene is not essential for early control of *Mycobacterium tuberculosis* in the murine lung. *Infect. Immun.* **68**, 6879–6882.

Cooper, A. M., Kipnis, A., Turner, J., et al. (2002). Mice lacking bioactive IL-12 can generate protective, antigen-specific cellular responses to mycobacterial infection only if the IL-12 p40 subunit is present. *J. Immunol.* **168**, 1322–1327.

Cunningham, J. A., Kellner, J. D., Bridge, P. J., et al. (2000). Disseminated bacille Calmette-Guerin infection in an infant with a novel deletion in the interferon-gamma receptor gene. *Int. J. Tuberc. Lung Dis.* **4**, 791–794.

Dalton, D. K., Pitts-Meek, S., Keshav, S., et al. (1993). Multiple defects of immune cell function in mice with disrupted interferon-gamma genes. *Science* **259**, 1739–1742.

de Jong, R., Altare, F., Haagen I. A., et al. (1998). Severe mycobacterial and *Salmonella* infections in interleukin-12 receptor-deficient patients. *Science* **280**, 1435–1438.

Dinarello, C. A. (1996). Biologic basis for interleukin-1 in disease. *Blood* **87**, 2095–2147.

Doffinger, R., Jouanguy, E., Dupuis, S., et al. (2000). Partial interferon-gamma receptor signaling chain deficiency in a patient with bacille Calmette-Guerin and *Mycobacterium abscessus* infection. *J. Infect. Dis.* **181**, 379–384.

Doherty, T. M. and Sher, A. (1997). Defects in cell-mediated immunity affect chronic, but not innate, resistance of mice to *Mycobacterium avium* infection. *J. Immunol.* **158**, 4822–4831.

Dorman, S. E. and Holland, S. M. (1998). Mutation in the signal-transducing chain of the interferon-gamma receptor and susceptibility to mycobacterial infection. *J. Clin. Invest.* **101**, 2364–2369.

Dorman, S. E. and Holland, S. M. (2000). Interferon-gamma and interleukin-12 pathway defects and human disease. *Cytokine Growth Factor Rev.* **11**, 321–333.

Dupuis, S., Doffinger, R., Picard, C., et al. (2000). Human interferon-gamma-mediated immunity is a genetically controlled continuous trait that determines the outcome of mycobacterial invasion. *Immunol. Rev.* **178**, 129–137.

Dupuis, S., Dargemont, C., Fieschi, C., et al. (2001). Impairment of mycobacterial but not viral immunity by a germline human STAT1 mutation. *Science* **293**, 300–303.

Dupuis, S., Jouanguy, E., Al-Hajjar, S., et al. (2003). Impaired response to interferon- alpha/beta and lethal viral disease in human STAT1 deficiency. *Nat. Genet.* **33**, 388–391.

Ehlers, S., Benini, J., Kutsch, S., et al. (1999). Fatal granuloma necrosis without exacerbated mycobacterial growth in tumor necrosis factor receptor p55 gene-deficient mice intravenously infected with *Mycobacterium avium. Infect. Immun.* **67**, 3571–3579.

Ehlers, S. and Richter, E. (2001). Differential requirement for interferon-gamma to restrict the growth of or eliminate some recently identified species of non-tuberculous mycobacteria *in vivo*. *Clin. Exp. Immunol.* **124**, 229–238.

Elloumi-Zghal, H., Barbouche, M. R., Chemli, J., et al. (2002). Clinical and genetic heterogeneity of inherited autosomal recessive susceptibility to disseminated *Mycobacterium bovis* bacille Calmette-Guerin infection. *J. Infect. Dis.* **185**, 1468–1475.

Emile, J. F., Patey, N., Altare, F., et al. (1997). Correlation of granuloma structure with clinical outcome defines two types of idiopathic disseminated BCG infection. *J. Pathol.* **181**, 25–30.

Erb, K. J., Holloway, J. W., Sobeck, A., et al. (1998). Infection of mice with *Mycobacterium bovis*-Bacillus Calmette-Guerin (BCG) suppresses allergen-induced airway eosinophilia. *J. Exp. Med.* **187**, 561–569.

Fang, F. C. (1997). Perspectives series: host/pathogen interactions. Mechanisms of nitric oxide-related antimicrobial activity. *J. Clin. Invest.* **99**, 2818–2825.

Fieschi, C., Dupuis, S., Catherinot, E., et al. (2003). Low penetrance, broad resistance, and favorable outcome of interleukin 12 receptor beta 1 deficiency: medical and immunological implications. *J. Exp. Med.* **197**, 527–535.

Fieschi, C., Dupuis, S., Picard, C., et al. (2001). High levels of interferon gamma in the plasma of children with complete interferon gamma receptor deficiency. *Pediatrics* **107**, E48.

Flynn, J. L., Chan, J., Triebold, K. J., et al. (1993). An essential role for interferon gamma in resistance to *Mycobacterium tuberculosis* infection. *J. Exp. Med.* **178**, 2249–2254.

Flynn, J. L., Goldstein, M. M., Chan, J., et al. (1995). Tumor necrosis factor-alpha is required in the protective immune response against *Mycobacterium tuberculosis* in mice. *Immunity* **2**, 561–572.

Frucht, D. M. and Holland, S. M. (1996). Defective monocyte costimulation for IFN-gamma production in familial disseminated *Mycobacterium avium* complex infection: abnormal IL-12 regulation. *J. Immunol.* **157**, 411–416.

Fujita, T., Reis, L. F., Watanabe, N., et al. (1989). Induction of the transcription factor IRF-1 and interferon-beta mRNAs by cytokines and activators of second-messenger pathways. *Proc. Natl. Acad. Sci. USA* **86**, 9936–9940.

Garcia, I., Duler, R., Vesin, D., et al. (2000). Lethal *Mycobacterium bovis* Bacillus Calmette Guerin infection in nitric oxide synthase 2-deficient mice: cell-mediated immunity requires nitric oxide synthase 2. *Lab Invest.* **80**, 1385–1397.

Gollob, J. A., Veenstra, K. G., Jyonouchi, H., et al. (2000). Impairment of STAT activation by IL-12 in a patient with atypical mycobacterial and staphylococcal infections. *J. Immunol.* **165**, 4120–4126.

Gomes, M. S., Florido, M., Pais, T. F., et al. (1999). Improved clearance of *Mycobacterium avium* upon disruption of the inducible nitric oxide synthase gene. *J. Immunol.* **162**, 6734–6739.

Holland, S. M., Dorman, S. E., Kwon, A., et al. (1998). Abnormal regulation of interferon-gamma, interleukin-12, and tumor necrosis factor-alpha in human interferon-gamma receptor 1 deficiency. *J. Infect. Dis.* **178**, 1095–1104.

Holscher, C., Atkinson, R. A., Arendse, B., et al. (2001). A protective and agonistic function of IL-12p40 in mycobacterial infection. *J. Immunol.* **167**, 6957–6966.

Jouanguy, E., Altare, F., Lamhamedi, S., et al. (1996). Interferon-gamma-receptor deficiency in an infant with fatal bacille Calmette-Guerin infection. *N. Engl. J. Med.* **335**, 1956–1961.

Jouanguy, E., Lamhamedi-Cherradi, S., Altare, F., et al. (1997). Partial interferon-gamma receptor 1 deficiency in a child with tuberculoid bacillus Calmette-Guerin infection and a sibling with clinical tuberculosis. *J. Clin. Invest.* **100**, 2658–2664.

Jouanguy, E., Lamhamedi-Cherradi, S., Lammas, D., et al. (1999). A human IFNGR1 small deletion hotspot associated with dominant susceptibility to mycobacterial infection. *Nat. Genet.* **21**, 370–378.

Jouanguy, E., Doffinger, R., Dupuis, S., et al. (1999). IL-12 and IFN-gamma in host defense against mycobacteria and salmonella in mice and men. *Curr. Opin. Immunol.* **11**, 346–351.

Jouanguy, E., Dupuis, S., Pallier, A., et al. (2000). In a novel form of IFN-gamma receptor 1 deficiency, cell surface receptors fail to bind IFN-gamma. *J. Clin. Invest.* **105**, 1429–1436.

Juffermans, N. P., Florquin, S., Camoglio, L., et al. (2000). Interleukin-1 signaling is essential for host defense during murine pulmonary tuberculosis. *J. Infect. Dis.* **182**, 902–908.

Kamijo, R., Le, J., Shapiro, D., et al. (1993). Mice that lack the interferon-gamma receptor have profoundly altered responses to infection with Bacillus Calmette-Guerin and subsequent challenge with lipopolysaccharide. *J. Exp. Med.* **178**, 1435–1440.

Kamijo, R., Nakayama, K., Penninger, J., et al. (1994). Requirement for transcription factor IRF-1 in NO synthase induction in macrophages. *Science* **263**, 1612–1615.

Kaneko, H., Yamada, H., Mizuno, S., et al. (1999). Role of tumor necrosis factor-alpha in *Mycobacterium*-induced granuloma formation in tumor necrosis factor-alpha-deficient mice. *Lab Invest.* **79**, 379–386.

Kindler, V., Sappino, A. P., Grau, G. E., et al. (1989). The inducing role of tumor necrosis factor in the development of bactericidal granulomas during BCG infection. *Cell* **56**, 731–740.

Lamhamedi, S., Jouanguy, E., Altare, F., et al. (1998). Interferon-gamma receptor deficiency: relationship between genotype, environment, and phenotype. *Int. J. Mol. Med.* **1**, 415–418.

Levin, M., Newport, M. J., D'Souza, S., et al. (1995). Familial disseminated atypical mycobacterial infection in childhood: a human mycobacterial susceptibility gene? *Lancet* **345**, 79–83.

MacMicking, J. D., North, R. J., La Course, R., et al. (1997). Identification of nitric oxide synthase as a protective locus against tuberculosis. *Proc. Natl. Acad. Sci. USA* **94**, 5243–5248.

McKusick, V.A. (1998). *Mendelian Inheritance in Man: A Catalog of Human Genes and Genetic Disorders.* Baltimore: Johns Hopkins University Press.

Micallef, M. J., Ohtsuki, T., Kohno, K., et al. (1996). Interferon-gamma-inducing factor enhances T helper 1 cytokine production by stimulated human T cells: Synergism with interleukin-12 for interferon-gamma production. *Eur. J. Immunol.* **26**, 1647–1651.

Murray, P. J., Young, R. A. and Daley, G. Q. (1998). Hematopoietic remodeling in interferon-gamma-deficient mice infected with mycobacteria. *Blood* **91**, 2914–2924.

Newport, M. J., Huxley, C. M., Huston, S., et al. (1996). A mutation in the interferon-gamma-receptor gene and susceptibility to mycobacterial infection. *N. Engl. J. Med.* **335**, 1941–1949.

Okamura, H., Tsutsi, H., Komatsu, T., et al. (1995). Cloning of a new cytokine that induces IFN-gamma production by T cells. *Nature* **378**, 88–91.

Olleros, M. L., Guler, R., Corazza, N., et al. (2002). Transmembrane TNF induces an efficient cell-mediated immunity and resistance to *Mycobacterium bovis* bacillus Calmette-Guerin infection in the absence of secreted TNF and lymphotoxin-alpha. *J. Immunol.* **168**, 3394–3401.

Picard, C., Fieschi, C., Altare, F., et al. (2002). Inherited interleukin-12 deficiency: IL12B genotype and clinical phenotype of 13 patients from six kindreds. *Am. J. Hum. Genet.* **70**, 336–348.

Pierre-Audigier, C., Jouanguy, E., Lamhamedi, S., et al. (1997). Fatal disseminated *Mycobacterium smegmatis* infection in a child with inherited interferon gamma receptor deficiency. *Clin. Infect. Dis.* **24**, 982–984.

Reichenbach, J., Rosenzweig, S., Doffinger, R., et al. (2001). Mycobacterial diseases in primary immunodeficiencies. *Curr. Opin. Allergy Clin. Immunol.* **1**, 503–511.

Roesler, J., Kofink, B., Wendisch, J., et al. (1999). *Listeria monocytogenes* and recurrent mycobacterial infections in a child with complete interferon-gamma-receptor (IFNgammaR1) deficiency: mutational analysis and evaluation of therapeutic options. *Exp. Hematol.* **27**, 1368–1374.

Roach, D. R., Bean, A. G., Demangel, C., et al. (2002). TNF regulates chemokine induction essential for cell recruitment, granuloma formation, and clearance of mycobacterial infection. *J. Immunol.* **168**, 4620–4627.

Rook, G. A., Taverne, J., Leveton, C., et al. (1987). The role of gamma-interferon, vitamin D3 metabolites and tumour necrosis factor in the pathogenesis of tuberculosis. *Immunology* **62**, 229–234.

Rosenzweig, S., Dorman, S. E., Roesler, J., et al. (2002). 561del4 defines a novel small deletion hotspot in the interferon-gamma receptor 1 chain. *Clin. Immunol.* **102**, 25–27.

Sakai, T., Matsuoka, M., Aoki, M., et al. (2001). Missense mutation of the interleukin-12 receptor beta1 chain-encoding gene is associated with impaired immunity against *Mycobacterium* avium complex infection. *Blood* **97**, 2688–2694.

Scanga, C. A., Mohan, V. P., Tanaka, K., et al. (2001). The inducible nitric oxide synthase locus confers protection against aerogenic challenge of both clinical and laboratory strains of *Mycobacterium tuberculosis* in mice. *Infect. Immun.* **69**, 7711–7717.

Senaldi, G., Shklee, C. L., Guo, J., et al. (1996). *Corynebacterium parvum-* and *Mycobacterium bovis* bacillus Calmette-Guerin-induced granuloma formation is inhibited in TNF receptor I (TNF-RI) knockout mice and by treatment with soluble TNF-RI. *J. Immunol.* **157**, 5022–5026.

Sugawara, I., Yamada, H., Kaneko, H., et al. (1999). Role of interleukin-18 (IL-18) in mycobacterial infection in IL-18-gene-disrupted mice. *Infect. Immun.* **67**, 2585–2589.

Sugawara, I., Yamada, H., Hua, S., et al. (2001). Role of interleukin (IL)-1 type 1 receptor in mycobacterial infection. *Microbiol. Immunol.* **45**, 743–750.

Verhagen, C. E., de Boer, T., Smits, H. H., et al. (2000). Residual type 1 immunity in patients genetically deficient for interleukin 12 receptor beta1 (IL-12Rbeta1): evidence for an IL-12Rbeta1-independent pathway of IL-12 responsiveness in human T cells. *J. Exp. Med.* **192**, 517–528.

Villella, A., Picard, C., Jouanguy, E., et al. (2001). Recurrent *Mycobacterium avium* osteomyelitis associated with a novel dominant interferon gamma receptor mutation. *Pediatrics* **107**, E47.

Wakeham, J., Wang, J., Magram, J., et al. (1998). Lack of both types 1 and 2 cytokines, tissue inflammatory responses, and immune protection during pulmonary infection by *Mycobacterium bovis* bacille Calmette-Guerin in IL-12-deficient mice. *J. Immunol.* **160**, 6101–6111.

Yamada, H., Mizumo, S., Horai, R., et al. (2000). Protective role of interleukin-1 in mycobacterial infection in IL-1 alpha/beta double-knockout mice. *Lab Invest.* **80**, 759–767.

CHAPTER 10

Mannose-binding lectin deficiency and susceptibility to infectious disease

Dominic L. Jack

Institute of Child Health, University College London, United Kingdom; Division of Genomic Medicine, University of Sheffield Medical School, Sheffield, United Kingdom

Nigel J. Klein and Malcolm W. Turner

Institute of Child Health, University College London, United Kingdom

1. INNATE IMMUNITY

The innate immune system is a set of cellular and humoral components which recognise the general features of microbes in order to clear these potentially damaging agents from the body. In contrast to the acquired immune system this does not require prior exposure to the infectious agent. The development of the innate immune system as a range of pattern recognition receptors (PRRs) designed to recognise pathogen-associated molecular patterns (PAMPs) is a response to the host's inability to store unique recognition molecules for every possible pathogen within the genome.

Mannose-binding lectin (MBL) is a part of the humoral innate immune system. It is a pattern-recognition molecule able to detect a wide range of microbial and altered self-targets and recruit a number of host immune effector systems to clear those targets (Turner, 1996). However, genetic deficiency of MBL is surprisingly common in most human populations (Turner and Hamvas, 2000).

The protein was first discovered through biochemical purification and the gene independently described after functional cloning by two research teams. The effects of human MBL deficiency were documented separately by these research efforts and led in due course to the discovery of the genetic polymorphisms that give rise to this deficiency. More recently, the details of disease susceptibilities and the mechanisms of these effects have been elucidated. This review will concentrate on the role of MBL in determining susceptibility to infectious disease, starting with the initial perturbations in immune function caused by MBL deficiency before considering some specific cases where MBL status apparently determines susceptibility to infectious disease.

2. THE COMMON OPSONIC DEFECT

The story of the discovery and characterisation of the common opsonic defect and eventually MBL deficiency is often traced back to 1968 when an article by Miller et al. described the case of a female patient, who had suffered from recurrent upper respiratory tract infections and diarrhoea in the first 2 years of life. The patient did not appear to have a deficiency of immunoglobulin but, in the presence of the patient's serum, neutrophils from human donors were unable to phagocytose *Saccharomyces cerevisiae* (Baker's yeast) efficiently. The patient's neutrophils were able to phagocytose efficiently when the serum of other donors was used in the same assay. There was a familial association of the opsonic defect, although the other family members were healthy. The patient was treated with plasma infusions, which ameliorated the condition. It was curious, therefore, that despite the deleterious condition described in this little girl, this opsonic defect could be found in a high proportion (5–8%) of apparently healthy populations (Soothill and Harvey 1976). The frequency of this defect was higher in children with recurrent unexplained infections, but the defect was also present in healthy adults with no history of persistent infection.

The complement system is one of the effector arms of the immune system. In the early 1980s, it was known to be activated by two or more immunoglobulin G molecules or one IgM molecule bound to a target surface (classical pathway) or by the binding of hydrolysed C3 to a permissive surface such as a bacterium (alternative pathway). Activation in this manner would lead to opsonisation of the target by the amplified deposition of multiple opsonic C3 fragments or the direct lysis of Gram-negative bacteria by the formation of a macromolecular complex, the membrane attack complex (for a review see Walport, 2001a, 2001b). Since a gross immunoglobulin defect did not appear to be the cause of the opsonic defect and in any case would be unlikely in such a large proportion of the population, a study of complement deposition was begun. Biochemical assays using the D-mannose polymer zymosan confirmed that the absence of a serum factor led to the poor deposition of C3b and iC3b on the yeast surface (Turner et al., 1981, 1986) without deficiency of any known component of either the alternative or classical pathways of complement activation (Turner et al., 1985).

3. MANNOSE-BINDING LECTIN

Lectins are sugar-binding proteins which are not immunoglobulins. In the late 1970s, proteins binding to the yeast cell wall components, mannan

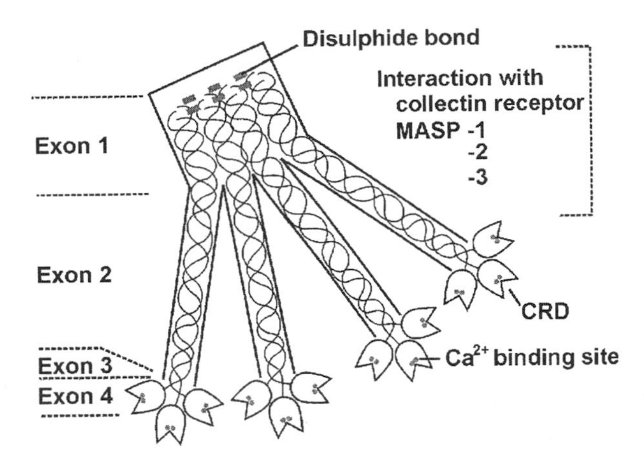

Figure 10.1. Major structural features of tetrameric human mannose-binding lectin. In each of the four subunits, three identical peptide chains of 32-kDa associate to give a collagenous triple helix. In the C-terminal region the three chains are independently folded to give C-type lectin domains. The MBL-associated serine protease (MASP) family of proteins interact with the collagenous region and MASP-2 appears to be the most important in promoting complement activation. The regions of the protein encoded by the four exons of the human MBL gene are indicated. All three known structural gene mutations are localised to the region encoded by exon 1.

or D-mannose, were found in the liver and serum of rabbits (Kawasaki et al., 1978; Kozutsumi et al., 1980) and later in rats (Mizuno et al., 1981) and humans (Kawasaki et al., 1983; Wild et al., 1983). This protein was variously called mannose- or mannan-binding protein (MBP), but is now more commonly described as mannose- or mannan-binding lectin or MBL. The protein is synthesised in the liver in humans and is found mainly in serum.

MBL has a bouquet-like structure reminiscent of C1q with various oligomeric structures (dimers, trimers, tetramers, hexamers) but it is still unclear which is the predominant circulating form (Lu et al., 1990). Full functional activity, including both binding to microbial surfaces and the activation of complement requires higher order structures such as tetramers (Yokota et al., 1995).

All higher order structures of MBL are based on trimeric subunits comprising three identical peptide chains of 32 kDa. Each chain is characterised by a carbohydrate recognition or lectin domain (CRD), a coiled-coil hydrophobic neck region, a collagenous region, and finally a cysteine rich N-terminal region (Sastry et al., 1989; Taylor et al., 1989). Three such chains interact to give a classical collagenous triple helix (Weis and Drickamer, 1994) (see Fig. 10.1).

MBL is a calcium-dependent (or C-type) lectin which makes coordination bonds with the 3- and 4-hydroxyl groups of various sugars, including mannose, N-acetyl-D-glucosamine, N-acetyl-mannosamine, fucose, and glucose. In contrast, the sugar D-galactose does not bind significantly (Drickamer, 1992; Weis et al., 1992). It is believed that the sugar group patterns decorating microbial surfaces make particularly appropriate targets for binding since the three sugar binding sites of one subunit array offer a flat platform with a constant distance between the sites [45Å in human (Sheriff et al., 1994); 54 Å separation in rat (Weis and Drickamer, 1994)]. Simultaneous, multiple binding is required because the K_d of each separate MBL–sugar interaction is relatively low (10^{-3}M) (Iobst et al., 1994). These features facilitate interaction with microbial surfaces but minimise the chances of disadvantageous self-recognition and emphasise the role of the protein as a pattern recognition molecule.

In the circulation, MBL is found in association with four proteins structurally related to each other. These are the MBL-associated serine proteases 1, 2, and 3 (MASP-1-3) (Dahl et al., 2001; Matsushita and Fujita, 1992; Thiel et al., 1997) and a truncated version of MASP-2 called MAp 19 (Stover et al., 1999; Takahashi et al., 1999). The stoichiometry of the MBL–MASP interaction is unclear, but it appears that MASP-2 appears to be the most important in complement activation (Thiel et al. 1997). The evidence available suggests that MBL–MASP2 complexes become activated to cleave C4 and C2 in an identical manner to C1 esterase when bound to appropriate sugar arrays on microbial surfaces (Thiel et al., 2000). The C4b fragments generated bind covalently either to the nearby microbial surface or possibly to the lectin itself and then act as a focus for C2 binding/activation. The resultant C4b2a complex has C3 convertase activity and cleaves C3 molecules in a similar manner to the C3 convertases of both the classical and alternative pathways of complement activation (see Fig. 10.2). In addition, there is evidence that MASP-1 is able to cleave C3 directly (Matsushita and Fujita, 1995), although apparently at low efficiency.

MBL may also interact directly with cell surface receptors and thereby promote opsonophagocytosis and other immune processes. This property was first reported by Kuhlman et al. (1989) in a study of MBL coated *Salmonella enterica* serovar *Montevideo* organisms. Subsequently, a number of putative MBL binding proteins/receptors have been proposed, including cC1qR/calreticulin (Malhotra et al., 1994), C1qR_p/CD93 (Steinberger et al., 2002; Tenner et al., 1995), and CR1/CD35 (Ghiran et al., 2000). However, it is unclear whether MBL is acting as a direct opsonin for microorganisms (Kuhlman et al., 1989) or is enhancing other pathways such as complement- or immunoglobulin-receptor-mediated phagocytosis (Tenner et al., 1995).

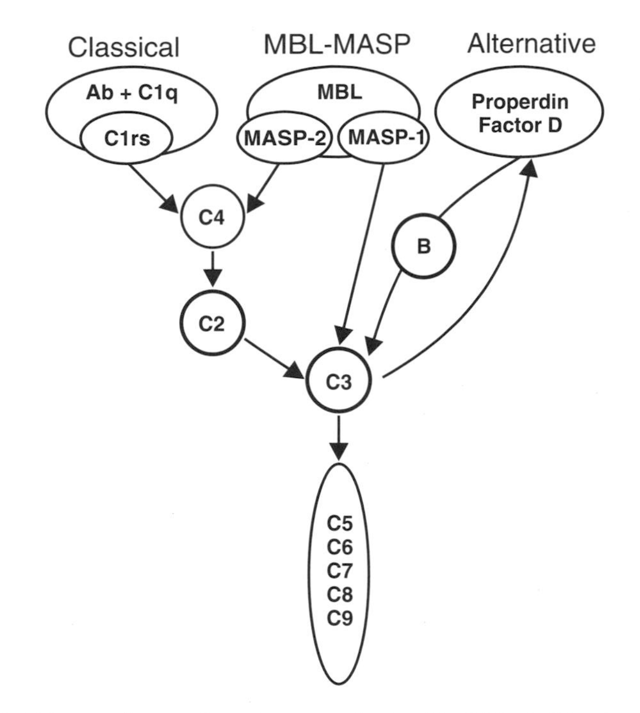

Figure 10.2. Three pathways of complement activation. The MBL-MASP pathways provide antibody and C1-independent mechanisms for generating C3 convertase enzymes. The major MBL initiated pathway involves the activation of MASP-2 and the sequential cleavage of C4 and C2. MASP-1 is believed to provide direct cleavage of C3 but probably at low efficiency. P, properdin; B, Factor B.

Indeed, it is still unclear whether MBL receptors are distinct from receptors for the structurally similar C1q molecule (Bajtay et al., 2000).

4. MBL DEFICIENCY AND GENETICS

MBL appeared to be an ideal candidate molecule for the serum factor deficient in patients with the common opsonic defect. Indeed, it was demonstrated these individuals were deficient in MBL and that the addition of purified MBL to the serum of such patients resulted in increased deposition of C3b, Factor B, and C4 on zymosan (Super et al., 1989). The basis of low MBL levels remained to be determined, but the familial associations that had been noted previously suggested a genetic cause.

The gene for human MBL was cloned by two groups using strategies similar to those used for the cloning of the two rat MBL genes (Drickamer

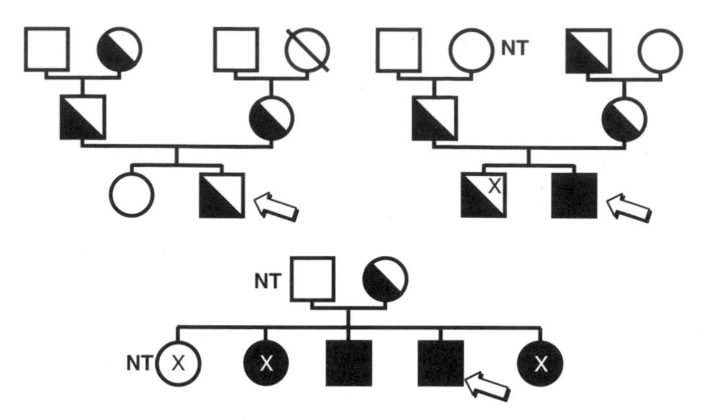

Figure 10.3. MBL B variant inheritance patterns in three families with an infant presenting with frequent unexplained infections. Probands are shown by arrows. Individuals wild type for the B variant (codon 54) allele are indicated by open symbols, those heterozygous for the B variant allele are shown as half symbols and those homozygous for the B variant are shown as filled symbols. It should be noted that this genotyping predated the discovery of the C and D variants and promotor polymorphisms. NT, not tested; X, individuals also affected by recurrent infections.

et al. 1986). In the rat, a deduced DNA sequence was made from the known polypeptide sequence of the CRD. This was used to probe a rat liver cDNA library and further probes based on the cDNA sequence were then used to interrogate a genomic library to identify the coding sequence.

The same rat cDNA was used by Ezekowitz et al. (1988) to probe a human liver plasmid cDNA library with the identification of the cDNA encoding human MBL, demonstrating that the rat MBL (MBL-C) carbohydrate recognition domain was highly homologous to human MBL. Probes were constructed by two groups using the cDNA of the CRD to identify genomic clones in a λphage library (Sastry et al., 1989) and a cosmid library (Taylor et al., 1989). The single human gene was localised to a region of chromosome 10 (10q11.2-q21) (Sastry et al., 1989), a region which also contains genes for the closely related molecules lung surfactant proteins- (SP) A and SP-D.

The human MBL gene consists of four exons (see Fig. 10.3). The first exon encodes the signal peptide, the N-terminal cysteine-rich region, and the first part of the collagenous domain (characterised by Gly-X-Y repeats). Exon 2 encodes the remainder of the collagenous domain and exon 3 encodes the α-coiled-coil neck region. The entire C-type lectin domain is contained within exon 4. Upstream, there are a number of regulatory elements. The TATAA and CAAT boxes are found at −38 and −79 bp respectively with a heat-shock consensus sequence at −592. There are glucocorticoid-responsive elements

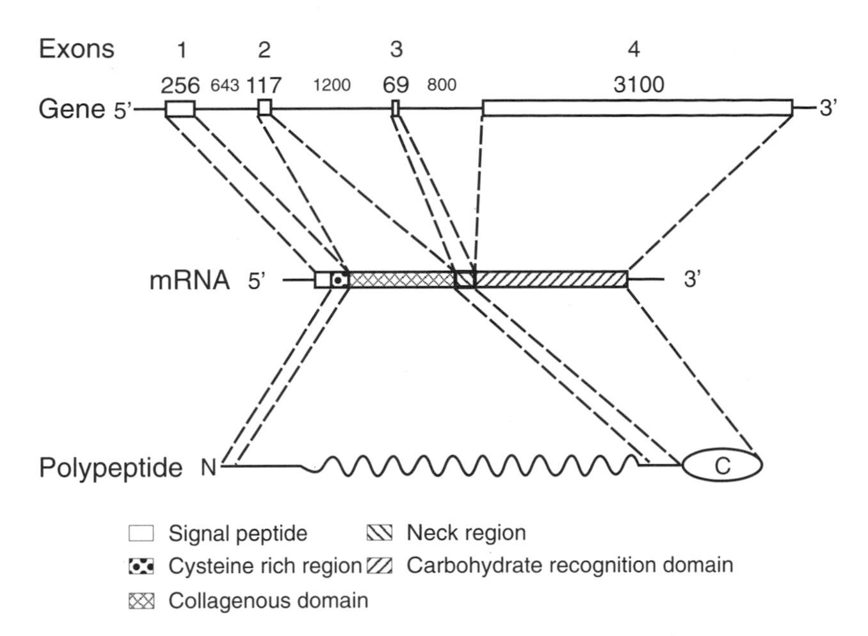

Figure 10.4. Structure of the human MBL gene, the corresponding mRNA and the translated protein domains. The single expressed functional human MBL gene is located at 10q11.2-q21 and consists of four exons interrupted by introns of 643, 1,200, and 800 bp respectively (see top of figure). The mRNA (centre) encodes for the various protein domains indicated at the bottom of the figure.

at −245, −656, and −736 bp and a region from −204 to −184 which has 90% homology to the gene for the acute-phase reactant, serum amyloid A. These regulatory elements suggest that MBL may be regulated as an acute phase reactant (Ezekowitz et al., 1988; Taylor et al., 1989). However, protein level data suggests that MBL levels increase by only 2–3 times in acute phase responses (Thiel et al., 1992).

To identify possible mutations in MBL deficient patients, the gene from three families affected by the opsonic defect was sequenced, starting at exon 4 and working backwards (see Fig. 10.4). The rationale for such an approach was that the CRD was thought to be the most likely mutated site. However, the mutation in these families was actually located in exon 1.

To date, three structural mutations resulting in amino acid substitutions (Lipscombe et al., 1992b; Madsen et al., 1994; Sumiya et al., 1991) have been identified in the human MBL gene. Mutations at codons 54 and 57 (B and C mutations) result in the substitution of glycine for a large dicarboxylic acid residue (Lipscombe et al., 1992b; Sumiya et al., 1991). Recombinant MBL molecules based on these mutations are thermodynamically less

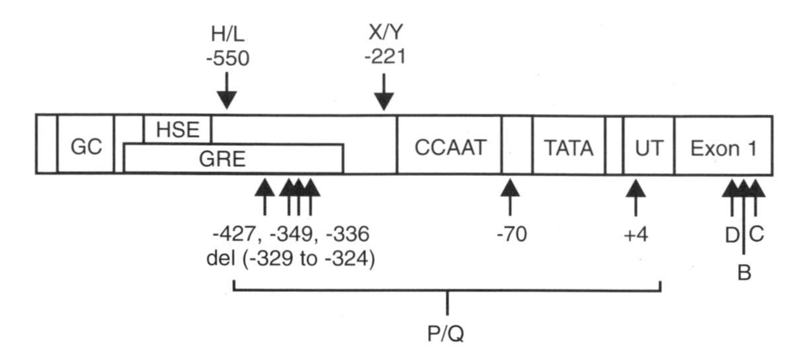

Figure 10.5. Polymorphic sites in the promoter region and exon 1 of the functional human MBL gene. The D, B, and C variants in exon 1 correspond to single point mutations in codons 52, 54, and 57. The H/L, X/Y, and P/Q polymorphic sites are in linkage disequilibrium with the three exon 1 variants and give rise to the haplotypes HYPD, LYPB, and LYQC respectively. The extended wild-type haplotype LXPA is also associated with low expression of MBL and together with heterozygous B, C, and D variants can lead to profoundly reduced levels of the protein. TATAA, CAAT, and GC represent the relevant box sequences; HSE, heat shock element; GRE, glucocorticoid-responsive element.

stable (Super et al., 1992), do not form higher order oligomers (Wallis and Cheng, 1999) and are more slowly secreted than the normal molecules (Heise et al., 2000; Wallis and Cheng, 1999). Additionally, these recombinant mutant forms of MBL are unable to interact with MASPs to activate complement (Super et al., 1992; Wallis and Cheng, 1999). An additional mutation at codon 52 (D mutation) has also been described, but this differs from the others in that a cysteine is inserted into the collagenous region (Madsen et al., 1994). Cysteine normally takes part in the formation of disulphide bonding and the insertion of an extra cysteine appears to result in the incorrect polymerisation of the MBL subunit (Wallis and Cheng, 1999).

These structural mutations appear to be inherited in a codominant fashion. Heterozygotes for MBL mutations have much reduced serum levels of MBL compared to homozygous normal wild-type individuals. Homozygosity for the structural gene mutations appears to be associated with almost absent serum MBL. Although MBL variant alleles are common throughout the world, the B mutation is found in populations of European and Asian origin, whereas the C mutation is found in sub-Saharan Africa. It appears likely that these MBL mutations have arisen independently and that both alleles have been positively selected in most populations (Turner and Hamvas, 2000).

In addition to the structural mutations, a number of polymorphisms within the gene promoter have also been identified which have an effect on protein levels (see Fig. 10.5) (Madsen et al., 1995, 1998).

The H/L (G or C) polymorphism is at −550 bp, the X/Y (C or G) polymorphism is at −221 bp, and the P/Q polymorphisms are at −427 (A or C), −349 (A or G), del (−329 to −324), −70 (C or T), and at +4 (C or T), respectively. Four promoter haplotypes are commonly found: LXP, LYP, LYQ, and HYP. The X/Y polymorphism is thought to have the most profound effect on protein levels, with the HYP haplotype, followed by LYQ and then LYP associated with the highest levels of MBL protein. The LXP haplotype is associated with the lowest levels. A homozygous HYP individual with normal MBL structural genotype has a protein level some six times higher than a homozygous LXP individual (Madsen et al., 1998).

5. MBL IN INFECTIOUS DISEASE

The significance of MBL in infectious disease was apparent from early reports in the literature detailing the role of the protein in recognising certain forms of *Salmonella enterica* serovar *typhimurium*. Mice are not normally susceptible to infection by *Salmonella* expressing Ra-chemotype lipopolysaccharide (LPS) and an Ra-reactive factor (RaRF) was identified in immunologically naïve mice which could bind to this chemotype. Ra-chemotype LPS are truncated forms which lack the O-antigen and are often termed 'rough' mutants due to their effect on microorganism colony morphology. RaRF was able to activate complement by the classical pathway in the absence of immunoglobulin and could be inhibited by mannose and *N*-acetyl-D-glucosamine (Ihara et al., 1982; Kawakami et al., 1982) and was subsequently shown to be identical to mouse mannose binding lectin (Matsushita et al., 1992).

More recently we have assayed the binding of MBL to a wide range of aerobic and anaerobic bacteria, some fungi, and protozoa by flow cytometry (Kelly et al., 2000; Neth et al., 2000; Townsend et al., 2001). As shown in Table 10.1, these studies and the work of others have shown that MBL binds to a wide variety of infectious bacteria, viruses, fungi, protozoa, and helminths (Emmerik et al., 1994; Hartshorn et al., 1993; Haurum et al., 1993; Klabunde et al., 2000; Schelenz et al., 1995).

6. MBL DISEASE ASSOCIATIONS

In the decade following the identification of the first polymorphism in the MBL gene, epidemiology has associated MBL deficiency with altered susceptibility to many significant infections (see Table 10.2). It should be noted that in a number of cases these associations have been controversial with other studies presenting conflicting evidence for the role of MBL. The

Table 10.1. *The binding of MBL to intact microorganisms*

	Organisms Positive for MBL Binding
Bacteria	*Actinomyces israelii, Bacteroides* sp., *Bifidobacterium bifidum, Burkholderia cepacia, Escherichia coli, Eubacterium* sp., *Fusobacterium* sp., *Haemophilus influenzae, Klebsiella* sp., *Listeria monocytogenes, Mycobacterium avium, Neisseria cinerea, Neisseria gonorrhoeae* and *Neisseria meningitidis* serogroup B and C (LOS non-sialylated), *Neisseria meningitidis* serogroup A, *Neisseria subflava, Proprionobacterium acnes, Salmonella montevideo, Salmonella typhimurium* (Ra chemotype), *Staphylococcus aureus,* β-haemolytic *Streptococcus group* A, *Streptococcus pneumoniae, Streptococcus suis, Veillonella* sp.
Yeasts and fungi	*Aspergillus fumigatus, Candida albicans, Cryptococcus neoformans* (unencapsulated)
Viruses	Influenza A, HIV-1
Other parasites	*Cryptosporidium parvum, Schistosoma mansoni*
	Organisms Negative or Low for MBL Binding
Bacteria	*Clostridium* sp., *Enterococcus, Neisseria meningitidis* serogroup B and C and *Neisseria gonorrhoeae* (LOS sialylated), *Neisseria mucosa, Pseudomonas aeruginosa, Salmonella typhimurium* (smooth chemotype), *Staphylococcus epidermidis,* β-haemolytic *Streptococcus* group B, *Streptococcus agalactiae, Streptococcus sanguis*
Fungi	*Cryptococcus neoformans* (encapsulated)

Note: Data from Anders et al. (1994), Devyatyarova-Johnson et al. (2000), Emmerik et al. (1994), Jack et al. (1998, 2001a), Kelly et al. (2000), Klabunde et al. (2000), Polotsky et al. (1997), Saifuddin et al. (2000), Schelenz et al. (1995), Townsend et al. (2001).

effect of MBL deficiency may also be more apparent in the presence of concomitant immunodeficiencies such as immunoglobulin subclass deficiency (Aittoniemi et al., 1998).

In combination with *in vitro* data we now have an ever-refining view of the role of this protein in infectious disease. We will consider three cases where epidemiology is matched by experimental data on the role of MBL in infectious disease.

Table 10.2. *Publications in which associations between MBL status and infectious disease susceptibility have been investigated*

Disease/causative agent	Increased risk associated with MBL deficiency	Decreased risk associated with MBL deficiency	No association
HIV	Garred et al. (1997a), Maas et al. (1998), Nielsen et al. (1995), Prohaszka et al. (1997)		
Mycobacterium tuberculosis	Selvaraj et al. (1999)	Garred et al. (1994), Garred et al. (1997b)	Bellamy et al. (1998)
Neisseria meningitidis	Bax et al. (1999), Hibberd et al. (1999)		Garred et al. (1993)
Generalised risk of infection	Garred et al. (1995), Summerfield et al. (1995), Summerfield et al. (1997)		
Malaria	Luty et al. (1998)		Bellamy et al. (1998)
Cryptosporidium parvum	Kelly et al. (2000)		
Leishmania		Santos et al. (2001)	
Chronic necrotising aspergillosis	Crosdale et al. (2001)		
Hepatitis B	Thomas et al. (1996)		Bellamy et al. (1998), Hohler et al. (1998)
Hepatitis C	Matsushita et al. (1998)		
Interferon treatment resistance			
Infections with cystic fibrosis	Garred et al. (1999)		

6.1 Meningococcal Disease

Neisseria meningitidis is the most common cause of bacterial meningitis in countries offering vaccination against *Haemophilus influenzae* type B. It is an exclusively human pathogen spread by close contact and is commonly present in the nasopharynx of up to 20% of the human population (Cartwright, 1995). In susceptible individuals, meningococci migrate from the nasopharynx to the vascular system, where they employ a number of mechanisms to avoid destruction by the immune system. Organisms may migrate to the central nervous system causing inflammation of the meninges (meningitis) or to synovial joints. In the absence of a positive blood culture for meningococcus, mortality is 2% in these focal infections. The major problem with meningococcal disease is when the organism is readily cultured from blood and the patient is affected by sepsis syndrome. In these cases, the average mortality rises to 12% or higher depending on the study population and is highly dependent on the speed with which the subject is hospitalised and receives antibiotics (Rosenstein et al., 1999). These patients present with excessive activation of the inflammatory cascade, disseminated intravascular coagulation, and hypovolaemia due to injury to the vascular endothelium (Deuren et al., 2000).

The main correlate of protection appears to be the bactericidal activity of serum, thought to be initiated by the presence of bactericidal antibodies. In contrast to many other organisms, *N. meningitidis* does not appear to be killed effectively by blood phagocytes (Fothergill and Wright, 1933). The major burden of meningococcal disease falls on children below the age of 1 year. This appears to correlate with the low prevalence in this population of antibodies able to activate the complement system following the decay of maternal antibodies specific for the organism (Cartwright, 1995). Despite low antibody protection and high population carriage, the incidence of meningococcal disease remains low (Rosenstein et al., 1999).

The importance of the complement system is underlined by the susceptibility to meningococcal infection of individuals deficient in complement components. Deficiencies in the terminal complement components (C6 to C9) results in increased susceptibility to infection but reduced severity of infection and is often suspected in individuals who have had multiple episodes of meningococcal infection (Figueroa et al., 1993). Individuals who are deficient in the alternative pathway protein properdin present later in life with meningococcal disease and appear less likely to survive the disease although some patients with recurrent disease have been reported (Fijen et al., 1999).

The rarity of complement deficiencies had suggested that they are unlikely to contribute significantly to the total incidence of meningococcal disease. MBL is present in serum soon after birth and rapidly reaches adult levels (Aittoniemi et al., 1996; Lau et al., 1995; Terai and Kobayashi, 1993; Thiel et al., 1995) and is therefore normally present in the age group of individuals with the highest incidence of meningococcal disease (Rosenstein et al., 1999). This suggested that MBL deficiency, which occurs at high frequency, might contribute significantly to the incidence of *N. meningitidis* disease.

A retrospective analysis of Norwegian army recruits and young adults (Garred et al., 1993) investigated MBL levels in individuals who had received an experimental vaccine against one serogroup of meningococcus, but had gone on to develop the disease. Convalescent serum from 99 survivors was compared to a control group of 40 healthy blood donor controls. There was no significant difference between the levels of MBL in patients and controls. There was also no increase in the frequency of individuals with very low levels of MBL (<0.1 μg/ml and therefore likely to be homozygous for variant alleles) and it was concluded that MBL was not associated with susceptibility to meningococcal disease.

A second population study was conducted in the UK (Hibberd et al., 1999). This study analysed MBL genotypes in a prospective hospital population of 194 patients and a retrospective population of 72 survivors of meningococcal disease. These were compared to 272 patients with noninfectious illnesses or 110 healthy controls. It was found that the overall frequency of individuals homozygous for MBL variant alleles in these populations was increased from 1.5 to 7.7% in the hospital study with an odds ratio of 6.5 (2.0–27.2, 95% confidence interval). In the community study the frequency of homozygous variants was increased from 2.7 to 8.3% with an odds ratio of 4.5 (0.9–29.1, 95% confidence interval). It was concluded that MBL variant alleles might be implicated in up to 32% of meningococcal cases. There was apparently a trend towards less severe disease in homozygous carriers of MBL mutant alleles, but this was not significant and remains to be confirmed in a larger study. Some support for these observations comes from the study of an extensive family pedigree in which MBL homozygosity was associated with susceptibility to meningococcal infection (Bax et al., 1999). However these observations are still controversial, since as pointed out by Tang and Kwiatkowski (1999), the codon 52 and 54 mutation frequencies were rather low in the noninfectious admission control group used by Hibberd et al. (1999). Further epidemiological studies will be required to define the role of MBL in meningococcal disease.

We have studied the molecular interactions of MBL and *Neisseria meningitidis* to define the mechanisms behind the possible association of MBL deficiency with meningococcal infection. We first addressed whether MBL was able to bind to the organism and what structural surface features determined binding. Using flow cytometry to detect MBL binding, we have examined the interplay between lipooligosaccharide (LOS) structure and capsule. The LOS of *Neisseria* are relatively small and lack O antigens (Schneider et al., 1991), but are often terminated in a single sialic acid residue which reduces the activity of the alternative pathway of complement activation (Jarvis and Vedros, 1987) by masking the terminal parts of the LOS (Estabrook et al., 1997).

We found that the absence of sialic acid from the LOS of *Neisseria meningitidis* serogroup B (Jack et al., 1998), serogroup C (Jack et al., 2001a), and the closely related organism *Neisseria gonorrhoeae* (Devyatyarova-Johnson et al., 2000) allowed MBL to bind to the bacterium. MBL appeared to bind very poorly or not at all to organisms with sialylated LOS. Sialic acid and the next carbohydrate exposed (galactose) are not MBL ligands (Weis et al., 1992). Removal of sugars within the LOS to which MBL is able to bind resulted in no change in binding to whole *N. gonorrhoeae* (Devyatyarova-Johnson et al., 2000). Such observations suggest that other factors must be important for MBL binding apart from simple composition and linear structure. This would be consistent with the role of MBL as a pattern recognition molecule rather than as a molecule simply recognising sugar composition.

We found that for *N. meningitidis* B1940, encapsulation appeared to have only a minor effect in reducing MBL binding (Jack et al., 1998). However, others have reported that encapsulation of two serogroup B meningococci (H44/76 and 2996) had an important role in reducing MBL binding, although the effect was less pronounced than with *Haemophilus influenzae* type b (Emmerik et al., 1994). Such discrepancies may reflect a difference in sensitivity between radioactive and flow cytometric protocols or may indicate a strain (and species) dependency of the effect of the capsule. Certainly, encapsulation of other organisms such as the yeast *Cryptococcus neoformans* does appear to reduce MBL binding (Schelenz et al., 1995). Further investigations will be required to identify whether encapsulation has a general or more specific effect on binding.

We next considered the influence of MBL on complement activation and killing. MBL is usually copurified in a complex with activated MASPs (Thiel et al., 2000). By incubating organisms with first MBL and then purified C4, we have detected activation of C4 on a number of different organisms, including *Neisseria meningitidis, Staphylococcus aureus, Burkholderia cepacia,* and *Cryptosporidium parvum,* suggesting that the lectin binding is functional (Davies et al., 2000; Jack et al., 1998; Kelly et al., 2000; Neth et al., 2000). More recently, we have assayed the effect of purified MBL on complement activation on *Neisseria meningitidis* in MBL-deficient serum and found that the protein increased the rate of activation, but not necessarily the total amount

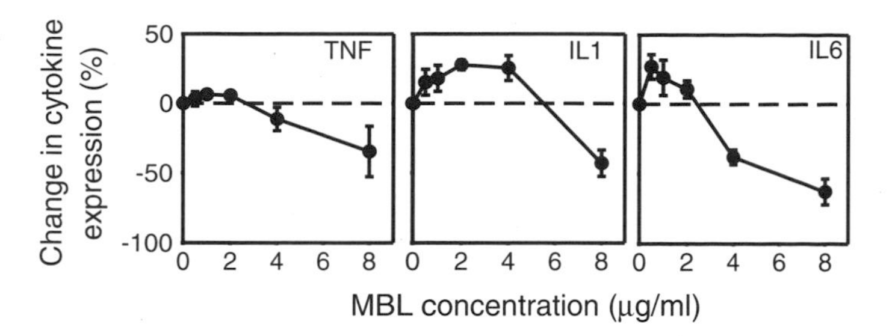

Figure 10.6. MBL-mediated modulation of TNF-α, IL1-β, and IL6 production by monocytes. The production of TNF-α, IL6 and IL1-β by monocytes in whole blood was assayed by flow cytometry and the effect of adding differing concentrations of purified MBL to MBL-deficient blood in response to meningococcus has been calculated as a percentage of the effect of no added MBL. Adapted from Jack et al. (2001b).

of complement activated. However, this did lead to significant increases in killing, which were greater when the organisms lacked LOS sialylation (Jack et al., 2001a).

Although neutrophils are not important in killing meningococci, these and other phagocytic cells are important in the clearance of killed organisms from the host and the activation of these cells is likely to be important in the pathogenesis of sepsis syndrome. With its potential role as a direct opsonin for Gram-negative bacteria, we have looked at the effect that MBL may have on the responses of phagocytic cells to meningococcus. We found that purified MBL enhanced the phagocytosis of meningococci by neutrophils, monocytes, and monocyte-derived macrophages (Jack et al., 2001b). In the case of macrophages, this appears to occur through an increase in the internalisation, but not the binding, efficiency. We presume that other receptors must be important in the initial attachment of these organisms to the phagocyte and that MBL is one of the signals involved in the internalisation.

The receptors involved in the internalisation of a particle are likely to have consequences in terms of the intracellular fate of that particle and the response of the cell. We have found that MBL down-regulates the normal changes in adhesion molecule expression by neutrophils and the concentration at which this occurs in whole blood, 8 μg/ml, was also associated with a reduction in the production by monocytes of the proinflammatory cytokines TNFα, interleukin (IL)-1β, and IL6 (Jack et al., 2001b). However, as shown in Fig. 10.6, lower concentrations of MBL in whole blood (<4 μg/ml) stimulate the production of both IL1β and IL6, suggesting that in the absence of an acute phase response, the lectin has a selective proinflammatory effect. In an acute phase response, MBL may depress the production of proinflammatory cytokines. It should be noted that other groups have not reported this change in the induction of proinflammatory cytokines and have observed consistent increases in TNFα with increasing levels of MBL in response to *Leishmania* (Santos et al. 2001), possibly indicating that MBL has different effects on different pathogens.

6.2 HIV Infection

There has been substantial interest in the role that MBL may play in infection by the human immunodeficiency virus (HIV) arising from a report early in the chronology of MBL publications (Ezekowitz et al., 1989), which showed that purified MBL was able to bind to HIV-infected cell lines and directly inhibit HIV infection of lymphoblasts. MBL binds to and activates complement on gp120 (Haurum et al., 1993) which is rich in mannose residues and is critical for interactions of the virus with the cell-surface marker, CD4, on the surface of the T cells targeted by the virus. MBL binds to both CCR5 and CXCR4 tropic primary isolates of whole virus (Saifuddin et al., 2000).

Despite this impressive *in vitro* data, epidemiological studies have been less conclusive. The complex pathogenesis of HIV infection means that two main questions have been addressed. First, does MBL deficiency lead to enhanced susceptibility to HIV infection; second, is there a correlation between the course/severity of HIV infection and MBL status?

There is broad agreement that MBL deficiency, assayed by genotype or phenotype, is associated with a greater susceptibility to HIV acquisition. MBL deficiency increased the acquisition of HIV infection by between three- and eightfold (Garred et al., 1997a, 1997b; Nielsen et al., 1995; Prohaszka et al., 1997) and increased the risk of vertical transmission from infected mothers to their offspring (Boniotto et al., 2000). However, a minority of studies have failed to demonstrate a role for MBL in HIV infection (McBride et al., 1998; Senaldi et al., 1995).

There is less clarity with regard to the role of MBL in HIV disease progression. Garred et al. (1997a) demonstrated that men with MBL variant alleles had a shorter survival time following the onset of acquired immune deficiency syndrome (AIDS) than did patients with wild-type MBL alleles. However, in a well-characterised cohort of homosexual men, variant MBL alleles had an insignificant effect on survival following the diagnosis of AIDS (Maas et al., 1998). In this latter study there appeared to be a protective effect of MBL variant alleles with a delay in the development of AIDS *from the time of HIV seroconversion*. Patients with MBL variant alleles had lower CD4 counts at the time of developing AIDS, indicating that MBL deficiency may influence the onset of AIDS for any given CD4 count. Furthermore, MBL mutations appeared to protect against the development of Kaposi sarcoma, a finding that was difficult to explain (Maas et al., 1998). Prohaszka et al. (1997) found that MBL levels were lower in asymptomatic HIV-positive individuals when compared with HIV-negative controls. However, the protective effect of MBL was lost in patients with an AIDS diagnosis; patients with high MBL levels had significantly lower numbers of CD4 cells. A possible explanation is that enhanced proinflammatory cytokine production in advanced HIV disease acts to increase MBL synthesis (Arai et al., 1993), elevating levels in patients with late-stage disease. In the light of studies indicating a role for MBL in inflammatory modulation, it is tempting to suggest that under some circumstances MBL may act to promote inflammatory cell activation, thereby accelerating the rate of CD4+ T-cell depletion.

A possible cause of the more rapid progression to death in AIDS observed in the study by Garred et al. (1997a) is a greater susceptibility to opportunistic infections. In Africa, persistent gut infections, such as cryptosporidiosis, lead to malnutrition and death (Blanshard et al. 1992). In a study of 72 Zambian AIDS patients with diarrhoea, we found that individuals homozygous for MBL structural gene mutations were at increased risk of developing cryptosporidiosis (Kelly et al. 2000). Furthermore, MBL was present in small intestinal fluids (presumably following transexudation) and could bind to *Cryptosporidium parvum* sporozoites and activate complement (Kelly et al. 2000).

It would appear that MBL is implicated as a determinant of HIV acquisition and progression but its involvement appears to be complex. All of the studies published were performed before the use of highly active antiretroviral therapy. Further studies are required to understand fully the role of MBL in HIV disease, particularly in the context of these new therapies.

6.3 Neutropenia

The importance of MBL in patients with other immunodeficiencies was demonstrated in patients receiving treatment for malignancy and rendered neutropenic by chemotherapy. In one study, MBL levels were measured in 54 adults treated with chemotherapy for a range of malignancies (Peterslund et al., 2001). No differences were observed between the distribution of MBL levels in this population and an apparently healthy population of local blood donors. However, in 16 patients who developed bacteraemia, pneumonia, or both within 3 weeks of starting chemotherapy, MBL levels were significantly lower than in patients without serious infections. Further analysis revealed

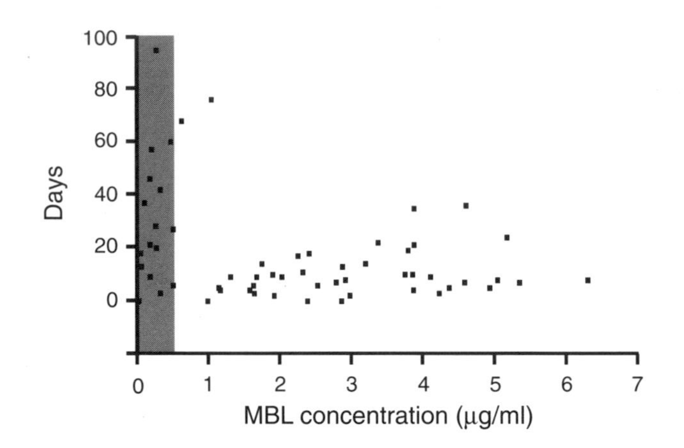

Figure 10.7. Plot of MBL serum levels against the total number of days of febrile neutropenia experienced by children with malignancy and receiving chemotherapy (modified from Neth et al., 2001). The shaded area at 0.5 μg/ml MBL and below represents the range of MBL concentrations which Peterslund et al. (2001) associated with a high risk of clinically significant infections.

that it was patients with an MBL concentration of 0.5 μg/ml or less that were particularly at risk of serious infection.

In another study, MBL phenotype and genotype were determined for 100 children receiving chemotherapy for malignancy (Neth et al., 2001). Their MBL status was then correlated with the causes, frequency, and duration of febrile neutropenic episodes. Children with variant MBL alleles suffered from twice as many days of febrile neutropenia compared to patients with a wild-type genotype (Fig. 10.7). As with the adult study, there appeared to be a level of MBL at diagnosis (0.5 μg/ml) which identified those patients most at risk. Taken together these two studies showed that MBL can play a role in protecting patients from infection in the context of neutropenia.

How MBL is operating in such patients is unclear. It had been assumed that the antimicrobial properties of MBL would be severely attenuated in the absence of neutrophils. A clue to its action may come from the analysis of MBL levels in the paediatric study (Neth et al., 2001). MBL levels were higher at diagnosis than those found in healthy adult Caucasians. Furthermore, levels increased still further during a febrile episode. This is consistent with the known acute phase properties of MBL and levels of the protein would be expected to increase following any infectious stimulus (Thiel et al., 1992). Patients with MBL deficiency did not show a significant rise in protein levels during febrile episodes, presumably reflecting the basic defect in protein assembly (Neth et al., 2001). This inability to increase MBL levels as required could be critical to the control of an infection during neutropenic episodes.

MBL can bind to a range of microorganisms, including those that particularly cause problem infections in chemotherapy patients (Neth et al., 2000). Increased concentrations of MBL led to enhanced complement activation which could result in improved complement-mediated immunity. Alternatively, MBL could be acting as a direct opsonin (Kuhlman et al., 1989) or by enhancing the effect of complement or other phagocytic receptors (Tenner et al., 1995) or through some, as yet uncharacterised, mechanism of phagocyte function. It was interesting that MBL levels were reported to affect only the likelihood or duration of infections but not the nature of the infecting organisms in these two chemotherapy studies (Neth et al., 2001; Peterslund et al., 2001).

These studies indicate that MBL deficiency is important in protecting both neutropenic adults and children from infection. This raises the possibility that MBL replacement therapy as described by Valdimarrson et al. (1998) could be a useful adjunct to cancer treatment in the future.

7. THE ROLE OF MBL DEFICIENCY

The high incidence of structural mutations in the MBL gene within most populations (Garred et al., 1992; Lipscombe et al., 1992a; Madsen et al., 1994) suggests that low levels of the protein may confer a selective advantage in certain situations (Turner and Hamvas, 2000). Two possible explanations, which are not mutually exclusive, have been proposed for the high frequencies of the codon 54 and 57 MBL mutations. Reduced MBL levels may reduce the ability to activate complement and this may limit the possibility of host damage through inflammation (Lipscombe et al., 1992b). This situation would be analogous to the reduced morbidity of meningococcal disease observed in patients deficient in terminal complement components.

Alternatively, a reduction in complement activation by MBL deficiency may protect the host against those parasites such as *Mycobacterium* and *Leishmania*, which use C3b coating and uptake by C3 receptors to gain entrance to cells (Garred et al., 1992). This is supported by higher median serum levels of MBL in lepromatous patients and patients with tuberculosis in Ethiopia compared to healthy individuals (Garred et al., 1994, 1997b). However, this evidence is controversial (Selvaraj et al., 1999), which may reflect the difficulty of analysing disease susceptibility in areas of high HIV prevalence. More recently, it has been shown that *Leishmania* infections occur more commonly in individuals with higher levels of MBL (Santos et al., 2001).

MBL deficiency has been associated with recurrent infection in children (Super et al., 1989) and profound deficiency may carry a lifelong risk of infections (Summerfield et al., 1995) especially when present with another mild immune defect (Garred et al., 1995; Super et al., 1989). Again, this has led to two not mutually exclusive theories of the action of MBL. First, it has been postulated that MBL may be important in early childhood after maternal antibody levels have fallen significantly and before the neonate is capable of producing its own immunoglobulins. This lack of antibody leaves a 'window of vulnerability' between the ages of approximately 6 months and 2 years in which MBL may provide important protection (Turner, 1996). Second, MBL could play an essential protective role in the initial stages of any infection, before immunoglobulin responses have occurred and the term ante-antibody has been coined to describe this activity (Ezekowitz, 1991). Certainly many of the studies of HIV infection susceptibility have been undertaken in adult populations.

8. CLINICAL PRESENTATION OF MBL DEFICIENCY

One of the cardinal features of severe immunodeficiencies is the association of a defect in host defences with infection caused by specific microorganisms. For example, patients with a profound deficiency of immunoglobulins are at a greatly increased risk of infection by encapsulated organisms. MBL deficiency does not have such a close relationship with specific susceptibility and is often present in the absence of any overt symptoms. Our experience of testing for MBL deficiency at the Great Ormond Street Hospital, London, over many years has helped to identify a cohort of patients whose clinical problems are likely to be caused by a deficiency of this protein. A typical history is of an infant who suffers from repeated infections, usually of the upper or lower respiratory tract. The onset of these infections is predominantly between 3 and 6 months and they occur once or twice a month until the acquisition of effective adaptive immune responses. Despite the frequency of infections, most children continue to thrive and rarely need intensive care but they will require multiple courses of antibiotics and frequent attention by health care professionals. Although the aetiology of the infections is often unknown, prophylactic antibiotics are usually beneficial, implying a bacterial component in many of the infections. Reassuringly, this pattern of frequent infections has usually declined or disappeared by the time children start formal education. This typical presentation is compatible with the predicted 'window of vulnerability' role of MBL and serves to highlight the importance of the innate immune system in protecting children prior to the maturation of more specific immune responses.

9. CONCLUSIONS

The past decade has seen a growing awareness of the importance of innate immune mechanisms regardless of the functional integrity of the adaptive immune system. In the case of mannose-binding lectin, extensive studies have been made of the structure, function, genetics, and clinical relevance of the protein. The recognition of the MBL–MASP pathways of complement activation is of particular significance and explains why the protein is of such importance in the immunological repertoire. Initially studies of MBL deficiency states suggested a major role for the protein in reducing susceptibility to a range of infectious diseases but there is, increasingly, a recognition that MBL may also be involved in modulating inflammation and disease severity. Future research will need to address the interplay between these two major roles.

ACKNOWLEDGMENTS

We would like to acknowledge the assistance of Dave Smithson for graphic design and Vania de Toledo in the preparation of this chapter. We are grateful for the financial support of the Wellcome Trust, Action Research and the Medical Research Council, UK for the research undertaken in our laboratories. Research at the Institute of Child Health and Great Ormond Street Hospital for Children NHS Trust benefits from RandD funding received from the NHS Executive.

REFERENCES

Aittoniemi, J., Baer, M., Soppi, E., et al. (1998). Mannan binding lectin deficiency and concomitant immunodefects. *Arch. Dis. Child.* **78**, 245–248.

Aittoniemi, J., Miettinen, A., Laippala, P., et al. (1996). Age-dependent variation in the serum concentration of mannan-binding protein. *Acta. Paediatr.* **85**, 906–909.

Anders, E. M., Hartley, C. A., Reading, P. C., et al. (1994). Complement-dependent neutralization of influenza virus by a serum mannose-binding lectin. *J. Gen. Virol.* **75**, 615–622.

Arai, T., Tabona, P. and Summerfield, J. A. (1993). Human mannose-binding protein gene is regulated by interleukins, dexamethasone and heat shock. *Q. J. Med.* **86**, 575–582.

Bajtay, Z., Jozsi, M., Banki, Z., et al. (2000). Mannan-binding lectin and C1q bind to distinct structures and exert differential effects on macrophages. *Eur. J. Immunol.* **30**, 1706–1713.

Bax, W. A., Cluysenar, O. J. J., Bartelink, A. K. M., et al. (1999). Association of familial deficiency of mannose-binding lectin and meningococcal disease. *Lancet* **354**, 1094–1095.

Bellamy, R., Ruwende, C., McAdam, K. P., et al. (1998). Mannose binding protein deficiency is not associated with malaria, hepatitis B carriage nor tuberculosis in Africans. *Q. J. Med.* **91**, 13–18.

Blanshard, C., Jackson, A. M., Shanson, D. C., et al. (1992). Cryptosporidiosis in HIV-seropositive patients. *Q. J. Med.* **85**, 813–823.

Boniotto, M., Crovella, S., Pirulli, D., et al. 2000. Polymorphisms in the MBL2 promoter correlated with risk of HIV-1 vertical transmission and AIDS progression. *Genes Immun.* **1**, 346–348.

Cartwright, K. (1995). Introduction and historical aspects. In: *Meningococcal Disease, 1st ed.*, Cartwright, K. (ed.). Chichester: John Wiley and Sons Ltd., pp. 1–19.

Crosdale, D. J., Poulton, K. V., Ollier, W. E., et al (2001). Mannose-binding lectin gene polymorphisms as a susceptibility factor for chronic necrotizing pulmonary aspergillosis. *J. Infect. Dis.* **184**, 653–656.

Dahl, M. R., Thiel, S., Willis, A. C., et al. (2001). Mannan-binding lectin associated serine protease 3 (MASP-3) – a new component of the lectin pathway of complement activation. *Immunity* **15**, 127–135.

Davies, J., Neth, O., Alton, E., et al. (2000). Differential binding of mannose-binding lectin to respiratory pathogens in cystic fibrosis. *Lancet* **355**, 1885–1886.

Deuren, M. V., Brandtzaeg, P. and van der Meer, J. W. M. (2000). Update on meningococcal disease with emphasis on pathogenesis and clinical management. *Clin. Microbiol. Rev.* **13**, 144–166.

Devyatyarova-Johnson, M., Rees, I. H., Robertson, B. D., et al. (2000). The lipopolysaccharide structures of *Salmonella enterica* serovar *Typhimurium* and *Neisseria gonorrhoeae* determine the attachment of human mannose-binding lectin to intact organisms. *Infect. Immun.* **68**, 3894–3899.

Drickamer, K. (1992). Engineering galactose-binding activity into a C-type mannose-binding protein. *Nature* **360**, 183–186.

Drickamer, K., Dordal, M. S. and Reynolds, L. (1986). Mannose-binding proteins isolated from rat liver contain carbohydrate-recognition domains linked to collagenous tails. *J. Biol. Chem.* **261**, 6878–6887.

Emmerik, L. C. V., Kuijper, E. J., Fijen, C. A., et al. (1994). Binding of mannan-binding protein to various bacterial pathogens of meningitis. *Clin. Exp. Immunol.* **97**, 411–416.

Estabrook, M. M., Griffiss, J. M. and Jarvis, G. A. (1997). Sialylation of *Neisseria meningitidis* lipooligosaccharide inhibits serum bactericidal activity by masking lacto-*N*-neotetraose. *Infect. Immun.* **65**, 4436–4444.

Ezekowitz, R. A. B. (1991). Ante-antibody immunity. *Curr. Opin. Immunol.* **1**, 60–62.

Ezekowitz, R. A. B., Day, L. and Herman, G. (1988). A human mannose-binding protein is an acute phase reactant that shares sequence homology with other vertebrate lectins. *J. Exp. Med.* **167**, 1034–1046.

Ezekowitz, R. A. B., Kuhlman, M., Groopman, J. E. and Byrn, R. A. (1989). A human serum mannose-binding protein inhibits *in vitro* infection by the human immunodeficiency virus. *J. Exp. Med.* **169**, 185–196.

Figueroa, J., Andreoni, J. and Densen, P. (1993). Complement deficiency states and meningococcal disease. *Immunol. Res.* **12**, 295–311.

Fijen, C. A., van den Bogaard, R., Schipper, M., et al. (1999). Properdin deficiency: molecular basis and disease association. *Mol. Immunol.* **36**, 863–867.

Fothergill, L. D. and Wright, J. (1933). Influenzal meningitis: The relationship of age incidence to the bactericidal power of blood against the causal organism. *J. Immunol.* **24**, 273–284.

Garred, P., Harboe, M., Oettinger, T., et al. (1994). Dual role of mannan-binding protein in infections: another case of heterosis? *Eur. J. Hum. Genet.* **21**, 125–131.

Garred, P., Madsen, H. O., Balsev, U., et al. (1997a). Susceptibility to HIV infection and progression of AIDS in relation to variant alleles of mannose-binding lectin. *Lancet* **349**, 236–240.

Garred, P., Madsen, H. O., Hofmann, B., et al. (1995). Increased frequency of homozygosity of abnormal mannan-binding protein alleles in patients with suspected immunodeficiency. *Lancet* **346**, 941–943.

Garred, P., Madsen, H. O., Kurtzhals, J. A., et al. (1992). Diallelic polymorphism may explain variations of the blood concentration of mannan-binding protein in Eskimos, but not in black Africans. *Eur. J. Immunogen.* **19**, 403–412.

Garred, P., Michaelsen, T. E., Bjune, G., et al. (1993). A low serum concentration of mannan-binding protein is not associated with serogroup B or C meningococcal disease. *Scand. J. Immunol.* **37**, 468–470.

Garred, P., Pressler, T., Madsen, H. O., et al. (1999). Association of mannose-binding lectin gene heterogeneity with severity of lung disease and survival in cystic fibrosis. *J. Clin. Invest.* **104**, 431–437.

Garred, P., Richter, C., Andersen, A. B., et al. (1997b). Mannan-binding lectin in the sub-Saharan HIV and tuberculosis epidemics. *Scand. J. Immunol.* **46**, 204–208.

Ghiran, I., Barbashov, S. F., Klickstein, L. B., et al. (2000). Complement receptor 1/CD35 is a receptor for mannan-binding lectin. *J. Exp. Med.* **192**, 1797–1807.

Hartshorn, K. L., Sastry, K., White, M. R., et al. (1993). Human mannose-binding protein functions as an opsonin for influenza A viruses. *J. Clin. Invest.* **91**, 1414–1420.

Haurum, J. S., Thiel, S., Jones, I. M., et al. (1993). Complement activation upon binding of mannan-binding protein to HIV envelope glycoproteins. *AIDS* **7**, 1307–1313.

Heise, C. T., Nicholls, J. R., Leamy, C. E., et al. (2000). Impaired secretion of rat mannose-binding protein resulting from mutations in the collagen-like domain. *J. Immunol.* **165**, 1403–1409.

Hibberd, M. L., Sumiya, M., Summerfield, J. A., et al. (1999). Association of variants of the gene for mannose-binding lectin with susceptibility to meningococcal disease. *Lancet* **353**, 1049–1053.

Hohler, T., Wunschel, M., Gerken, G., et al. (1998). No association between mannose-binding lectin alleles and susceptibility to chronic hepatitis B virus infection in German patients. *Exp. Clin. Immunogenet.* **15**, 130–133.

Ihara, I., Harada, Y., Ihara, S., et al. (1982). A new complement-dependent bactericidal factor found in nonimmune mouse sera: Specific binding to polysaccharide of Ra chemotype *Salmonella*. *J. Immunol.* **128**, 1256–1260.

Iobst, S. T., Wormald, M. R., Weis, W. I., et al. (1994). Binding of sugar ligands to Ca^{2+}-dependent animal lectins. I. Analysis of mannose binding by site-directed mutagenesis and NMR. *J. Biol. Chem.* **269**, 15505–15511.

Jack, D. L., Dodds, A. W., Anwar, N., et al. (1998). Activation of complement by mannose-binding lectin on isogenic mutants of *Neisseria meningitidis* serogroup B. *J. Immunol.* **160**, 1346–1353.

Jack, D. L., Jarvis, G. A., Booth, C. L., et al. (2001a). Mannose-binding lectin accelerates complement activation and increases serum killing of *Neisseria meningitidis* serogroup C. *J. Infect. Dis.* **184**, 836–845.

Jack, D. L., Read, R. C., Tenner, A. J., et al. (2001b). Mannose-binding lectin regulates the inflammatory response of human professional phagocytes to *Neisseria meningitidis* serogroup B. *J. Infect. Dis.* **184**, 1152–1162.

Jarvis, G. A. and Vedros, N. A. (1987). Sialic acid of group B *Neisseria meningitidis* regulates alternative complement pathway activation. *Infect. Immun.* **55**, 174–180.

Kawakami, M., Ihara, I., Suzuki, A., et al. (1982). Properties of a new complement-dependent bactericidal factor specific for Ra chemotype *Salmonella* in sera of conventional and germ-free mice. *J. Immunol* **129**, 2198–2201.

Kawasaki, N., Kawasaki, T. and Yamashina, I. (1983). Isolation and characterization of a mannan-binding protein from human sera. *J. Biochem.* **94**, 937–947.

Kawasaki, T., Etoh, R. and Yamashina, I. (1978). Isolation and characterization of mannan-binding proteins from rabbit liver. *J. Biochem.* **210**, 167–174.

Kelly, P., Jack, D. L., Mandanda, B., et al. (2000). Mannose binding lectin is a component of innate mucosal defence against *Cryptosporidium parvum* in AIDS. *Gastroenterol.* **119**, 1236–1242.

Klabunde, J., Berger, J., Jensenius, J. C., et al. (2000). *Schistosoma mansoni*: adhesion of mannan-binding lectin to surface glycoproteins of cercariae and adult worms. *Exp. Parasitol.* **95**, 231–239.

Kozutsumi, Y., Kawasaki, T. and Yamashina, I. (1980). Isolation and characterization of a mannan-binding protein from rabbit serum. *Biochem. Biophys. Res. Commun.* **95**, 658–664.

Kuhlman, M., Joiner, K. and Ezekowitz, R. A. (1989). The human mannose-binding protein functions as an opsonin. *J. Exp. Med.* **169**, 1733–1745.

Lau, Y. L., Chan, S. Y., Turner, M. W., et al. (1995). Mannose-binding protein in preterm infants – developmental profile and clinical-significance. *Clin. Exp. Immunol.* **102**, 649–654.

Lipscombe, R. J., Lau, Y. L., Levinsky, R. J., et al. (1992a). Identical point mutation leading to low levels of mannose binding protein and poor C3b mediated opsonisation in Chinese and Caucasian populations. *Immunol. Lett.* **32**, 253–257.

Lipscombe, R. J., Sumiya, M., Hill, A. V., et al. (1992b). High frequencies in African and non-African populations of independent mutations in the mannose binding protein gene. *Hum. Mol. Genet.* **1**, 709–715.

Lu, J. H., Thiel, S., Wiedemann, H., et al. (1990). Binding of the pentamer/hexamer forms of mannan-binding protein to zymosan activates the proenzyme C1r2C1s2 complex, of the classical pathway of complement, without involvement of C1q. *J. Immunol.* **144**, 2287–2294.

Luty, A. J. F., Kun, J. F. J. and Kremsner, P. G. (1998). Mannose-binding lectin plasma levels and gene polymorphisms in *Plasmodium falciparum* malaria. *J. Infect. Dis.* **178**, 1221–1224.

Maas, J., de Roda, H., Brouwer, M., et al. (1998). Presence of the variant mannose-binding lectin alleles associated with slower progression to AIDS. Amsterdam Cohort Study. *AIDS* **12**, 2275–2280.

Madsen, H. O., Garred, P., Kurtzhals, J. A., et al. (1994). A new frequent allele is the missing link in the structural polymorphism of the human mannan-binding protein. *Immunogenetics* **40**, 37–44.

Madsen, H. O., Garred, P., Thiel, S., et al. (1995). Interplay between promotor and structural gene variants control basal serum level of mannan-binding protein. *J. Immunol.* **155**, 3013–3020.

Madsen, H. O., Satz, M. L., Hogh, B., et al. (1998). Different molecular events result in low protein levels of mannan-binding lectin in populations from southeast Africa and South America. *J. Immunol.* **161**, 3169–3175.

Malhotra, R., Haurum, J. S., Thiel, S., et al. (1994). Binding of human collectins (SP-A and MBP) to influenza virus. *Biochem. J.* **304**, 455–461.

Matsushita, M. and Fujita, T. (1992). Activation of the classical complement pathway by mannose-binding protein in association with a novel C1s-like serine protease. *J. Exp. Med.* **176**, 1497–1502.

Matsushita, M. and Fujita, T. (1995). Cleavage of the third component of complement (C3) by mannose-binding protein-associated protease (MASP) wilth subsequent complement activation. *Immunobiology* **194**, 443–448.

Matsushita, M., Hijikata, M., Ohta, Y., et al. (1998). Hepatitis C virus infection and mutations of mannose-binding lectin gene MBL. *Arch. Virol.* **143**, 645–651.

Matsushita, M., Takahashi, A., Hatsuse, H., et al. (1992). Human mannose-binding protein is identical to a component of Ra-reactive factor. *Biochem. Biophys. Res. Commun.* **183**, 645–651.

McBride, M. O., Fischer, P. B., Sumiya, M., et al. (1998). Mannose-binding protein in HIV-seropositive patients does not contribute to disease progression or bacterial infections. *Int. J. STD AIDS* **9**, 683–688.

Miller, M. E., Seals, J., Kaye, R., et al. (1968). A familial, plasma-associated defect of phagocytosis. *Lancet* **ii**, 60–63.

Mizuno, Y., Kozutsumi, Y., Kawasaki, T., et al. (1981). Isolation and characterization of a mannan-binding protein from rat liver. *J. Biol. Chem.* **256**, 4247–4252.

Neth, O., Hann, I., Turner, M. W., et al. (2001). Deficiency of mannose-binding lectin and burden of infection in children with malignancy: A prospective study. *Lancet* **358**, 614–618.

Neth, O., Jack, D. L., Dodds, A. W., et al. (2000). Mannose-binding lectin binds to a range of clinically relevant microorganisms and promotes complement deposition. *Infect. Immun.* **68**, 688–693.

Nielsen, S. L., Andersen, P. L., Koch, C., et al. (1995). The level of the serum opsonin, mannan-binding protein in HIV-1 antibody-positive patients. *Clin. Exp. Immunol.* **100**, 219–222.

Peterslund, N. A., Koch, C., Jensenius, J. C., et al. (2001). Association between deficiency of mannose-binding lectin and severe infections after chemotherapy. *Lancet* **358**, 637–638.

Polotsky, V. Y., Belisle, J. T., Mikusova, K., et al. (1997). Interaction of human mannose-binding protein with *Mycobacterium avium*. *J. Infect. Dis.* **175**, 1159–1168.

Prohaszka, Z., Thiel, S., Ujhelyi, E., et al. (1997). Mannan-binding lectin serum concentrations in HIV-infected patients are influenced by the stage of the disease. *Immunol. Lett.* **58**, 171–175.

Rosenstein, N. E., Perkins, B. A., Stephens, D. S., et al. (1999). The changing epidemiology of meningococcal disease in the United States 1992–1996. *J. Infect. Dis.* **180**, 1894–1901.

Saifuddin, M., Hart, M. L., Gewurz, H., et al. (2000). Interaction of mannose-binding lectin with primary isolates of human immunodeficiency virus type 1. *J. Gen. Virol.* **81**, 949–955.

Santos, I. K., Costa, C. H., Krieger, H., et al. (2001). Mannan-binding lectin enhances susceptibility to visceral leishmaniasis. *Infect. Immun.* **69**, 5212–5215.

Sastry, K., Herman, G. A., Day, L., et al. (1989). The human mannose-binding protein gene: Exon structure reveals its evolutionary relationship to a human pulmonary surfactant gene and localization to chromosome 10. *J. Exp. Med.* **170**, 1175–1189.

Schelenz, S., Malhotra, R., Sim, R. B., et al. (1995). Binding of host collectins to the pathogenic yeast *Cryptococcus neoformans*: Human surfactant protein D acts as an agglutinin for acapsular yeast cells. *Infect. Immun.* **63**, 3360–3366.

Schneider, H., Griffiss, J. M., Boslego, J. W., et al. (1991). Expression of paragloboside-like lipooligosaccharides may be a necessary component of gonococcal pathogenesis in men. *J. Exp. Med.* **174**, 1601–1605.

Selvaraj, P., Narayanan, P. R. and Reetha, A. M. (1999). Association of functional mutant homozygotes of the mannose binding protein gene with susceptibility to pulmonary tuberculosis in India. *Tuberc. Lung Dis.* **79**, 221–227.

Senaldi, G., Davies, E. T., Mahalingam, M., et al. (1995). Circulating levels of mannose binding protein in human immunodeficiency virus infection. *J. Infect.* **31**, 145–148.

Sheriff, S., Chang, C. Y. and Ezekowitz, R. A. B. (1994). Human mannose-binding protein carbohydrate recognition domain trimerizes through a triple α-helical coiled-coil. *Nat. Struct. Biol.* **1**, 789–794.

Soothill, J. F., and Harvey, B. A. M. (1976). Defective opsonisation: A common immunity deficiency. *Arch. Dis. Child.* **91**, 91–99.

Steinberger, P., Szekeres, A., Wille, S., et al. (2002). Identification of human CD93 as the phagocytic C1q receptor (C1qRp) by expression cloning. *J. Leukoc. Biol.* **71**, 133–140.

Stover, C. M., Thiel, S., Thelen, M., et al. (1999). Two constituents of the initiation complex of the mannan-binding lectin activation pathway of complement are encoded by a single structural gene. *J. Immunol.* **162**, 3481–3490.

Sumiya, M., Super, M., Tabona, P., et al. (1991). Molecular basis of opsonic defect in immunodeficient children. *Lancet* **337**, 1569–1570.

Summerfield, J. A., Ryder, S., Sumiya, M., et al. (1995). Mannose-binding protein gene-mutations associated with unusual and severe infections in adults. *Lancet* **345**, 886–889.

Summerfield, J. A., Sumiya, M., Levin, M., et al. (1997). Mannose-binding protein gene mutations are associated with childhood infections in a consecutive hospital series. *Br. Med. J.* **314**, 1229–1232.

Super, M., Gillies, S. D., Foley, S., et al. (1992). Distinct and overlapping functions of allelic forms of human mannose binding protein. *Nat. Genet.* **2**, 50–55.

Super, M., Thiel, S., Lu, J., et al. (1989). Association of low levels of mannan-binding protein with a common defect of opsonisation. *Lancet* **ii**, 1236–1239.

Takahashi, M., Endo, Y., Fujita, T., et al. (1999). A truncated form of mannose-binding lectin-associated serine protease (MASP)-2 expressed by alternative polyadenylation is a component of the lectin complement pathway. *Int. Immunol.* **11**, 859–863.

Tang, C. and Kwiatkowski, D. (1999). Mannose-binding lectin and meningococcal disease. *Lancet* **354**, 336.

Taylor, M. E., Brickell, P. M., Craig, R. K., et al. (1989). Structure and evolutionary origin of the gene encoding a human serum mannose-binding protein. *Biochem. J.* **262**, 763–771.

Tenner, A. J., Robinson, S. L. and Ezekowitz, R. A. B. (1995). Mannose-binding protein (MBP) enhances mononuclear phagocyte function via a receptor that contains the 126,000 M(r) component of the C1q receptor. *Immunity* **3**, 485–493.

Terai, I. and Kobayashi, K. (1993). Perinatal changes in serum mannose-binding protein (MBP) levels. *Immunol. Lett.* **38**, 185–187.

Thiel, S., Bjerke, B. T., Poulsen, L. K., et al. (1995). Ontogeny of human mannan-binding protein, a lectin of the innate immune system. *Pediatr. Allergy Immunol.* **6**, 20–23.

Thiel, S., Holmskov, U., Hviid, L., et al. (1992). The concentration of the C-type lectin, mannan-binding protein, in human plasma increases during an acute phase response. *Clin. Exp. Immunol.* **90**, 31–35.

Thiel, S., Petersen, S. V., Vorup-Jensen, T., et al. (2000). Interaction of C1q and mannan-binding lectin (MBL) with C1r, C1s, MBL-associated serine proteases 1 and 2, and the MBL-associated protein MAp19. *J. Immunol.* **165**, 878–887.

Thiel, S., Vorup-Jensen, T., Stover, C. M., et al. (1997). A second serine protease associated with mannan-binding lectin that activates complement. *Nature* **386**, 506–510.

Thomas, H. C., Foster, G. R., Sumiya, M., et al. (1996). Mutation of gene for mannose-binding protein associated with chronic hepatitis B viral infection [published erratum appears in Lancet 1997; 349(22 Feb):578]. *Lancet* **348**, 1417–1419.

Townsend, R., Read, R. C., Klein, N. J., et al. (2001). Differential binding of mannose-binding lectin to obligate anaerobic bacteria. *Clin. Exp. Immunol.* **124**, 223–228.

Turner, M. W. (1996). Mannose-binding lectin: the pluripotent molecule of the innate immune system. *Immunol. Today* **17**, 532–540.

Turner, M. W., Grant, C., Seymour, N. D., et al. (1986). Evaluation of C3b/C3bi opsonization and chemiluminescence with selected yeasts and bacteria using sera of different opsonic potential. *Immunology* **58**, 111–115.

Turner, M. W. and Hamvas, R. M. (2000). Mannose-binding lectin: Structure, function, genetics and disease associations. *Rev. Immunogenet.* **2**, 305–322.

Turner, M. W., Mowbray, J. F. and Roberton, D. R. (1981). A study of C3b deposition on yeast surfaces by sera of known opsonic potential. *Clin. Exp. Immunol.* **46**, 412–419.

Turner, M. W., Seymour, N. D., Kazatchkine, M. D., et al. (1985). Suboptimal C3b/C3bi deposition and defective yeast opsonization. II. Partial purification and preliminary characterisation of an opsonic co-factor able to correct sera with the defect. *Clin. Exp. Immunol.* **62**, 435–441.

Valdimarsson, H., Stefansson, M., Vikingsdottir, T., et al. (1998). Reconstitution of opsonizing activity by infusion of mannan-binding lectin (MBL) to MBL-deficient humans. *Scand. J. Immunol.* **48**, 116–123.

Wallis, R. and Cheng, J. T. (1999). Molecular defects in variant forms of mannose-binding protein associated with immunodeficiency. *J. Immunol.* **163**, 4953–4959.

Walport, M. J. (2001a). Complement: first of two parts. *N. Engl. J. Med.* **344**, 1058–1066.

Walport, M. J. (2001b). Complement: second of two parts. *N. Engl. J. Med.* **344**, 1140–1144.

Weis, W. I. and Drickamer, K. (1994). Trimeric structure of a C-type mannose-binding protein. *Structure* **2**, 1227–1240.

Weis, W. I., Drickamer, K. and Hendrickson, W. A. (1992). Structure of a C-type mannose-binding protein complexed with an oligosaccharide. *Nature* **360**, 127–134.

Wild, J., Robinson, D. and Winchester, B. (1983). Isolation of mannose-binding proteins from human and rat liver. *Biochem. J.* **210**, 167–174.

Yokota, Y., Arai, T. and Kawasaki, T. (1995). Oligomeric structures required for complement activation of serum mannan-binding proteins. *J. Biochem.* **117**, 414–419.

CHAPTER 11

Blood group phenotypes and infectious diseases

C. Caroline Blackwell

Discipline of Immunology and Microbiology and Hunter Immunology Unit, University of Newcastle, Newcastle, Australia; Institute for Scientific Evaluation of Naturopathy, University of Cologne, Cologne, Germany

Donald M. Weir, Abdulhamid M. Alkout, Omar R. El Ahmer, Doris A. C. Mackenzie, and Valerie S. James

Department of Medical Microbiology, The Medical School, University of Edinburgh

J. Matthias Braun

Department of Medical Microbiology, The Medical School, University of Edinburgh; Institute for Scientific Evaluation of Naturopathy, University of Cologne, Cologne, Germany

Osama M. Almadani

Department of Medical Microbiology, The Medical School University of Edinburgh; Forensic Medicine Unit, Edinburgh, Scotland, United Kingdom

Anthony Busuttil

Forensic Medicine Unit, The Medical School University of Edinburgh, Edinburgh, Scotland, United Kingdom

1. INTRODUCTION

The first observations on associations between blood groups and infectious diseases were made in the 1950s, but the underlying mechanisms were not elucidated for many years. This could have been due to limited explanations for the epidemiological findings or to conflicting reports of associations between different blood groups with the same disease. An example of the latter is the large numbers of papers on *Helicobacter pylori* and ABO or Lewis blood groups/secretor status during the past few years which have reported inconsistent or conflicting results. Because determination of blood groups is a relatively simple and inexpensive procedure, many investigators have used it for quick "simple" studies without consideration of possible confounding factors. For all studies on blood groups and infection, the following points (gained with the experience of hindsight) need to be considered in planning or assessment of surveys:

1. The disease or organism under investigation needs to be clearly defined. Severity of the symptoms should be also be considered, *e.g.*, differentiation of cases of *Escherichia coli* O157 infection between patients with uncomplicated diarrhoeal disease and those that develop haemolytic uraemic syndrome (HUS) (Blackwell et al., 2002).
2. It should be made clear that the investigation examined an outbreak or defined epidemic due to a particular strain in contrast to sporadic cases which could be due to strains with different antigenic characteristics or virulence factors.
3. Different populations express different quantities of antigens such as H, Lewis^{a}, or Lewis^{b}. This can be due to developmental status (age) (Issit, 1986) or genetic (Race and Sanger, 1975) or environmental factors (smoking, fasting, acute illness) (Alkout et al., 1997; Weinmeister and Dal Nogare, 1994).
4. Use of serological evidence of infection, particularly qualitative not quantitative data, can be misleading (see below).
5. In relation to serological surveys, the incidence of the asymptomatic infection or condition among the control population needs to be assessed as this can mask significant differences (Alkout et al., 2000).

2. HYPOTHESES

The first hypothesis proposed to explain the association between blood groups and diseases was that isohaemagglutinins were protective against microorganisms with antigens cross-reactive with the A or B blood groups. The advent of studies on lectin interactions between microorganisms and their hosts provided new approaches to investigating these associations, particularly the possibility that the carbohydrate structures of the blood group antigens could act as receptors for lectinlike adhesins on microorganisms. The blood group antigens are histocompatibility antigens and new evidence indicates that blood group is associated with inflammatory responses to some organisms. This review will concentrate on mechanisms proposed to explain associations between blood groups and secretor status identified in various epidemiological studies (Tables 11.1 and 11.2). There are two major stages in pathogenesis of infection to be considered: (1) colonisation of mucosal surfaces and (2) the host's immune or inflammatory responses to the microorganism. In recent years, the majority of studies on the role of blood groups, secretor status, and susceptibility to infection have used *H. pylori* as a model, and this is reflected in the number of times the studies on *H. pylori* will be cited in the sections below.

Table 11.1. *Examples of associations between blood groups and susceptibility to infectious agents*

Infectious agent/disease	ABO	Reference
Genitourinary tract		
Escherichia coli	B/AB	Kinane et al. (1982)
Pseudomonas. aeruginosa	B	Ratner et al. (1986)
Neisseria gonorrhoeae	B	Kinane et al. (1983)
Staphylococcus saprophyticus	A/AB	Beuth, Stoffel and Pulverer (1996)
Oral/gastrointestinal tract		
Candida albicans	O	Aly et al. (1991, 1992)
		Ben-Aryeh et al. (1995)
Salmonella and *E. coli*	B/AB	Socha, Belinska, and Kaczera (1969)
		Robinson, Tolchin, and Halpern (1971)
		Black et al. (1987)
Vibrio cholerae	O	Barua and Paguio (1977)
		Chaudhuri and De (1977)
		Swerdlow et al. (1994)
Helicobacter pylori (peptic ulcers)	O	Aird et al. (1954)
		Mentis et al. (1990)
		Luman et al. (1996)
		Hein et al. (1997)
		Lin et al. (1998)
E. coli O157	O	Blackwell et al. (2002)
	P^-	Blackwell et al. (2002)
	not-B	Shimazu et al. (2000)
Respiratory tract		
Streptococcus pyogenes	not-O	Haverkorn and Goslings (1969)
Streptococcus pneumoniae	not-B	Reed et al. (1974)
Influenzae A	O/B	MacDonald and Zuckerman (1962)
Influenzae A2	O	Potter (1969)
tuberculosis	O	Viskum (1975)
Acute otitis externa		
Ps. aeruginosa	A	Beuth, Stoffel, and Pulverer (1996)

(*cont.*)

Table 11.1. (*cont.*)

Infectious agent/disease	ABO	Reference
Plasmodium falciparum (protection)	O	Migot-Nabius et al. (2000)
Infection associated?		
Ischaemic heart disease	O	Suadicuni, Hein, and Gyntelberg (2000)
Gastric cancer	A	Mourant, Kopec, and Domaniewska-Sobczak (1978)

3. BLOOD GROUP ANTIGENS AS RECEPTORS FOR MICROORGANISMS OR THEIR TOXINS

The ABO, Lewis, P, S, Anton, and Duffy blood group antigens have all been identified as receptors for microorganisms (Table 11.3). Usually, the blood group antigen is not the only receptor for the organism and the contribution of the antigen to colonisation in relation to other receptors needs to be assessed by quantitative methods (see below). Recent studies also indicate that some adhesins can use more than one blood group antigen as receptors, *e.g.*, a 61-kDa component of *H. pylori* binds H type 2, Lewis^b, and Lewis^a (Alkout et al., 1997). Evidence that blood group antigens can act as receptors for microorganisms comes from various sources as follows: assays to determine binding of microorganisms to cells expressing the candidate antigen, inhibition of binding by blocking the putative receptor with monoclonal or polyclonal antibodies, anti-idiotypic reagents, inhibition of binding of microorganism to epithelial cells by synthetic blood group oligosaccharides, and isolation of adhesins by affinity methods using synthetic blood group antigens.

3.1 Binding of Microorganisms to Host Cells

Binding of microorganisms to cells expressing the blood group antigen under investigation has been assessed by several methods as follows: light microscopy, enzyme-linked immunosorbent assays (ELISA), isotope labelling, and flow cytometry. The advantages of the flow cytometry method used by our group have been summarised elsewhere (Raza et al., 1993; Saadi et al., 1993). A major advantage of the flow cytometry method is that

Table 11.2. *Secretor status/Lewis phenotype and susceptibility to disease or carriage of microorganisms*

Infectious agent/disease	Secretor status	Reference
Genitourinary tract		
E. coli		
Recurrent infections	Nonsecretor	Kinane et al. (1982)
Postmenopausal women	Nonsecretor	Raz et al. (2000)
HIV (heterosexual transmission) (protection)	Secretor	Blackwell et al. (1991)
	Secretor	Ali et al. (2000)
Respiratory tract		
Neisseria meningitidis	Nonsecretor	Blackwell et al. (1986a, 1990)
Stretp. pyogenes	Nonsecretor	Haverkorn and Goslings (1969)
Haemophilus influenzae type b	Nonsecretor	Blackwell et al. (1986b)
Streptococcus pneumoniae	Nonsecretor	Blackwell et al. (1986a)
Influenza virus	Secretor	Raza et al. (1991)
Rhino virus	Secretor	Raza et al. (1991)
Respiratory syncytial virus	Secretor	Raza et al. (1991)
Oral/gastrointestinal tract		
C. albicans	Nonsecretor	Aly et al. (1991, 1992)
		Ben-Aryeh et al. (1995)
V. cholerae	Nonsecretor	Chaudhuri and Das Adhikary (1978)
Infection associated?		
H. pylori/peptic ulcers	Nonsecretor	Clarke et al. (1956)
	Nonsecretor	Yang et al. (2001)
Insulin dependent diabetes	Nonsecretor	Blackwell et al. (1987)
	Nonsecretor	Aly et al. (1991, 1992)
Non-insulin-dependent diabetes	No association	Blackwell et al. (1987)
		Aly et al. (1991, 1992)
Graves disease	Nonsecretor	Collier et al. (1988)
Ischaemic heart disease	$Lewis^{a-b-}$	Chaudhary and Shukla (1999)

Table 11.3. *Examples of blood group antigens that act as receptors for microorganisms*

Organism	Receptor antigen	Reference
Escherichia coli	P	Kallenius et al. (1980)
	S	
Haemophilus influenzae	Anton	Van Alphen, Poole, and Overbeeke (1986)
	Lewis[a]	Essery et al. (1994b)
Neisseria meningitidis	Lewis[a]	Essery et al. (1994b)
Bordetella pertussis	Lewis[a], Lewis[x]	Van t'Wout et al. (1992)
		Saadi et al. (1996)
Helicobacter pylor	Lewis[b]	Boren et al. (1993)
	H type 2, Lewis[a], Lewis[b]	Alkout et al. (1997)
Staphylococcus aureus	Lewis[a], Lewis[x]	Saadi et al. (1993)
Streptococcus salivarius	B	Ciopraga, Motas, and Doyle (1995)
Streptococcus mutans	H, B	Ciopraga, Motas, and Doyle (1995)
Streptococcus cricetus	H, A, B	Ciopraga, Motas, and Doyle (1995)
Candida albicans	H, Lewis[a], Lewis[b]	Cameron and Douglas (1996)
Plasmodium knowlesi	Duffy	Barnwell et al. (1989)
Plasmodium vivax		Chitnis et al. (1996)
B19 parvovirus	P	Brown, Anderson, and Young (1993)

monoclonal antibodies to the blood group antigens on the cell population under investigation can be used to assess correlations between expression of the antigen (reflected in level of binding of the antibodies) and binding of the bacteria. This is important as environmental factors can alter expression of the blood group antigens and this can result in wide variations in results for individual donors or even cell lines.

When using epithelial cells from donors, a number of factors which can affect binding of microorganisms need to be considered in addition to blood group or secretor status. The amount of the antigens under investigation can vary with environmental factors such as hormonal changes (Schaeffer et al., 1994), smoking, fasting, virus infection, or underlying diseases (Alkout et al.,1997; Raza et al., 1999; El Ahmer et al., 1999a, 1999b; Weinmeister and Dal Nogare, 1994). Although epithelial cells of group O individuals were found to bind significantly more monoclonal anti-H type 2 than cells from donors of other blood groups, both smoking and fasting affected the levels of antibody bound (Alkout et al., 1997). It is generally accepted that nonsecretors have high levels of Lewis[a]; however, we have found a wide range of binding of monoclonal anti-Lewis[a] to cells from individual secretors (Saadi et al., 1993). Under these circumstances, examining binding only in relation to secretor status and not expression of Lewis[a] could result in false conclusions.

3.2 Studies on Inhibition of Binding of Microorganisms

Both polyclonal and monoclonal antibodies have been used in inhibition of binding studies to examine the role of blood group and other antigens as receptors for microorganisms. There is, however, the criticism that the actual receptor is adjacent to the antigen to which the antibody is directed and inhibition is due to stearic hindrance. Other methods are needed to confirm the inhibition of binding by treatment of the cells with antibodies, e.g., inhibition of binding by synthetic blood group antigen(s) or agglutination by anti-idiotypic antibodies produced by immunisation with a monoclonal antibody to the putative receptor (Essery et al., 1994). Production of anti-idiotypic reagents can be unreliable and the availability of synthetic blood group antigens makes them the preferable method (Saadi et al., 1999; Gordon et al., 1999).

3.3 Binding of Synthetic Blood Group Antigens

Attempts to use blood group antigens derived from secretions or cells in inhibition studies could be criticised on the grounds that there were also other substances such as proteins, lipids, or secretory antibodies present in these preparations, and these substances, not the blood group oligosaccharide, were contributing to inhibition of binding. In the past few years, synthetic blood group oligosaccharides labelled with biotin or fluorochromes have become available in sufficient quantities and at reasonable costs. These have greatly enhanced our ability to assess directly the binding of these putative receptors. They also allow comparison of level of binding of different antigens and tests can be carried out by relatively simple spectrophotometric methods (Alkout et al., 1997).

Table 11.4. *Blood group antigens that bind bacterial toxins*

Toxin	Antigen	Reference
Cholera	A	Monferran et al. (1990)
E. coli heat labile enterotoxin	A, B	Barra et al. (1992)
Pertussis	Le^a, Le^x	Van t'Wout et al. (1992)
Staphylococcal	Le^a, Le^b	Essery et al. (1994a)
SEB, TSST-1		Saadi et al. (1996)
E. coli O157 shiga toxin	P	Bitzan et al. (1994)

3.4 Isolation of Adhesins By Affinity Adsorption with Synthetic Antigens

Methods for affinity purification of Gram-positive (Saadi et al., 1994) and Gram-negative (Alkout et al., 1997) bacterial surface components with synthetic blood group antigens attached to an inert matrix have been described. The relative efficiency of binding of the adhesins to the receptor can be estimated by the amount of protein eluted from the different oligosaccharides (Alkout et al., 1997) and the adhesin can be used to block bacterial binding to epithelial cells.

3.5 Blood Group Antigens as Receptors or Inhibitors for Bacterial Toxins

A variety of bacterial toxins bind to blood group antigens (Table 11.4). These include the enterotoxins of *E. coli*, cholera toxin, pertussis toxin, and some of the pyrogenic staphylococcal toxins. The P blood group is a receptor on endothelial surfaces for the shiga toxin (ST) produced by enterohaemorrhagic strains of *E. coli* O157 (Bitzan et al., 1994). There is a growing body of evidence that glycoconjugates with blood group antigen activity obtained from mucosa (Bennun et al., 1989; Montferran et al., 1990; Barra et al., 1992) and human milk (Saadi et al., 1996a, 1999) or even synthetic antigens (Saadi et al., 1996a) can inhibit the activity of some bacterial toxins.

4. FREQUENCY AND DENSITY OF COLONISATION OF EPITHELIAL SURFACES

4.1 Frequency of Colonisation

Studies on carriage of potentially pathogenic bacteria found that nonsecretors were overrepresented among individuals from whom *Streptococcus pyogenes* (reviewed by Haverkorn and Goslings, 1969) or *Neisseria meningitidis* (Blackwell et al., 1990) were isolated. In longitudinal studies on rectal or vaginal colonisation of women with recurrent urinary tract infection, nonsecretors were significantly more likely to be colonised by F fimbriated strains of *E. coli* (Stapleton et al., 1995).

Frequency of colonisation might be related to developmental factors associated with expression of oligosaccharides on mucosal surfaces. For example, the susceptibility of newborn calves to enterotoxigenic strains of *E. coli* expressing the K99 adhesin reflects developmental changes in oligosaccharide composition. As this oligosaccharide receptor for the K99 adhesin disappears with age, so does susceptibility to disease caused by these bacteria. An analogue of the receptor fed to the calves will prevent the colonisation of the calves by these bacteria and subsequent disease due to the toxin (Mouricourt et al., 1990).

Expression of the Lewis^{a} and Lewis^{b} antigen in early life is associated with developmental control of expression of the fucosyl transferase coded for by the secretor gene. This has been investigated in relation to the role of infectious agents in cot deaths. During the age range in which most cases of Sudden Infant Death Syndrome (SIDS) occur, 80–90% of infants express the Lewis^{a} antigen on their red blood cells. The proportion of infants expressing this antigen declines with age; and by 18–24 months, the antigen is usually found on red cells of approximately 20–25% of children, a proportion similar to that observed in adults (Issit, 1986). Lewis^{a} was identified in 71% of SIDS infants examined (Blackwell et al., 1992; Saadi et al., 1993).

Toxigenic bacteria have been implicated in SIDS either by direct identification of their toxins in SIDS infants (*S. aureus*) (Newbould et al., 1989; Malam et al., 1992; Zorgani et al., 1999; Blackwell et al., 2001, 2002) or by indirect evidence in epidemiological studies (*B. pertussis*) (Nicholl and Gardner, 1988; Lindgren, Milerad and Lagercrantz, 1997; Heininger et al., 1996). The Lewis^{a} and structurally related Lewis^{x} antigens were found to be receptors for *S. aureus* and *B. pertussis* (Saadi et al., 1993, 1996), and an adhesin was obtained from *S. aureus* by affinity purification with synthetic Lewis^{a} (Saadi et al., 1994).

Longitudinal studies of infants found that during the first 3 months of life, the predominant species in the normal nasopharyngeal flora of infants was *S. aureus*; 57% of healthy babies were carriers of these bacteria. The proportion of *S. aureus* isolates declined significantly with age to be replaced with other species (Blackwell et al., 1999). Similar results were obtained in two independent studies (Harrison et al., 1999; Anianson et al., 1992).

The predominant species isolated from 37 SIDS infants during the period in which the healthy infants were studied was also *S. aureus*. These

Table 11.5. *Distribution of H and Lewis antigens on cells and in body fluids of secretors and nonsecretors*

	Secretors		Nonsecretors	
	Cells	Fluids	Cells	Fluids
H type 2	+	−	+	−
H type 1	+	+	−	−
Lewis[a]	±*	± *	+	+
Lewis[b]	+	+	−	−

* Lewis[a] can be present in highly variable amounts in secretors. When assessing the effect of secretor status on binding of microorganisms to epithelial cells, it is highly recommended that the amount of both Lewis[a] and Lewis[b] be determined by flow cytometry as controls for the experiments.

bacteria were isolated from 19/22 (86.4%) SIDS infants 3 months or younger compared with 143/253 (57%) of the healthy infants sampled in this age range ($\chi^2 = 5.32$, $P = 0.02$) and 7/15 (46.6%) of those over 3 months of age compared with 235/672 (35.2%) specimens obtained from healthy infants at the second, third, and fourth samplings ($\chi^2 = 0.41$) (Blackwell et al., 1999).

4.2 Density of Colonisation

Density of colonisation is an important factor in development of disease. The more organisms present on a mucosal surface, the greater the probability that the host will develop disease in contrast to asymptomatic infection (Ofek and Kahane, 1996).

4.3 Secretor Status and Density of Colonisation

Two hypotheses based on the distribution of ABO and Lewis blood groups on cells and in body fluids of secretors and nonsecretors (Table 11.5) were proposed to explain susceptibility of nonsecretors to disease or carriage of some bacteria and yeasts. First, the H type 2 antigen present on epithelial cells of all individuals except the extremely rare Bombay phenotype (Race and Sanger, 1975) is a receptor for adhesins on the microorganism, and the terminal fucose of H type 1 or Lewis[b] in body fluids binds to the adhesin and reduces or inhibits colonisation. Second, the Lewis[a] antigen usually found in higher levels on epithelial cells of nonsecretors is a major receptor for the microorganism, and despite the presence of the Lewis[a] antigen in body fluids, their epithelial cells are more densely colonised than those of secretors (Blackwell, 1989).

Lomberg et al. (1986) found uroepithelial cells of nonsecretors bound more *E. coli* expressing the P fimbriae. This was explained by a third observation, that the P antigen was more accessible on epithelial cells of nonsecretors. Other groups found *E. coli* was bound to two components in the uroepithelial cells of nonsecretors that was absent from cells of secretors (sialosyl galactose-globoside and disialosyl galactose-globoside). The authors suggested these are expressed only by cells of nonsecretors because the galactose-globoside precursor glycolipid which in secretors is fucosylated is sialylated in nonsecretors (Stapleton et al., 1992, 1995).

5. BINDING OF BACTERIA TO BLOOD GROUP ANTIGENS: ASSESSMENT OF EXPERIMENTAL STUDIES IN RELATION TO EPIDEMIOLOGY

Evidence from the studies of Boren et al. (1993) indicated that Lewis[b] was the only fucose containing blood group antigen to which *H. pylori* bound and molecular studies on the gene for the adhesin that binds Lewis[b] have identified it in a variety of patient populations (Gerhard et al., 1999). Since nonsecretors are genetically incapable of producing Lewis[b], this could not explain their reported increased susceptibility to ulceration (Table 11.2). We used several of the methods outlined above to demonstrate that H type 2, Lewis[b], and Lewis[a] are all receptors for *H. pylori*. Binding of the bacteria could be inhibited by pretreatment of buccal epithelial cells or the Kato III cell line with monoclonal antibodies to H type 2, Lewis[b], or Lewis[a]. Binding of *H. pylori* to epithelial cells of different donors was correlated with binding of monoclonal antibodies to H type 2 ($P < 0.005$) and Lewis[b] ($P < 0.001$) but not Lewis[a]. Biotinylated H type 2, Lewis[b], and Lewis[a] (but not Lewis[x]) each bound to the NCTC 11637 strain of *H. pylori* and 51 clinical isolates obtained from patients attending local gastroscopy clinics (Alkout et al., 1997). A 61-kDa protein was obtained by affinity purification with each of the three synthetic blood group antigens. H type 2 appears to be the most effective of the three antigens for binding the adhesin. The highest yield of the adhesin was obtained with the H type 2 affinity matrix and the majority of the strains bound more biotinylated H type 2 than the other two oligosaccharides. Pretreatment of epithelial cells with the 61-kDa protein significantly reduced binding of the bacteria to epithelial cells.

These results help explain the increased susceptibility of group O to peptic ulceration. Density of colonisation is related to inflammatory responses and duodenal ulcer among patients infected with these bacteria (Atherton et al., 1996). H type 2, which appears to be the most efficient receptor for the bacteria, is present in significantly higher levels on cells of group O. The second most effective receptor was Lewis^b which has a terminal fucose that in secretions might block the adhesins that bind the fucose containing antigens. Nonsecretors will lack Lewis^b in their body fluids and this could contribute to their being more densely colonised than secretors.

Density of *H. pylori* assessed by number of colonies on the primary isolation plate was significantly higher for patients with duodenal ulcers compared with patients in whom no ulcers were present. The numbers of bacteria identified in patients with duodenal ulcers was also significantly greater than those found among patients with gastric ulcers (Mentis et al., 1990). Reassessment of the data found no significant differences between the proportions of patients of group A compared with those of group O who were considered to be heavily colonised (>100 colonies). There was no difference between A secretors and A nonsecretors, but the proportion of O secretors with >100 colonies (21%) was lower than that for O nonsecretors (35%). If Lewis^b in secretions binds the adhesin more effectively than Lewis^a, the levels of Lewis^a and Lewis^b expressed on the cells of an individual, not simply secretor status, need to be considered in future studies.

5.1 Environmental Factors Affect Expression of Blood Group Antigens and Bacterial Binding

Environmental factors such as smoking are associated with susceptibility to ulcers (McCarthy, 1984). Binding of *H. pylori* to buccal epithelial cells from 13 smokers and 13 nonsmokers attending gastroscopy clinics was assessed, and the cells from smokers bound significantly greater numbers of bacteria. Additional experiments with cells from 8 pairs of smokers and nonsmokers, healthy individuals with no symptoms of peptic ulcer or gastritis, also showed higher levels of binding of *H. pylori* to cells of smokers, but the difference was not significant. The patient group had no significant medication to account for the differences observed; however, each had fasted for approximately 12 h before the buccal cells were obtained at the clinic. The effects of fasting on binding of *H. pylori* were further investigated with cells from 30 healthy Muslim men, 15 smokers and 15 nonsmokers, during Ramadan, when they were fasting 12–16 h per day, and after the fast when they were eating and drinking normally. The effects of smoking were reversed in this group; there was significantly lower binding to cells of smokers (Alkout et al., 1997).

For both the patient and control groups, the levels of H type 2 expressed on the epithelial cells correlated with binding of *H. pylori*; they were increased among the smokers in the patient group and decreased among the smokers in the control group. Expression of the Lewis antigens were not significantly affected by smoking if the donors were not fasting but were significantly higher among nonsmokers who were fasting. For healthy individuals, this was the first example we found in which cells of smokers bound significantly fewer bacteria (Alkout et al., 1997; El Ahmer et al., 1999a). A water-soluble cigarette smoke extract (CSE) was used to treat cells from nonsmokers to assess the effects of smoking on bacterial binding. The undiluted CSE masked some blood group antigens, suggesting that that increased bacterial binding observed with cells of smokers is due to mechanisms not linked to expression of blood group antigens; however, the CSE also contained material that was cross-reactive with the H type 2 antigen (El Ahmer et al., 1999a).

6. IMMUNE/INFLAMMATORY RESPONSES AND BLOOD GROUPS

6.1 Immune Responses

6.1.1 Protection

The possibility that isohaemagglutinins might act as "natural" opsonising or bactericidal antibodies was one of the first explanations for associations between ABO blood groups and infectious diseases. Antigens cross-reactive with A, B, and H had been identified on microorganisms (Springer, 1970; Springer et al., 1971; Reed et al., 1974). The human immunodeficiency virus (HIV) incorporates blood group antigens into its envelope and monoclonal antibody to the blood group antigen inactivated the virus (Arendrup et al., 1991). If an individual were exposed to virus that had grown in the presence of an incompatible blood group, the isohaemagglutinin might exert a protective effect against infection. Similar mechanisms might underlie the observation that nonsecretors are less susceptible to acquiring HIV by heterosexual intercourse (Blackwell et al., 1991; Ali et al., 2000). If the infected partner is a secretor, incorporation of Lewis^b into the viral envelope might result in anti-Lewis^b antibodies in the uninfected partner acting as effective opsonins.

There also appear to be subtle differences in humoral responses of secretors and nonsecretors. Although it was reported that nonsecretors had lower levels of secretory and serum IgA (Waissbluth and Langman, 1971; Grundbacher, 1972), these observations did not control for carriage of potential pathogens, current or recurrent infection. We found IgA levels were significantly increased among nonsecretor women with recurrent urinary tract infections (Blackwell et al., 1989). Assessment of secretory IgA levels for individuals who were carriers of meningococci compared with noncarriers found that there was no difference associated with secretor status; higher levels were associated with carriage of the bacteria. Both total salivary IgM and salivary IgM binding to meningococci in a whole-cell ELISA were significantly lower in nonsecretors (Zorgani et al., 1992). Consistent correlations between bactericidal activity against capsulate strains of meningococci with IgG antibodies to *Neisseria lactamica* were found for secretors but not for nonsecretors (Zorgani et al., 1994, 1996).

6.1.2 Tolerance

The other possibility is that some antigens on microorganisms similar to A, B, or H do not induce an immune response in hosts of the corresponding blood group and this tolerance makes the host more susceptible to disease or to reinfection. Individuals of blood group O are at greater risk of severe cholera (Swerdlow et al., 1994), and in Bangladesh, killed oral vaccines protected individuals of blood group O less well than those of other blood groups (Clemmens et al., 1989). In contrast, more recent studies in Chile found that there was a higher vibriocidal response among children of blood group O compared with non-O individuals (Lagos et al., 1995).

Breakdown of tolerance and consequent induction of antibodies to antigens on *H. pylori* similar to Lewis^x and Lewis^y has been suggested to play a role in pathogenesis of peptic ulcer disease (Appelmelk et al., 1996). This needs to be assessed in relation to recent findings in a Mexican population that the Lewis^x antigen is identified significantly more often on *H. pylori* isolates obtained from children compared with isolates from adults and there is a significant trend for loss of strains expressing Lewis^x with increasing age (Munoz et al., 2001).

In a study of an Irish population, no coincidence between expression of Lewis^x or Lewis^y on the host and bacteria was noted. There were no significant differences between the quantity of these antigens expressed on isolates from patients with ulcers compared with patients with chronic gastritis (Heneghan et al., 2000). In a study in Sweden, strains expressing Lewis^x or Lewis^y were isolated more often from patients with duodenal ulceration than from patients who were infected but were asymptomatic (Thoreson et al., 2000).

There were significantly higher levels of antibodies to Lewis^x and Lewis^y in sera of patients with gastric cancer and other pathological conditions associated with *H. pylori* compared with controls who were not infected with this bacteria. The antibody response to Lewis^x or Lewis^y was unrelated to the host phenotype but was significantly associated with expression of the antigens on the bacterial isolates (Heneghan et al., 2001). Expression of these antigens on *H. pylori* has been associated with density of colonisation by the bacteria, neutrophil infiltration and lymphocyte infiltration (Heneghan et al., 2000).

6.2 Inflammatory Responses

Inflammatory responses play an important role in pathogenesis of many infections. The contribution of antigens on the microorganism (Braun et al., 2002) and host responses need to be assessed when evaluating severity and outcome of the disease.

Greater inflammatory responses of nonsecretors to urinary tract infections have been noted (Lomberg et al., 1992). Higher levels of interleukin-6 (IL-6), tumor necrosis factor-α (TNFα), and nitric oxide (NO) were produced in response to whole cells of *H. pylori* or its antigens by monocytes of group O blood donors compared with responses by cells of other ABO groups (Alkout et al., 2000). These higher levels of inflammatory mediators might contribute to tissue damage leading to ulceration. Histological studies found that lymphocyte infiltration was significantly greater in *H. pylori*-infected nonsecretors. Group O nonsecretors had a higher grade of lymphocyte infiltration compared to nonsecretors of other ABO groups (Heneghan et al., 1998).

Inflammatory responses play a major role in the pathogenesis of disease due to *E. coli* O157 (Tesh, 1998). Blood donors who lack the P blood group antigen had higher TNF responses to culture filtrates of toxigenic *E. coli* O157. Although there were no significant differences between *in vitro* TNF or IL-10 responses associated with ABO group, there was a significant excess of group O among patients affected in the Scottish outbreak of this disease and the mortality rate was significantly higher for group O (Blackwell et al., 2002).

7. DISEASES TRIGGERED BY INFECTIOUS AGENTS?

7.1 *H. pylori* and Peptic Ulcers

Thirty years before the association among *H. pylori*, gastritis, and ulcers was identified (Marshall et al., 1985), the associations between peptic ulceration with blood group O and nonsecretion were reported (Aird et al., 1954; Clarke et al., 1956). The associations between O and ulceration have been confirmed in more recent studies (Mentis et al., 1991; Luman et al., 1996), and men of group O had significantly higher risk of hospitalisation associated with ulcers than others (Hein et al., 1997). In a Taiwanese population, the prevalence and relative risk of *H. pylori* infection and development of ulcers was significantly higher in blood group O patients than other ABO groups (Lin et al., 1998). Infection with *H. pylori* was significantly higher in patients expressing Lewis^{a} and secretors expressing Lewis^{b} had a significantly lower rate of gastroduodenal ulcers (Yang et al., 2001).

A number of surveys failed to find any association between evidence for infection with *H. pylori* and blood group or secretor status. Some studies were based on serological tests (Hook-Nikanne et al., 1990; Loffeld and Stobberingh, 1991; Chesner et al., 1992) and others on culture or microscopic identification of the bacteria (Dickey et al., 1993).

7.2 *H. pylori* and Gastric Carcinoma

The mechanisms proposed for increased susceptibility of group O to peptic ulceration do not appear to be relevant to the association between group A and gastric cancer (Mourant et al., 1978). *H. pylori* infection is now recognised as a factor related to development of gastric carcinoma. Chronic infection and inflammation have been recognised as risk factors for a variety of human cancers, and it has been proposed that active oxygen species such as superoxide anion, hydrogen peroxide, and hydroxyl radicals generated in inflamed tissue can cause injury to target cells and also damage DNA (reviewed by Ohshima and Bartsch, 1994). There is increasing evidence that nitric oxide and its derivatives produced by activated phagocytes can also contribute to the multistage carcinogenesis process (Ohshima and Bartsch, 1994). NO produced in response to *H. pylori* or its antigens by monocytes of group O blood donors were higher compared with responses by cells of group A donors. In contrast to findings for TNF and IL-6, NO production by monocytes exposed to *H. pylori* or its antigens showed an inverse relationship with the numbers of bacteria per cell. The highest levels of NO were induced by the lowest numbers of bacteria (Blackwell et al., 1997).

7.3 *H. pylori* and Ischaemic Heart Disease (IHD)

Infection-induced inflammatory responses are implicated in pathophysiological events associated with IHD (Vallance, Collier, and Bhagat, 1997).

There is some support for the hypothesis that *H. pylori* infection might be a significant factor (Mendall et al., 1994; Miragliotta et al., 1994; Martin-de-Argila et al., 1995; Morgando et al., 1995; Murray et al., 1995; Danesh et al., 1998, 1999). The association of socioeconomic status with risk of IHD is only partly explained by the uneven distribution of conventional risk factors. A Danish study found that only among men of group O was socioeconomic status associated with a significant excess risk of IHD (Suadicani, Hein, and Gyntelberg, 2000). Lower socioeconomic status, particularly during childhood, and group O are both risks for acquisition of *H. pylori* (Glynn, 1994).

Most studies examined sera of patients and comparison groups for presence of antibodies to *H. pylori*. The assays, as in many other studies on these bacteria, were qualitative rather than quantitative and the presence of gastritis, peptic ulcer, or heart disease was not always assessed in the comparison groups (e.g., blood donors). By enzyme-linked immunoassay we compared IgG levels to *H. pylori* outer membrane extract in sera obtained from autopsy investigations of individuals who died of IHD ($n = 40$) with those from age- and sex-matched individuals who died in accidents ($n = 40$). Although the IgG levels to *H. pylori* were higher in the IHD group, the difference was not significant. Detailed examination of the autopsy findings found that eight subjects had evidence of heart disease or evidence of ulcers. Reassessment of the results found that the differences between the IHD and accident groups were significant.

Additional sera from the following four groups were examined: (1) men who survived one myocardial infarction (MI), (2) men matched for age and socioeconomic background to group 1, (3) individuals who died suddenly of IHD, and (4) accidental deaths with no evidence of IHD based on autopsy findings matched for age and sex to group 3. Levels of IgG to *H. pylori* increased with age ($P < 0.005$) but were not associated with smoking or socioeconomic groups. There was a correlation between IgG to the bacteria and decreasing socioeconomic levels only among group 1 ($P < 0.01$). Antibodies to *H. pylori* were higher for subjects who died of IHD (median = 151 ng ml^{-1}) compared with survivors (median = 88 ng ml^{-1}) ($P = 0.034$) and higher for survivors compared with their controls (median = 58 ng ml^{-1}) ($P = 0.039$). If the antibody levels reflect density of colonisation or inflammatory activity, future serological studies of *H. pylori* in relation to IHD must take into account severity of disease and serological analyses must be quantitative. Larger studies need to be carried out to determine if the association between group O, socioeconomic group and IHD are related to infection with *H. pylori* (Alkout et al., 2000).

8. CONCLUSIONS AND APPLICATIONS

Antiadhesin therapy is an obvious application of the studies on blood group phenotype and susceptibility to infection. The use of synthetic blood group antigens and other host cell receptors for prevention of infection are under development for *H. pylori* and respiratory pathogens (Zopf and Roth, 1996; Zopf et al., 1996). These methods are being applied to problems associated with the development of resistance to antifungal agents, particularly among patients on long term prophylaxis for prevention of superficial mucosal infections. Development of resistance is emerging not only among *C. albicans* but also among other yeasts causing superficial or systemic infection among immunosuppressed patients (Willocks et al., 1991; Johnson et al., 1995; Rex et al., 1995; Odds, 1996). Our epidemiological studies indicated that blood group O and nonsecretors were at increased risk of carriage or oral disease due to candida (Aly et al., 1991, 1992; Ben-Aryeh et al., 1995), and *C. albicans* adhesins that bind fucose have been characterised (Cameron and Douglas, 1996). Strains of both serotypes A and B of *C. albicans* bind biotinylated H type 2, Lewis^a and Lewis^b and binding of the yeasts to epithelial cells could be reduced by the synthetic oligosaccharides (Saadi et al., 1998).

An additional application of the antiadhesin approach is based on the protective effect of blood group oligosaccharides in human milk. Oligosaccharides with blood group antigen activity reduced binding of toxigenic strains of *S. aureus* and also *Cl. perfringens* (Saadi et al., 1999; Gordon et al., 1999); however, much more work could be done on their role in neutralisation of toxins. These studies have implications not only for understanding the beneficial effects of breastfeeding, but also for development of infant formula preparations or perhaps supplements to rehydration therapies for treatment of diarrhoeal diseases.

It is becoming clear that the blood group antigens are recognition molecules of the immune/inflammatory systems. Although they have been studied in relation to transplantation and transfusion incompatibilities, very little is known about their role in immune and inflammatory responses to microorganisms. We predict that this is an area in which there will be significant interest and advances in elucidating the associations between blood groups and susceptibility to infections.

REFERENCES

Aird, I., Bentall, H. H., Mehigan, J. A., et al. (1954). The blood groups in relation to peptic ulceration and carcinoma of colon, rectum, breast and bronchus:

an association between the ABO groups and peptic ulceration. *Br. Med. J.* **ii**, 315–321.

Ali, S., Niang, M. A. P., N'doye, I., et al. (2000). Secretor polymorphism and human immune deficiency virus infection in Senegalese women. *J. Infect. Dis.* **181**, 737–739.

Alkout, A. H., Blackwell, C. C., Weir, D. M., et al. (1997). Isolation of a cell surface component of *Helicobacter pylori* that binds H type 2, Lewis[a] and Lewis[b] antigens. *Gastroenterology* **112**, 1179–1187.

Alkout, A. M., Blackwell, C. C., and Weir, D. M. (2000). The inflammatory response to *Helicobacter pylori* in relation to ABO blood group antigens. *J. Infect. Dis.* **181**, 1364–1369.

Alkout, A. M., Ramsay, E., Blackwell, C. C., et al. (2000). IgG levels to *Helicobacter pylori* among individuals who died of ischaemic heart disease compared with patients who experienced a first heart attack. *FEMS Immunol. Med. Microbiol.* **29**, 271–274.

Aly, F. Z., Blackwell, C. C., MacKenzie, D. A. C., et al. (1991). Chronic atrophic oral candidiasis among patients with diabetes mellitus – the role of secretor status. *Epidemiol. Infect.* **106**, 355–363.

Aly, F. Z. M., Blackwell, C. C., MacKenzie, D. A. C., et al. (1992). Factors influencing oral carriage of yeasts among individuals with diabetes mellitus. *Epidemiol. Infect.* **109**, 507–518.

Aniansson, G., Alm, B., Anderson, B., et al. (1992). Nasopharyngeal colonization during the first year of life. *J. Infect. Dis.* **165**(Suppl.), s38–s42.

Appelmelk, B. J., Simmons-Smit, I., Negrini, R., et al. (1996). Potential role of molecular mimicry between *Helicobacter pylori* lipopolysaccharide and host Lewis blood group antigens in autoimmunity. *Infect. Immun.* **64**, 2031–2040.

Arendrup, M., Hansen, J. E., Clausen, H., et al. (1991). Antibody to histo-blood group A antigen neutralizes HIV produced by lymphocytes from blood group A donors but not from blood group B or O donors. *AIDS* **5**, 441–444.

Atherton, J. C., Tham, K. T., Peek, R. M., et al. (1996). Density of *Helicobacter pylori* infection *in vivo* as assessed by quantitative culture and histology. *J. Infect. Dis.* **174**, 552–556.

Barnwell, J. W., Nicholas, M. E. and Rubenstein, P. (1989). *In vitro* evaluation of the role of the Duffy blood group in erythrocyte invasion by *Plasmodium vivax*. *J. Exp. Med.* **169**, 1795–1802.

Barra, J. L, Monferran, C. G., Balanzino, L. E., et al. (1992). *Escherichia coli* heat-labile enterotoxin preferentially interacts with blood group A-active glycolipids from pig intestinal mucosa and A- and B-active glycolipids from human red cells compared to H-active glycolipids. *Mol. Cell. Biochem.* **115**, 63–70.

Barua, A, D., and Paguio, A. S. (1977). ABO blood groups and cholera. *Ann. Hum. Biol.* **4**, 489–492.

Ben-Aryeh, H., Blumfield, E., Szargel, R., et al. (1995). Oral *Candida* carriage and blood group antigen secretor status. *Mycoses* **38**, 355–358.

Bennun, F. R., Roth, G. A., Monferrran, C. G., et al. (1989). Binding of cholera toxin to pig intestinal mucosa glycosphingolipids: relationship with the ABO blood group system. *Infect. Immun.* **57**, 969–974.

Beuth, J., Stoffel, B., and Pulverer, G. (1996). Inhibition of bacterial adhesion and infections by lectin blocking. In: *Toward Anti-Adhesin Therapy of Microbial Diseases*. Ofek, I., and Kahane, I. (eds.). New York: Plenum Press, pp. 51–56.

Bitzan, M., Richardson, S., Huang, C., et al. (1994). Evidence that verotoxins (Shiga-like toxins) bind to P blood group antigens of human erythrocytes. *Infect. Immun.* **62**, 3337–3347.

Black, R. E., Levine, M. M., Clements, M. L., et al. (1987). Association between O blood group and occurrence and severity of diarrhoea due to *Escherichia coli*. *Trans. Roy. Soc. Trop. Med. Hyg.* **81**, 120–123.

Blackwell, C. C. (1989). The role of ABO blood groups and secretor status in host defences. *FEMS Microbiol. Immunol.* **47**, 341–350.

Blackwell, C. C., Dundas, S., James, V. S., et al. (2002). Blood group and susceptibility to disease caused by *Escherichia coli* O157. *J. Infect. Dis.* **185**, 393–396.

Blackwell, C. C., Gordon, A. E., James, V. S., et al. (2002). The role of bacterial toxins in Sudden Infant Death Syndrome (SIDS). *Int. J. Med. Microbiol.* **291**, 561–570.

Blackwell, C. C., Gordon, A. E., James, V. S., et al. (2001). Making sense of the risk factors for Sudden Infant Death Syndrome (SIDS): infection and inflammation. *Rev. Med. Microbiol.* **12**, 1–11.

Blackwell, C. C., James, V. S., Davidson, S., et al. (1991). Secretor status and heterosexual transmission of HIV. *Br. Med. J.* **303**, 825–826.

Blackwell, C. C., James, V. S. Gemmill, J. D., et al. (1987). Secretor state of patients with type 1 or type 2 diabetes mellitus. *Br. Med. J.* **295**, 1024–1025.

Blackwell, C. C., Jonsdottir, K., Hanson, M. F., et al. (1986a). Non-secretion of ABO blood group antigens and susceptibility to infection with *Neisseria meningitidis* and *Streptococcus pneumoniae*. *Lancet* **ii**, 284–285.

Blackwell, C. C., Jonsdottir, K., Weir, D. M., et al. (1986b). Non-secretion of ABO blood group antigens predisposing to infection by *Haemophilus influenzae*. *Lancet* **ii**, 687.

Blackwell, C.C., MacKenzie, D. A. C., James, V. S., et al. (1999). Toxigenic bacteria and SIDS: nasopharyngeal flora in the first year of life. *FEMS Immunol. Med. Microbiol.* **25**, 51–58.

Blackwell, C.C., May, S.J., MacCallum, C.J., et al. (1989). Secretor state and susceptibility to recurrent urinary tract infections. In: *Host–Parasite Interactions in Urinary Tract Infections.* Kass, E.H., and Svanborg-Eden, C. (eds.). Chicago/London: University of Chicago Press, pp. 234–240.
Blackwell, C. C., Saadi, A. T., Raza, M. W., et al. (1992). Susceptibility to infection in relation to sudden infant death syndrome. *J. Clin. Pathol.* **45**(Suppl.), 20–24.
Blackwell, C. C., Weir, D. M., Alkout, A. M., et al. (1997). Bacterial infection and blood group phenotype. *Nova Acta Leopoldina* **301**, 67–88.
Blackwell, C. C., Weir, D. M., James, V. S., et al. (1990). Secretor status, smoking and carriage of *Neisseria meningitidis. Epidemiol. Infect.* **104**, 203–209.
Boren, T., Falk, P., Roth, K. A., et al. (1993). Attachment of *Helicobacter pylori* to human gastric epithelium mediated by blood group antigens. *Science* **262**, 1892–1895.
Braun, J. M., Blackwell, C. C., Poxton, I. R., et al. (2002). Pro-inflammatory responses to lipoologosaccharide *of Neisseria meningitidis* immunotype strains in relation to virulence and disease. *J. Infect. Dis.* **185**, 1431–1438.
Brown, K. E., Anderson, S. M. and Young, N. S. (1993). Erythrocyte P antigen: cellular receptor for B19 parvovirus. *Science* **262**, 114–117.
Cameron, B. and Douglas, L. J. (1996). Blood group glycolipids as epithelial receptors for *Candida albicans. Infect. Immun.* **64**, 891–896.
Chaudhary, R. and Shukla, J. S. (1999). Association of Lewis blood group with ischaemic heart disease. *Ind. J. Med. Res.* **109**, 103–104.
Chaudhuri, A. and Das Adhikary, C. R. (1978). Possible rolc of blood group secretory substances in the aetiology of cholera. *Trans. Roy. Soc. Trop. Med. Hyg.* **72**, 664–665.
Chaudhuri, A. and De, S. (1977). Cholera and blood groups. *Lancet* **ii**, 404.
Chesner, I. M., Nicholson, G., Ala, F., et al. (1992). Predisposition to gastric antral infection by *Helicobacter pylori*: an investigation of any association with ABO or Lewis blood group and secretor status. *Eur. J. Gastroenterol. Hepatol.* **4**, 377–379.
Chitnis, C. E., Chaudhuri, A., Horuk, R., et al. (1996). The domain on the Duffy blood group antigen for binding *Plasmodium vivax* and *P. knowlesi* malarial parasites to erythrocytes. *J. Exp. Med.* **184**, 1531–1536.
Ciopraga, J., Motas, C. and Doyle, R. J. (1995). Inhibition of saliva-induced oral streptococcal aggregation by blood group glycoproteins. *FEMS Immunol. Med. Microbiol.* **10**, 145–149.
Clarke, C. A., Edwards, J. W., Haddock, D. R. W., et al. (1956). ABO blood groups and secretor character in duodenal ulcer. *Br. Med. J.* **ii**, 725–731.

Clemmens, J. D., Sack, D. A., Harris, J. R., et al. (1989). ABO blood groups and cholera: new observations of specificity of risk and modification of vaccine efficacy. *J. Infect. Dis.* **159**, 770–773.

Collier, A., Patrick, A. W., Toft, A. D., et al. (1988). Increased prevalence of non-secretors in patients with Graves' disease – evidence for an infective aetiology? *Br. Med. J.* **296**, 1162.

Danesh, J. and Peto, R. (1998). Risk factors for coronary heart disease and infection with *Helicobacter pylori*: meta-analysis of 18 studies. *Br. Med. J.* **316**, 1130–1132.

Danesh, J., Youngman, L., Clark, S., et al. (1999). *Helicobacter pylori* infection and early onset myocardial infarction: case control and sibling pair study. *Br. Med. J.* **319**, 1157–1162.

Dickey, W., Collins, J. S. A., Watson, R. G. P., et al. (1993). Secretor status and *Helicobacter pylori* infection are independent risk factors for gastroduodenal disease. *Gut* **34**, 351–353.

El Ahmer, O. R., Essery, S. D., Saadi, A. T, et al. (1999a). The effect of cigarette smoke on adherence of respiratory pathogens to buccal epithelial cells. *FEMS Immunol. Med. Microbiol.* **23**, 27–36.

El Ahmer, O. R., Raza, M. W., Ogilvie, M. M, et al. (1999b). Binding of bacteria to HEp-2 cells infected with influenza A virus. *FEMS Immunol. Med. Microbiol.* **23**, 331–341.

Essery, S. D., Saadi, A. T., Twite, S. J., et al. (1994a). Lewis antigen expression on human monocytes and binding of pyrogenic toxins. *Agents Actions* **41**, 108–110.

Essery, S. D., Weir, D. M., James, V. S., et al. (1994b). Detection of microbial surface antigens that bind Lewis[a] blood group antigen. *FEMS Immunol. Med. Microbiol.* **9**, 15–22.

Gerhard, M., Lehn, N., Neumayer, N., et al. (1999). Clinical relevance of the *Helicobacter pylori* gene for blood-group antigen-binding adhesin. *Proc. Natl. Acad. Sci. USA* **96**, 12778–12783.

Glynn, J. R. (1994). *Helicobacter pylori* and the heart. *Lancet* **344**, 146.

Gordon, A. E., Saadi, A. T., MacKenzie, D. A. C., et al. (1999). The protective effect of breast feeding in relation to Sudden Infant Death Syndrome (SIDS). II. The effect of human milk and infant formula preparations on binding of *Clostridium perfringens* to epithelial cells. *FEMS Immunol. Med. Microbiol.* **25**, 167–173.

Grundbacher, F. J. (1972). Immunoglobulin, secretor status and the incidence of rheumatic heart disease. *Hum. Hered.* **25**, 399–404.

Harrison, L. M., Morris, J. A., Telford, D. R., et al. (1999). The nasopharyngeal bacterial flora in infancy: effects of age, gender, season, viral upper respiratory tract infections and sleeping position. *FEMS Immunol. Med. Microbiol.* **25**, 19–28.

Haverkorn, M. J. and Goslings, W. R. O. (1969). Streptococci, ABO blood groups and secretor status. *Am. J. Hum. Genet.* **21**, 360–375.

Hein, H. O., Suadicani, P. and Gyntelberg, F. (1997). Genetic markers for peptic ulcer. A Study of 3387 men aged 54 to 74 years: The Copenhagen Male Study. *Scand. J. Gastroenterol.* **32**, 16–21.

Heininger, U., Stehr, K., Schmidt-Schlapfer, G., et al. (1996). *Bordetella pertussis* infections and sudden unexpected deaths in children. *Eur. J. Pediatr.* **155**, 551–553.

Heneghan, M. A., McCarthy, C. F. and Moran, A. P. (2000). Relationship of blood group determinants on *Helicobacter pylori* lipopolysaccharide with host Lewis phenotype and inflammatory response. *Infect. Immun.* **68**, 937–934.

Heneghan, M. A., McCarthy, C. F., Janulaityte, D., et al. (2001). Relationship of anti-Lewis^x and anti-Lewis^y antibodies in serum samples from gastric cancer and chronic gastritis patients to *Helicobacter pylori*-mediated autoimmunity. *Infect. Immun.* **69**, 4774–4778.

Heneghan, M. A., Moran, A. P., Feeley, K. M., et al. (1998). The effect of host Lewis and ABO blood group antigen expression on *Helicobacter pylori* colonisation density and the consequent inflammatory response. *FEMS Immunol. Med. Microbiol.* **20**, 257–266.

Hook-Nikanne, J., Sistonen, P. and Kosunen, T. U. (1990). Effect of ABO blood group and secretor status on the frequency of *Helicobacter pylori* antibodies. *Scand. J. Gastorenterol.* **25**, 815–818.

Issit, P.D. (1986). *Applied Blood Group Serology, 3rd ed.* Miami: Montgomery, pp. 169–191.

Johnson, E. M., Warnock, D. W., Lucker, J., et al. (1995). Emergence of azole drug resistance in *Candida* species from HIV-infected patients receiving prolonged fluconazole therapy for oral candidosis. *J. Antimicrob. Chemother.* **35**, 103–114.

Kallenius, G., Mollby, R., Svensson, S. B., et al. (1980). The pk antigen as receptor for the haemagglutinin of pyelonephritic *Escherichia coli. FEMS Microbiol. Lett.* **7**, 297–302.

Kinane, D. F., Blackwell, C. C., Brettle, R. P., et al. (1982). Blood group, secretor state and susceptibility to recurrent urinary tract infection in women. *Br. Med. J.* **285**, 7–9.

Kinane, D. F., Blackwell, C. C., Winstanley, F. P., et al. (1983). Blood group, secretor status and susceptibility to infection by *Neisseria gonorrhoeae. Br. J. Vener. Dis.* **59**, 44–46.

Lagos, R., Avendano, A., Prado, V., et al. (1995). Attenuated live cholera vaccine strain CVD 103-HgR elicits significantly higher serum vibriocidal antibody titres in persons of blood group O. *Infect. Immun.* **63**, 707–709.

Lin, C. W., Chang, Y. S., Wu, S. C., et al. (1998). *Helicobacter pylori* in gastric biopsies of Taiwanese patients with gastroduodenal diseases. *Jpn. J. Med. Sci. Biol.* **51**, 13–23.

Lindgren, C., Milerad, J. and Lagercrantz, H. (1997). Sudden infant death and prevalence of whooping cough in the Swedish and Norwegian communities. *Eur. J. Paediatr.* **156**, 405–409.

Loffeld, R. J. and Stobberingh, E. (1991). *Helicobacter pylori* and ABO blood groups. *J. Clin. Pathol.* **44**, 516–517.

Lomberg, H., Cedergren, B., Leffler, H., et al. (1986). Influence of blood group on the availability of receptors for attachment of uropathogenic *Escherichia coli*. *Infect. Immun.* **51**, 919–926.

Lomberg, H., Jodal, U., Leffler, H., et al. (1992). Blood group non-secretors have an increased inflammatory response to urinary tract infections. *Scand. J. Infect. Dis.* **24**, 77–83.

Luman, W., Alkout, A. M., Blackwell, C. C., et al. (1996). *Helicobacter pylori* in the mouth – negative isolation from dental plaque and saliva. *Eur. J. Gastroenterol. Hepatol.* **8**, 11–14.

MacDonald, J. C. and Zuckerman, A. J. (1962). ABO blood groups and acute respiratory virus disease. *Br. Med. J.* **ii**, 89–90.

Malam, J. E., Carrick, G. F., Telford, D. R., et al. (1992). Staphylococcal toxins and sudden infant death syndrome. *J. Clin. Pathol.* **445**, 716–721.

Marshall, B. J., Armstrong J. A., McGhechie, D. B., et al. (1985). *Campylobacter* infection and gastroduodenal disease. *Med. J. Austral.* **142**, 436–439.

Martin-de-Argila, C., Boixeda, D., Canton, R., et al. (1995). High seroprevalence of *Helicobacter pylori* infection in coronary heart disease. *Lancet* **346**, 310.

Mendall, M. A., Goggin, P. A. M., Molineaux, N., et al. (1994). Relation of *Helicobacter pylori* infection and coronary heart disease. *Br. Heart J.* **71**, 437–439.

McCarthy, D. (1984). Smoking and ulcer-time to quit. *New Eng. J. Med.* **13**, 726–727.

Mentis, A., Blackwell, C. C., Weir, D. M., et al. (1991). ABO blood group, secretor status and detection of *Helicobacter pylori* among patients with gastric or duodenal ulcers. *Epidemiol. Infect.* **106**, 221–229.

Migot-Nabias, F., Mombo, L. E., Luty, A. J., et al. (2000). Human genetic factors related to susceptibility to mild malaria in Gabon. *Genes Immun.* **1**, 4335–4341.

Miragliotta, G., Fimmarola, D., Mosca, A., et al. (1994). *Campylobacter pylori*-associated gastritis and procoagulant activity production. *Am. Soc. Microbiol. Ann. Meet.* B222.

Monferran, C. G., Roth, G. A., and Cumar, F. A. (1990). Inhibition of cholera toxin binding to membrane receptors by pig gastric mucin-derived glycopeptides: differential effect depending on the ABO blood group antigenic determinants. *Infect. Immun.* **58**, 3966–3972.

Morgando, A., Sanseverina, P., Perotto, C., et al. (1995). *Helicobacter pylori* seropositivity in myocardial infarction. *Lancet* **345**, 1380.

Mourant, A. E., Kopec, A. C. and Domaniewska-Sobczak, K. (1978). *Blood Groups and Diseases.* Oxford: Oxford University Press.

Mouricourt, M., Petit, J. M., Carias, J. R., et al. (1990). Glycoprotein glycans that inhibit adhesion of *Escherichia coli* mediated by K99 fimbriae: Treatment of experimental colibacillosis. *Infect. Immun.* **58**, 98–106.

Munoz, L., Gonzalez-Valencia, G., Perez-Perez, G. I., et al. (2001). A comparison of Lewis^x and Lewis^y expression in *Helicobacter pylori* obtained from children and adults. *J. Infect. Dis.* **183**, 1147–1151.

Murray, L. J., Bamford, K. B., O'Reilly, D. P., et al. (1995). *Helicobacter pylori* infection: relation with cardiovascular risk factors, ischaemic heart disease, and social class. *Br. Heart. J.* **74**, 497–501.

Newbould, M. J., Malam, J., McIllmurray, J. M., et al. (1989). Immunohistological localisation of staphylococcal toxic shock syndrome toxin (TSST-1) in sudden infant death syndrome. *J. Clin. Pathol.* **42**, 935–939.

Nicholl, A. and Gardner, A. (1988). Whooping cough and unrecognized post-perinatal mortality. *Arch. Dis. Child.* **63**, 41–47.

Odds, F. C. (1996a). Resistance of yeasts to azole derivative antifungals. *J. Antimicrob. Chemother.* **31**, 463–471.

Ofek, I. and Kahane, I. (1996). *Toward Anti-Adhesin Therapy of Microbial Diseases.* New York: Plenum.

Ohshima, H. and Bartsh, H. (1994). Chronic infections and inflammatory processes as cancer risk factors: possible role of nitric oxide in carcinogenesis. *Mutation Res.* **305**, 253–264.

Potter, C. W. (1969). Haemagglutination inhibition antibody to various influenza viruses and adenoviruses in individuals of blood groups A and O. *J. Hyg.* **67**, 67–74.

Race, R.R. and Sanger, R. (1975). *Blood Groups in Man, 6th ed.* Oxford: Blackwell Scientific.

Ratner, J. J., Thomas, V. L. and Forland, M. (1986). Relationship between human blood groups, bacterial pathogens, and urinary tract infections. *Am. J. Med. Sci.* **292**, 87–91.

Raz, R., Gennensin, Y., Wasser, J., et al. (2000). Recurrent urinary tract infections in post menopausal women. *Clin. Infect. Dis.* **30**, 152–156.

Raza, M. W., Blackwell, C. C., Molyneaux, P., et al. (1991). Association between secretor status and respiratory viral illness. *Br. Med. J.* **303**, 815–818.

Raza, M. W., Ogilvie, M. M., Blackwell, C. C., et al. (1999). Infection with respiratory syncytial virus enhances expression of native receptors for non-pilate *Neisseria meningitidis* on HEp-2 cells. *FEMS Immunol. Med. Microbiol.* **23**, 115–124.

Raza, M. W., Ogilvie, M. M., Blackwell, C. C., et al. (1993). Effect of respiratory syncytial virus infection on binding of *Neisseria meningitidis* and type b *Haemophilus influenzae* to human epithelial cell line (HEp-2). *Epidemiol. Infect.* **110**, 339–347.

Reed, W. P., Drach, F. W. and Williams, R. C., Jr. (1974). Antigens common to human and bacterial cells. IV. Studies of human pneumococcal disease. *J. Lab. Clin. Med.* **93**, 599–610.

Rex, J. H., Rinaldi, M. G. and Pfuller, M. A. (1995). Resistance of *Candida* species to fluconazole. *Antimicrob. Agents Chemother.* **39**, 1–8.

Robinson, M. G., Tolchin, D. and Halpern, C. (1971). Enteric bacteria and the ABO blood groups. *Am. J. Hum. Genet.* **23**, 135–145.

Saadi, A. T., Blackwell, C. C., Essery, S. D., et al. (1996). Comparison of human milk and infant formula on inhibition of bacterial binding and neutralisation of toxins. *Fourth SIDS International* A166.

Saadi, A. T., Blackwell, C. C., Essery, S. D., et al. (1996a). Developmental and environmental factors that enhance binding of *Bordetella pertussis* to human epithelial cells in relation to sudden infant death syndrome. *FEMS Immunol. Med. Microbiol.* **16**, 81–89.

Saadi, A. T., Blackwell, C. C., Raza, M. W., et al. (1993). Factors enhancing adherence of toxigenic staphylococci to epithelial cells and their possible role in sudden infant death syndrome. *Epidemiol. Infect.* **110**, 507–517.

Saadi, A. T., Gordon, A. E., MacKenzie, D. A. C., et al. (1999). The protective effect of breast feeding in relation to Sudden Infant Death Syndrome (SIDS). I. The effect of human milk and infant formula preparations on binding of toxigenic *Staphylococcus aureus* to epithelial cells. *FEMS Immunol. Med. Microbiol.* **25**, 155–165.

Saadi, A. T., Mackenzie, D.A. C, Weir, D.M., et al. (1998). Detection of fucose binding surface components on *Candida* species. In: *Fourth Congress of the European Confederation of Medical Mycology, Glasgow, Scotland.*

Saadi, A. T., Weir, D. M., Poxton, I. R., et al. (1994). Isolation of an adhesin from *Staphylococcus aureus* that bind Lewis[a] blood group antigen and its relevance to sudden infant death syndrome. *FEMS Immunol. Med. Microbiol.* **8**, 315–320.

Schaeffer, A. J., Bavas, E. L., Venegas, M. F., et al. (1994). Variation of blood group antigen expression on vaginal cells and mucus in secretor and nonsecretor women. *J. Urol.* **152**, 859–864.

Shimazu, T., Shimaoka, M., Sugimoto, H., et al. (2000) Does blood group B protect against haemolytic uraemic syndrome? An analysis of the 1996 Sakai outbreak of *Escherichia coli* O157H7 (VETC O157) infection. The Osaka Critical Care Study Group. *J. Infect.* **41**, 45–49.

Socha, W., Belinska, M. and Kaczera, A. (1969). *Escherichia coli* and ABO blood groups. *Folia Biol (Krakova)* **17**, 259–269.

Springer, G. F. (1970). Importance of blood-group substances in interactions between man and microbes. *Ann. N. Y. Acad. Sci.* **169**, 134–152.

Springer, G. F., Williamson, P. and Brandes, W. C. (1971). Blood group activity of Gram-negative bacteria. *J. Exp. Med.* **113**, 1077–1093.

Stapleton, A., Hooton, T. M., Fennell, C., et al. (1995). Effect of secretor status on vaginal and rectal colonization with fimbriated *Escherichia coli* in women with and without recurrent urinary tract infection. *J. Infect. Dis.* **171**, 717–720.

Stapleton, A., Nudelman, E., Clausen, H., et al. (1992). Binding of uropathogenic *Escherichia coli* R45 to glycolipids extracted from vaginal epithelial cells is dependent on histo-blood secretor status. *J. Clin. Invest.* **90**, 965–972.

Suadicani, P., Hein, H. O. and Gyntelberg, F. (2000). Socioeconomic status, ABO phenotypes and risk of ischaemic heart disease: An 8-year follow-up study in the Copenhagen Male Study. *J. Cardiovasc. Risk.* **7**, 2777–283.

Swerdlow, D. L., Mintz, E. D., Rodriguez, M., et al. (1994). Severe life-threatening cholera associated with blood group O in Peru: Implications for the Latin American epidemic. *J. Infect. Dis.* **170**, 468–472.

Tesh, V. L. (1998). Virulence of enterohemorrhagic *Escherichia coli*: Role of molecular cross talk. *Trends Microbiol.* **6**, 228–233.

Thoreson, A. C., Hamlet, A., Celik, J., et al. (2000). Differences in surface-exposed antigen expression between *Helicobacter pylori* strains isolated from duodenal ulcer patients and from asymptomatic subjects. *J. Clin. Microbiol.* **38**, 3436–3441.

Vallance, P., Collier, J. and Bhagat, K. (1997). Infection, inflammation, and infarction: Does acute endothelial dysfunction provide a link? *Lancet* **349**, 1391–1392.

Van Alphen, L., Poole, J. and Overbeeke, M. (1986). The Anton blood group antigen is the erythrocyte receptor for *Haemophilus influenzae. FEMS Microbiol. Lett.* **37**, 69–71,

Van t'Wout, J., Burnette, W. N., Mar, V. L., et al. (1992). Role of carbohydrate recognition domains of pertussis toxin in adherence of *Bordetella pertussis* to human macrophages. *Infect. Immun.* **60**, 3303–3308.

Viskum, K. (1975). The ABO and Rhesus blood groups in patients with pulmonary tuberculosis. *Tubercle* **56**, 329–334.

Waissbluth, J. G. and Langman, J. S. (1971). ABO blood group, secretor status, salivary protein and serum and salivary immunoglobulin concentrations. *Gut* **12**, 646–649.

Weinmeister, K. D. and Dal Nogare, A. R. (1994). Buccal cell carbohydrates are altered during critical illness. *Am. J. Respir. Crit. Care Med.* **150**, 131–134.

Willcocks, L., Leen, C. L. S., Breettle, R. P., et al. (1991). Increase in *Candida krusei* infection among patients with bone marrow transplantation and neutropenia treated prophylactically with fluconazole. *N. Eng. J. Med.* **325**, 1274–1277.

Yang, H. B., Sheu, B. S., Chen, R. G., et al. (2001). Erythrocyte Lewis antigen phenotypes of dyspeptic patients in Taiwan – Correlation of host factor with *Helicobacter pylori* infection. *J. Formos. Med. Assoc.* **100**, 227–232.

Zopf, D. and Roth, S. (1996). Oligosaccharide anti-infective agents. *Lancet* **347**, 1017–1021.

Zopf, D., Simon, P., Barthelson, R., et al. (1996). Development of anti-adhesion carbohydrate drugs for clinical use. In: *Toward Anti-Adhesin Therapy of Microbial Diseases*. Ofek, I., and Kahane, I. (eds.). New York: Plenum, pp. 35–39.

Zorgani, A. A., Essery, S. D., Al Madani, O., et al. (1999). Detection of pyrogenic toxins of *Staphylococcus aureus* in cases of sudden infant death syndrome. *FEMS Immunol. Med. Microbiol.* **25**, 103–108.

Zorgani, A. A., James, V. S., Stewart, J., et al. (1996). Serum bactericidal activity in a secondary school population following an outbreak of meningococcal disease: effects of carriage and secretor status. *FEMS Immunol. Med. Microbiol.* **14**, 73–82.

Zorgani, A. A., Stewart, J., Blackwell, C. C., et al. (1994). Inhibitory effect of saliva from secretors and non-secretors on binding of meningococci to epithelial cells. *FEMS Immunol. Med. Microbiol.* **9**, 135–142.

Zorgani, A. A., Stewart, J., Blackwell, C. C., et al. (1992). Secretor status and humoral immune responses to *Neisseria lactamica* and *Neisseria meningitidis*. *Epidemiol. Infect.* **109**, 445–452.

CHAPTER 12

Genetics of human susceptibility to infection and hepatic disease caused by schistosomes

Alain J. Dessein, Sandrine Marquet, Carole Eboumbou Moukoko, Hélia Dessein, Laurent Argiro, Sandrine Henri, Dominique Hillaire, and Christophe Chevillard

Immunologie et Génétique des Maladies Parasitaires, Institut National de la Santé et de la Recherche Médicale, Laboratoire de Parasitologie-Mycologie, Faculté de Médecine, Marseille, France

Nasureldin El Wali and Mubarak Magzoub

Institute of Nuclear Medicine and Molecular Biology, University of Gezira, Wad Medani, Sudan

Laurent Abel

Génétique Humaine des Maladies Infectieuses, Institut National de la Santé et de la Recherche Médicale, Faculté de Médecine Necker, Paris

Virmondes Rodrigues, Jr. and Aluizio Prata

Faculty of Medicine do Triangulo Mineiro, Ubéraba, Brazil

Gachuhi Kimani

Kenya Medical Research Institute, Biomedical Sciences Research Centre, Nairobi, Kenya

Schistosome infections cause much suffering in millions of people living in tropical regions of Africa, Asia, and South America (Prata, 1987; Chitsulo et al., 2000). The most severe clinical symptoms affect the kidneys and urinary tract. However, schistosomes also cause various other disorders such as heart failure and neurological diseases. Three species of schistosome are responsible for most human infections (*Schistosoma mansoni, Schistosoma japonicum,* and *Schistosoma haematobium*). These species are found in different geographical locations, have different vectors, and cause different symptoms. Schistosomes are multicellular parasites that are disseminated as free swimming larvae (cercariae) in ponds, lakes, and rivers by snails. Humans become infected when they stay in contaminated water for a few minutes. The cercariae penetrate the human skin and develop into male or female adult schistosomes within 5 or 6 weeks. These small worms (Fig. 12.1A) can live in the vascular system of their vertebrate host for 2 to 5 years. Schistosomes do not multiply within their vertebrate host. The female worms, however, lay

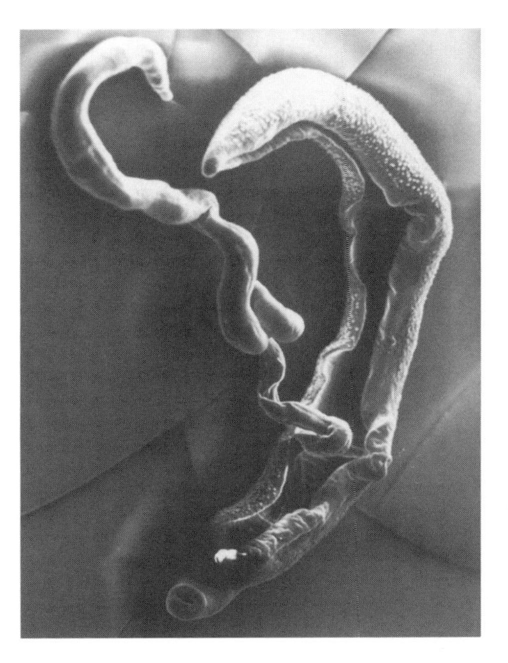
A

B
C

hundreds of eggs per day in the mesenteric or vesical veins of their host. Most of the symptoms associated with these infections are caused by the inflammation that is induced by the immunogenic and toxic substances produced by the eggs. The chronic cellular reaction that develops around the eggs is organised in a granuloma (Von Lichtenberg, 1962; Warren et al., 1967). Various organs can be affected depending on the tissues in which the eggs are trapped: intestine, liver, gall bladder, ureter, lungs, and the central nervous system.

As for many infectious diseases, schistosomes cause mild clinical symptoms in most subjects, whereas a small proportion of individuals (up to 15% in certain endemic areas) present severe clinical disease that may lead to death. Various hypotheses have been put forward to explain the heterogeneous distribution of severe cases of schistosomiasis in endemic populations: existence of pathogenic 'variants' of schistosomes, heavy infections, co-infections with viruses such as the hepatitis B virus or genetic predisposition of the host.

This report summarises the observations made by our group, showing that the genetic background of the host plays a major role in controlling infection and disease in schistosome infections. We also have observed that hepatitis viruses aggravate schistosomiasis and probably play an important role in the high prevalence of severe schistosomiasis in certain endemic areas. Our studies have failed, however, to detect more pathogenic or more virulent 'variants' among various schistosomes isolated from subjects with different clinical statuses.

1. HUMAN RESISTANCE TO INFECTION BY SCHISTOSOMES IN A BRAZILIAN POPULATION

1.1 Evidence that a Major Gene is Involved in the Control of Infection

We began our studies in a Brazilian village located in an area endemic for schistosomiasis. This village, Caatinga do Moura (CM), had already been studied by A. Prata and his team. Our study included the subjects living on the west bank of a small river that passed through the village, because this

Figure 12.1. (*facing page*). (A) Male and female adult schistosomes. The female lodges in the gynaecophoral canal of the male. Each worm measures approximately 0.5 to 1.5 cm in length. Hepatic vascular trees of a liver from a healthy subject who died in a car accident (1B) and of a liver from a subject who died with severe PPF due to *Schistosoma mansoni* infection (1C). The hepatic vascular tree was injected with acrylic blue liquid resin. After polymerisation, tissues were digested. Courtesy of Dr. E. Chapadeiro, Faculty of Medicine Uberaba, Brazil. (See color plate.)

Figure 12.2. Infection levels were taken as the mean parasite egg count in stool samples. When three to five stool samples are taken on different days this provides an accurate evaluation of the parasite load. Exposure was calculated from the water contact index as described previously (Dessein et al., 1988; Abel et al., 1991).

group had not been studied previously and therefore had not been treated with antihelminthic drugs for a long time. Furthermore, these subjects were geographically clustered and came into contact with the river at a few major sites that were easy to survey. A large number of these individuals belonged to large families (Dessein et al., 1992).

We were interested in the factors that determine infection levels, as it was thought that high infection could be a major cause of clinical disease. A critical epidemiological factor in infectious diseases is 'exposure' as exposure differences might have significant effects on infection. Another important factor is the age of the subjects, as age may influence the subject's comportment or immunity. In schistosome-infected populations, both age and exposure have important effects on infection levels (Fig. 12.2 and Butterworth et al., 1985; Dessein et al., 1988; Abel et al., 1991).

Age, exposure, and gender only explained a third of the variance of infection levels in this population, indicating that other unidentified factors are involved (Dessein et al., 1988; Abel et al., 1991). As high infection levels were clustered in certain families, we postulated that some inherited factors or some inherited but not genetic factors such as cultural habits might also have a critical effect on infection (Abel et al., 1991). Therefore, we tested whether infection levels were subject to genetic control in CM families. First,

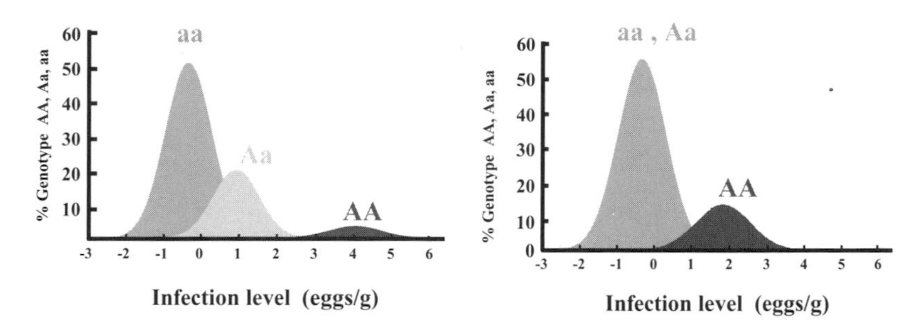

Figure 12.3. Distribution of predicted genotypes according to the results of segregation analysis in Caatinga (Brazil) and Ulillinzi (Kenya). Infection levels were adjusted for covariates that have significant effects on infection: age, gender, and exposure.

we looked for a major genetic effect at a single locus. Segregation analysis allowed us to test this hypothesis taking into account the effects of exposure, age, and gender on the phenotype (Abel et al., 1991).

Therefore, infection levels were adjusted for exposure, age, and gender. We tested whether the distribution of the trait (infection levels) in CM families (20 pedigrees, 269 individuals) supported the hypothesis of a single locus controlling the trait. This analysis provided clear and convincing evidence that infection levels are controlled by a major codominant locus. This means that the distribution of infection levels in families is best explained by a model that includes age, gender, exposure, and a major codominant gene effect. Importantly, the effects of this genetic control account for half of the variance in infection levels indicating that gene(s) at this locus exerts a strong genetic control on infection. Figure 12.3 shows the distribution of the genotypes based on the hypothesis that two alleles, resistant (a) and susceptible (A), of a single major gene account for the effects of the major locus. The frequency of the deleterious allele was estimated to be between 0.20 and 0.25; about 5% of the population was predisposed to a high level of infection, 60% were resistant and 35% had an intermediate level of resistance. This model is likely to be an oversimplification of the real situation. Susceptibility alleles are more likely to be haplotypes that segregate as a block in the families. Nevertheless, the important conclusion was that the genetic control of infection can best be explained by the segregation of a single genetic region. These findings provided the first genetic explanation for the earlier observations on the predisposition of certain individuals living in areas endemic for schistosomiasis to reinfection.

It is important to know whether this conclusion can be extended to other populations living in different epidemiological conditions. It seems that this might be the case: we provided evidence for a major genetic control of infection in a Kenyan population (Fig. 12.3; Dessein and Gachuhi, unpublished data). Furthermore, as we shall see below, Horstmann and co-workers identified a major locus involved in the control of *S. mansoni* infection in a Senegalese population (Muller-Myhsok et al., 1997). Nevertheless, these results, providing strong evidence that one or several genes at a major locus are involved in the control of infection levels by *S.mansoni*, do not exclude the possibility that other minor susceptibility loci that could not be detected with the strategies described above are involved.

1.2 Localisation of *SM1* on Chromosome 5q31-q33

Linkage analysis using data from a whole genome scan was used to map the locus that controls infection levels in the CM population. The genetic model was refined in a second segregation analysis, which took the covariates (exposure, age, gender) into account together with the major gene effect. The codominant major gene model was obtained with similar parameters as those described above. According to this model, the effect of the locus accounts for 66% of the infection intensity residual variance from covariate effects and the frequency of the allele predisposing to high infection levels is 0.16 (Marquet et al., 1996).

The linkage analysis was conducted on 142 Brazilian subjects belonging to 11 informative families (2 large pedigrees, 5 smaller pedigrees, and 4 nuclear families) (Marquet et al., 1998). The genome was scanned with 246 polymorphic microsatellites, corresponding to a 15-cM map. Fifty-four markers provided maximum lod scores of above 0.1, but only one region on chromosome 5 (5q31-q33) showed suggestive linkage. Two adjacent markers, D5S393 and D5S410, provided a maximum lod score of 3.18 and 3.06 for a recombination fraction (θ) of 0.09 and 0.15, respectively. To investigate this region, 11 additional markers were analysed and significant linkage (lod score >3.3) was observed for two close markers. The maximum 2-point lod score was 4.74 ($\theta : 0.07$) for D5S636 and 4.52 ($\theta : 0.04$) for the colony stimulating factor (CSF1R) according to the estimated marker allele frequencies. This successful mapping confirmed the existence of a major gene, denoted *SM1*, controlling the intensity of *S. mansoni* infection and showed that this gene is localised on chromosome 5q31-q33. Multipoint linkage analysis including five markers for this region indicated that the most likely location of *SM1* was close to the CSF1R marker, with a maximum lod score of 5.45.

Horstmann and co-workers also analysed genetic predisposition to *S. mansoni* infection in Richard-Toll in Senegal (Muller-Myhsok et al., 1997).

Unlike the subjects that we studied in Caatinga, the subjects in this study were not born in an endemic area, as this area was only infected by *S. mansoni* recently (less than 7 years ago) following a major outbreak of schistosomiasis in the region. The Senegalese study confirmed the presence of a major *S. mansoni* susceptibility locus on chromosome 5q31-q33 (Muller-Myhsok et al., 1997). Thus, one or several genes in the 5q31-q33 region exert a major control on infection. This control acts in two different populations that have been immunised against schistosomes in very different epidemiological conditions. Furthermore, this finding suggests that this locus also plays an important role in the control of other infections as susceptibility alleles at this locus could not have been selected by schistosomes in the Senegalese population. This view is also supported by (1) the observations made in our laboratory that gene(s) in 5q31-q33 also control(s) blood parasitaemia in *P. falciparum* infections (Garcia et al., 1998, Rihet et al., 1998) and (2) the presence in the 5q31-q33 region of a large number of genes that encode molecules such as IL-4, IL-5, IL-12p40, and IL-13 that play key functions in T-cell-dependent immunity to helminths, protozoa, bacteria, and viruses. Finally, this locus has also been linked to the control of blood IgE levels and to eosinophilia (Marsh et al., 1994; Martinez et al., 1998; Rioux et al., 1998). Both eosinophils and IgE are critical effectors of human protective immunity against helminths.

1.3 Search for Additional Loci Controlling the Levels of *S. mansoni* Infection

The strategy used to map the major gene could not detect other genetic effects as the linkage analysis was based on the major gene model provided by the segregation analysis. To determine whether additional loci are involved in the control of the intensity of *S. mansoni* infection, a model-free weighted pairwise correlation analysis was applied to the genome scan data for the Brazilian pedigrees. The most significant linkage was observed in the 5q31-q33 region, confirming that this locus has a major effect (Zinn-Justin et al., 2001). When *SM1* was taken into account, two additional regions (1p21-q23 and 6p21-q21) showed linkage with significance levels of about 0.001. The result for the 1p21-q23 region was independent of *SM1*, whereas the locus in the 6p21-q21 region interacted with *SM1*. The 1p21-q23 region, which yielded the higher multipoint results, contains the CSF1 gene. Interestingly, the gene encoding the CSF1 receptor is located in the 5q31-q33 region. The locus in 6p21-q21 is located between the MHC locus and IFNGR1 (see below). Additional studies on larger populations are required to confirm the existence of these minor loci involved in susceptibility to *S. mansoni* infections.

Thus, we can conclude that the intensity of *S. mansoni* infection in humans is markedly influenced by host genetics. One major susceptibility locus has been identified in the 5q31-q33 region. This region contains various genes encoding molecules that are critical in human defences against parasites, including helminths. A model-free linkage analysis in one population suggested that two other susceptibility loci also exist. These loci are located in 6p21-q21 and in 1p21-q23. The existence of these minor loci must now be confirmed by studies in other populations.

2. HEPATIC FIBROSIS IN SCHISTOSOME-INFECTED SUBJECTS

2.1 The Egg Granuloma

Adult *Schistosoma mansoni* live in the portal vein of their human host. Female worms lay large numbers of eggs, which cross the intestinal wall and are eliminated with the faeces. Some eggs, however, enter the blood flow in the hepatic portal vein and remain trapped in the hepatic sinusoids, where they trigger an inflammatory reaction. Antigenic materials released by adult worms also contribute to this immune reaction in the periportal space. This inflammation causes the major clinical manifestations of the disease: portal hypertension and hepatosplenomegaly. Portal hypertension results from an excessive accumulation of extracellular matrix protein (ECMP) in the periportal space. This excess ECMP forms a network of cross-linked proteins that is referred to as periportal fibrosis (PPF). PPF occludes small and large vessels, leading to the degeneration of the hepatic vascular tree (Figs. 12.1B and 12.1C). Hepatosplenomegaly is partly caused by portal hypertension (congestive splenomegaly) and partly caused by the multiplication of cells like Kupffer cells and Ag-stimulated lymphocytes (Grimaud et al., 1977). It was long believed that hepatosplenomegaly developed only in subjects with severe PPF. Recent ultrasonic observations have shown that splenomegaly can also occur in schistosomes-infected subjects with mild PPF (Mohamed-Ali et al., 1999).

Experimental studies have shown that the immune reaction caused by schistosome eggs in the liver is a manifestation of cell-mediated immunity that is regulated by egg-specific CD4+ T lymphocytes (Warren, 1977). Eggs possess a shell that resists digestion by phagocytes such as eosinophils, macrophages, and neutrophils. Therefore, the anti-egg immune reaction tends to organise in a granuloma where the inflammatory infiltrate is progressively replaced by fibroblasts and a dense network of ECMP. The changing properties of the T lymphocytes that participate in the granuloma have been studied. In the first weeks following oviposition, CD4 T lymphocytes producing Th1-like cytokines are produced, later on the granuloma depends on a Th2-type response involving IL-4, IL-5, and IL-13 (Chensue et al., 1994; Wynn et al., 1997; Kaplan et al., 1998). In mice, the granulomatous reaction is down-regulated 5 or 6 weeks after the beginning of oviposition. In humans, both Th1 cytokines and Th2 cytokines are released by egg-stimulated PBMC in subjects with chronic infections. No studies have looked at the inflammatory cells in the human hepatic granuloma.

2.2 Susceptibility to Periportal Fibrosis Caused by *Schistosoma mansoni* in a Sudanese Population

In regions endemic for schistosomiasis, 5 to 10% of infected subjects develop a severe hepatic disease characterised by hepatosplenomegaly, portal hypertension, ascites, and haematemesis. It has been suggested that high infection levels are a major cause of disease because advanced schistosomiasis is more frequent in populations in which infection is highly prevalent and infection levels are high (World Health Organization, 1974). To evaluate the risk factors for severe hepatic disease, we studied a population living in an endemic area of Gezira, Sudan, that had not been treated with antihelminthic drugs for a long time. Killing the worms is known to interrupt disease development. *Schistosoma mansoni* has been endemic in the irrigated region of Gezira for more than 50 years. An attempt was made to control transmission 20 years ago by treating infected subjects. Since this time, however, drugs have not been available to prevent human infections or to cure infections. As most agricultural activities depend on water from irrigation channels, most inhabitants of the poor villages have been exposed to infection since they were born or since they moved to this irrigated area. We selected one village in which the prevalence of schistosome infection was elevated. Ultrasonography was used to test the whole village population for PPF (Mohamed-Ali et al., 1999). PPF was observed in 11.3% of subjects ($n = 792$) and was 4.5 times more frequent in males than in females. PPF was graded in four stages (none, mild, advanced, and severe PPF) according to the WHO guidelines. A few of the advanced cases and almost all of the severe cases showed evidence of portal hypertension. As shown in Fig. 12.4 a significant fraction (>20%) of the population exhibited no signs of fibrosis (F0), regardless of age class; the prevalence of PPF (FII and FIII) was considerably elevated in the 20/25/30-years age classes of males; finally no more than a quarter of the subjects with advanced PPF developed severe PPF (FII or FIII with portal hypertension). The data suggested that certain subjects rapidly developed severe PPF,

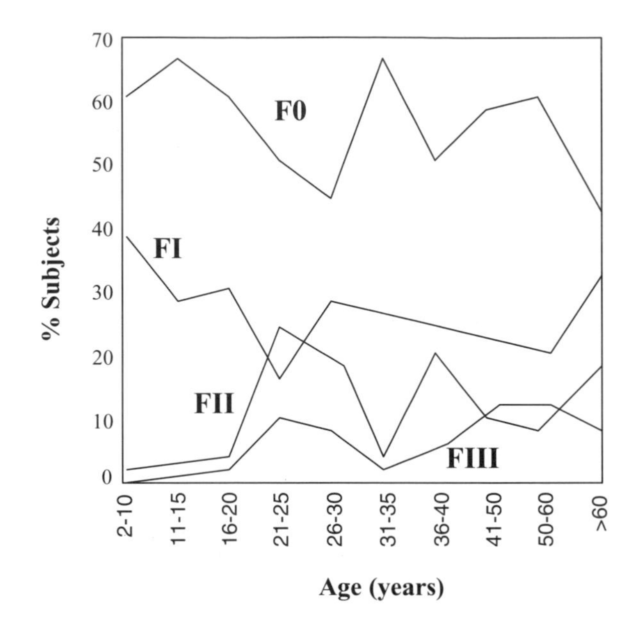

Figure 12.4. Prevalence of fibrosis by age in the study population: F0, no fibrosis; FI, mild fibrosis; FII, moderate fibrosis; FIII, advanced fibrosis.

whereas others either did not develop fibrosis at all or progressed slowly to severe grades. This is similar with the conclusion of a study by Poynard and co-workers (Poynard et al., 1997), who described rapid, intermediate, and slow fibrosers in a cohort of subjects infected by HCV. This raises questions about the nature of the factors that determine the rate of fibrosis progression. Gender is clearly an important factor in hepatic fibrosis. The low prevalence of PPF in females in Taweela (Mohamed-Ali et al., 1999) is probably related to the mechanisms of pathogenesis and not only to exposure as we first thought. Our data also showed that the intensity of infection is not a major factor in disease progression in this population.

PPF was frequent in certain families and absent in others. An analysis of the living habits of the families did not reveal any differences that could explain this and suggested that some inherited factors are involved (Dessein et al., 1999). To test for the existence of inherited genetic factors we performed a segregation analysis on 65 families [5 large pedigrees (>30 members), 29 smaller pedigrees, and 31 nuclear families]. The studied phenotype was defined as 'advanced (FII) or severe PPF (FIII) associated with portal hypertension'. The results indicated that PPF adjusted for environmental covariates segregated in families as a Mendelian trait and that a gender dimorphism might be involved in the control of the trait. A co-dominant gene named *SM2* accounted for the familial distribution of the phenotype. The frequency of the deleterious allele was estimated to be 0.16 and the respective proportions of homozygous susceptible, heterozygous, and homozygous resistant subjects were 0.03, 0.27, and 0.70. The most important predictions of the genetic model were (1) the frequency of the disease-causing allele (A) is elevated and (2) 50% penetrance is reached after 9, 14, and 19 years of residency in the area for AA males, AA females, and Aa males, respectively, whereas the penetrance remains lower than 2% after 20 years of exposure for other subjects. Nevertheless, with a sufficiently long time of residence, all heterozygote males are likely to present the disease. Consequently, in this population, 30% (3% of homozygotes and 27% heterozygotes) of males could potentially develop severe schistosomiasis if left untreated. The estimated penetrance of *SM2* strongly depends on gender, which explains the lower prevalence of fibrosis in females than in males. Gender differences in the prevalence of fibrosis have also been reported by other groups working in Sudan (Homeida et al., 1988a, 1988b) and in Egypt (Abdel-Wahab et al., 1990).

2.3 Severe Hepatic Fibrosis is Controlled by *SM2* Located in the 6q22-q23 Region

To localize *SM2*, we first tested the linkage of *SM2* with four chromosomal regions that contain good candidates genes (Dessein et al., 1999): (1) the 5q31-q33 region that contains *SM1* and several of the candidate genes, such as those coding for the granulocyte-macrophage colony stimulating factor (*CSF2*), several interleukins (*IL-13, IL-3, IL-4, IL-5, IL-9*), the interferon regulatory factor 1 (*IRF1*), the colony stimulating factor-1 receptor (*CSF-1R*), and gene(s) controlling total serum IgE levels and familial hypereosinophilia; (2) the HLA-TNF region (6p21) containing the HLA locus and the *TNF-α* and *TNF-β* genes; (3) the 12q15 region, including the *IFN-γ* gene and a gene controlling total serum IgE levels; and (4) the 6q22-q23 region containing the *IFN-γR1* gene. Genotyping was carried out on all informative families with multiple cases of severe fibrosis (8 families including 112 individuals) for 19 markers located in the four candidate regions. Two-point linkage analysis was performed and significant maximum lod scores (Z_{max}) were obtained in the 6q22-q23 region, with both D6S310 ($Z_{max} = 2.81$ at $\theta = 0$) and the FA1 intragenic marker ($Z_{max} = 1.80$ at $\theta = 0$) (Dessein et al., 1999). Similar results were observed using equal marker allele frequencies for D6S310 ($Z_{max} = 2.77$ at $\theta = 0$) and FA1 ($Z_{max} = 2.14$ at $\theta = 0$). No Z_{max} values of above 0.1 were observed with any markers from regions 5q31-q33 and 6p21. In the 12q15 region, D12S83 provided a Z_{max} of 0.44 at $\theta = 0.08$, but multipoint analysis showed a lod score <-4 at a location corresponding to the gene coding for IFNγ. Combined segregation–linkage analysis was performed with the D6S310 microsatellite from the 6q22-q23 region. The maximum lod score was 3.11 at $\theta = 0$. The parameters of *SM2* estimated using this combined analysis were very close to those obtained with the previous segregation analysis. For AA males and AA females, the penetrance is almost complete after 12 and 17 years of exposure, respectively, whereas for Aa males the penetrance is only 73% after 20 years of exposure. For other subjects, the penetrance was less than 2% after 20 years of residency in the area. The multipoint analysis using four markers simultaneously provided a Z_{max} of 3.12 with the D6S310 microsatellite and a Z_{max} of 2.49 with the FA1 intragenic marker.

These results indicate that *SM2* is located in the 6q22-q23 region, close to the IFNGR1 gene, which encodes the β-chain of the IFN-γ receptor and could be a good candidate gene according to animal model studies (Czaja et al., 1987, 1989). The fact that *SM1* and *SM2* are located on two different chromosomes (5q31-q33 and 6q22-23, respectively) indicates that the genetic control of infection is separate from the genetic control of severe fibrosis. Our results do not provide any indications of an interaction between *SM1* and *SM2*. To evaluate this possibility, it will be necessary to search for *SM1* in the Sudanese pedigrees. Unfortunately, the low and rather uniform levels of infection in this population do not make this possible. We must also remember that the data do not exclude the possibility that additional genes are involved in susceptibility/resistance to PPF. Complementary studies, such as genome scanning and appropriate association studies, are necessary to evaluate this hypothesis.

Our linkage results obtained with two microsatellite markers from the HLA-TNF region also indicate that the major SM2 locus is unlikely to be located within this region, whereas associations have been reported between (1) some HLA class I alleles (A1 and B5) and hepatosplenomegaly in Egypt (Salam et al., 1979; Abaza et al., 1985) and (2) an HLA class II allele (DQB1*0201) and biopsy-confirmed hepatic schistosomiasis in Brazil (Secor et al., 1996).However, it should be stressed that these results do not exclude the possibility that additional polymorphisms are involved, such as specific HLA antigens. Such polymorphisms could be detected by an appropriate association study in this population.

The next step was to identify the genes involved in the control and the characterisation of the pathways in which they are involved. The most successful applications of Mendelian genetics have characterised the inheritance of traits in which single genotypic changes result in large or discrete phenotypic differences. However, the segregation of phenotypic traits in a Mendelian manner is the exception rather than the rule. Most phenotypic differences between individuals are quantitative in nature, exhibiting a continuous nearly normal phenotype distribution. These complex traits arise from the interaction between multiple segregating genetic variants and the environment.

3. HOW DOES (DO) THE GENE(S) IN 5q31-q33 CONTROL *S. MANSONI* INFECTION?

Studies by Hagan et al., by our laboratory, and by Butterworth and Dunne's group have identified the principal immune effector mechanisms associated with resistance to *S.mansoni* or *S. haematobium* in humans living in endemic areas. Hagan and colleagues (working in Gambia) and our group (working in Brazil) demonstrated that resistance to schistosomes is positively associated with the production of antilarval and antiadult worm IgE antibodies (Hagan et al., 1991; Rihet et al., 1991). The positive association between IgE and protection is counterbalanced by a negative association between antilarval and antischistosome IgG4 antibodies and resistance (Hagan et al., 1991; Demeure et al., 1993). IgG4 probably competes with IgE for binding sites on antigens (Rihet et al., 1992). The data suggest that protection against schistosomes depends on a balance between these two isotypes. These results were later confirmed by Dunne and colleagues working in a Kenyan population (Dunne et al., 1992). Capron et al. showed that rats are protected against *S. mansoni* infections by the passive transfer of schistosome-specific IgE antibodies (Capron et al., 1980), which then activate macrophages, eosinophils, and platelets for the killing of helminth larvae (Capron et al., 1975, 1981, 1984; Joseph et al., 1983). Finally, our group has shown that neonatal suppression of IgE increases the susceptibility of rats to *S. mansoni* infection (Kigoni et al., 1986). Taken together these data indicate that the IgE/IgG4 balance probably controls the degree of human resistance to infection in endemic areas. This conclusion was further strengthened by reports that parasite-specific T-cell clones derived from resistant subjects from CM were Th2 or Th0/2, whereas those derived from susceptible subjects were Th1 or Th0/1 (Couissinier et al., 1995; Rodriguez et al., 1999). Butterworth and colleagues demonstrated that eosinophils were also a critical cytotoxic effector in immunity to schistosomes (Butterworth, 1977; Kimani et al., 1991).

The analysis of the 5q31-q33 region revealed a number of genes encoding cytokines that are critical in the production of IgE, in the eosinophil response or in the regulation of the Th1/Th2 balance. More specifically, IL-4, IL-5,

IL-13, and the β-chain of IL-12 are all encoded by genes located in this region. We have found many polymorphisms in most genes, except the IL-5 gene. None of these polymorphisms, however, showed a clear segregation pattern between resistant and susceptible subjects. A major limitation of our study was the small population size (because the model based linkage studies used to test the existence of and to map this locus does not require a large number of susceptible subjects). This sample is too small if control involves more than one gene at the locus or if the control genes differ among families. This may explain why we have not yet been able to identify the gene(s) in 5q31-q33 that are responsible for the control of infection. We are currently recruiting further subjects, although this is difficult as subjects living in most endemic areas are regularly treated with antihelminthic drugs, making it difficult to define resistant and susceptible phenotypes. When enough subjects have been recruited, we will study the levels of gene expression of the 300 or 400 known and unknown genes in the 5q31-q33 region by use of a microarray method, with the aim of identifying genes that are differentially expressed in resistant and susceptible subjects, and in infected and in treated subjects. These genes will be good candidates for the control of infection.

4. ROLE OF INTERFERON-γ AND TNF-α IN SUSCEPTIBILITY TO HEPATIC FIBROSIS CAUSED BY *S. MANSONI*

We are currently trying to confirm the results that mapped the SM2 locus involved in the control of PPF to the 6q22-q23 region in another population. In parallel, we are also carrying out immunological studies on patients with advanced PPF and evaluating some of the candidate genes within this genetic region and in other regions. The candidate genes were chosen on the basis of immunophysiological studies in experimental models of infection and in humans. PPF results from an abnormal deposition of ECMP in the periportal spaces due to a chronic inflammation triggered by eggs and schistosome antigens. ECMP are mainly produced by myofibroblasts. These myofibroblasts differentiate from stellate cells (or Ito cells) following stimulation by various molecules that are released by platelets and sinusoidal endothelial and Kupffer cells in response to irritation caused by certain molecules secreted by the eggs (Fig. 12.5a). ECMP production is regulated by a number of cytokines (Fig. 12.5b), including TGF-β, IL-1β, IL-6, IL-4, IL-5, IL-10, TNF-α, and IFN-γ (Gressner et al., 1995; Poli, 2000). In particular, *in vitro* work has shown that IFN-γ is a strong antifibrogenic cytokine. It inhibits the production of ECMP by stellate cells and increases the collagenase activity of the liver by stimulating metalloprotease (MP) synthesis and by inhibiting the synthesis

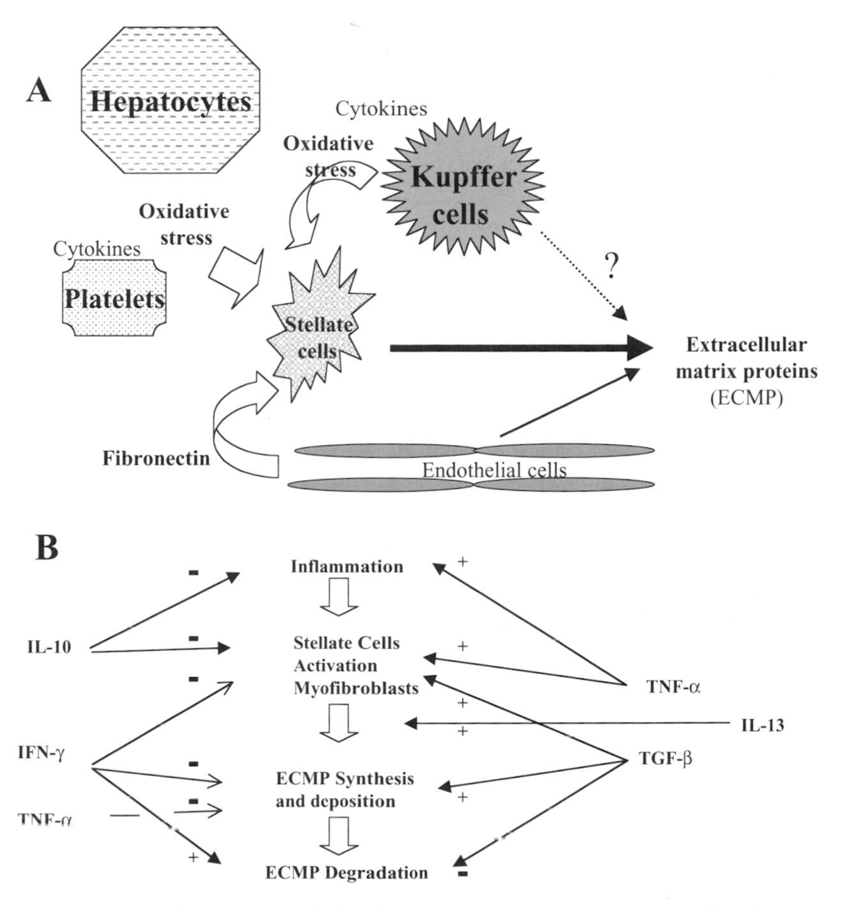

Figure 12.5. (A) The principal cells involved in ECMP production. Periportal fibrosis is caused by the deposition of extracellular matrix proteins (ECMP) in the periportal space. ECMP are produced mainly by myofibroblasts that differentiate from stellate (Ito) cells. This differentiation is initiated by substances released by damaged tissues, Kupffer cells, hepatocytes, platelets activated/damaged by products released by schistosome eggs. (B) Principal cytokines involved in the regulation of ECMP production and degradation. The amount of ECMP in tissues is determined by the rates of ECMP synthesis and degradation. A number of cytokines regulate this process. Some, such as TGF-β and IL-13, increase the amount of ECMP in tissues, whereas others, such as IL-10 and IFN-γ, decrease the ECMP concentration. The role of TNF-α is dual and not fully understood.

of tissue inhibitors of metalloprotease (TIMP) (Jimenez et al., 1984; Duncan et al., 1985; Rockey et al., 1994; Mallat et al., 1995; Tamai et al., 1995). TGF-β, IL-1, and IL-4 are fibrogenic and stimulate the differentiation of stellate cells into myofibroblasts. They also exert the opposite effects on the synthesis of ECMP and TIMPs to IFN (Roberts et al., 1986; Postlethwaite et al., 1988; Tiggelman et al., 1995). The roles of IL-4, IL-5, IL-10, TNF-α, and IFN-γ in the granuloma and in fibrosis have been evaluated in experimental models of schistosomiasis. It was confirmed that IFN-γ is the major down-regulator (Czaja et al., 1989), whereas IL-4 is strongly proinflammatory (Cheever et al., 1994; Chensue et al., 1994; Kaplan et al., 1998). Recent observations suggest that IL-10 has a key regulatory role in controlling excessive Th1 and Th2 polarisation during the granulomatous response (Wynn et al., 1997; Hoffmann et al., 2000). The administration of IL-12 together with egg antigens protects against fibrosis by increasing IFN-γ and TNF-α (Wynn et al., 1997). Finally TNF-α may have protective (Hoffmann et al., 1998), proinflammatory and profibrogenic effects (Berkow et al., 1987; Slungaard et al., 1990; Amiri et al., 1992).

As cytokines play an important role in the regulation of hepatic granuloma and in ECMP deposition and turnover, we hypothesised that PPF may result from the altered production of certain cytokines. We have initiated a study on a population from Taweela to test this hypothesis. These types of study must evaluate nonimmunological covariates that are cofounders in the analysis; these covariates cannot be evaluated in hospital-based studies. Though sinusoidal Kupffer, stellate, and endothelial cells are important players in hepatic fibrosis, there were good reasons to believe that evaluating cytokine production by blood leukocytes was a useful approach: (1) experimental studies carried out with lymphocytes from various tissues have identified the principal cytokines involved in the hepatic granuloma and (2) the regulation of mouse schistosome egg granuloma is dependent on CD4+ T lymphocytes.

Thus, we evaluated the cytokines produced by egg antigen-stimulated blood mononuclear cells from subjects with advanced and severe PPF and from subjects with no or mild hepatic fibrosis (Henri et al., 2002). Of all cytokines tested (TNF-α, IFN-γ, IFN-α, IL-12, IL-10, IL-1β, IL-4, IL-5, IL-6), only IFN-γ and TNF-α were significantly associated with fibrosis: the best model included IFN-γ, TNF-α, age, and gender as covariates. Thus, the associations of IFN-γ and TNF-α were adjusted for age and gender as the association between these cytokines and the disease phenotype varies (confounder effect) in the different age and gender classes. This model does not adjust for the intensity of infection because this covariate does not improve the likelihood of the model.

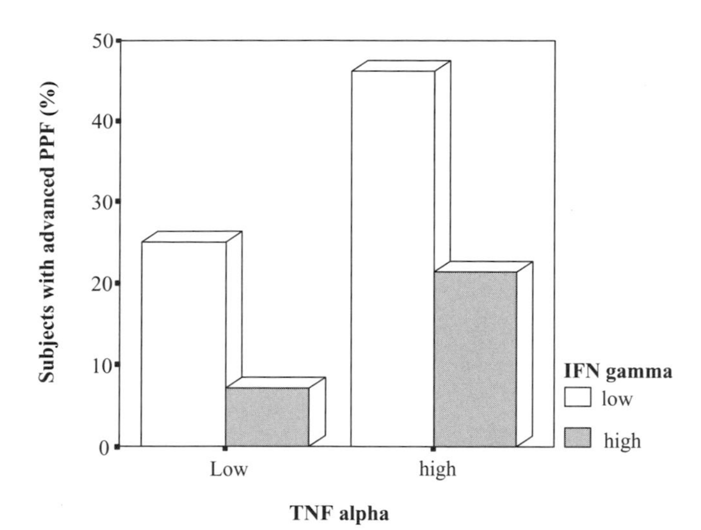

Figure 12.6. Proportion of subjects with advanced and severe PPF among INF-γ low/high and TNF-α low/high producers. PPF is more frequent in subjects who produce low amounts of IFN-γ and high amounts of TNF-α.

Thus, IFN-γ was significantly (inversely) associated ($P < 0.01$) with periportal fibrosis; high levels of IFN-γ are associated with a reduced risk of periportal fibrosis. The odds ratio that measures the strength of the association between IFN-γ and fibrosis after adjustment for other covariates with significant association with disease (age, gender, and TNF-α) was 0.11 (confidence interval: 0.03–0.6). This indicates that the risk of developing severe disease is on average 9 times higher among individuals who produce low amounts of IFN-γ than among those who produce high amounts of IFN-γ (Fig. 12.6).

TNF-α was found to be positively associated with PPF (Fig. 12.6): the risk of FII–III is on average 4 times higher in individuals who produce high amounts of TNF-α than in those who produce low amounts of TNF-α ($P = 0.05$, OR $= 4.6$, CI : 1–22). The odds ratio for TNF-α was adjusted for age, gender, and IFN-γ levels to take into account other covariates significantly associated with advanced fibrosis. No other cytokines were significantly associated with disease.

These results together with the mapping of *SM2* near the gene encoding the β-chain of the IFN-γ receptor indicate that the genes encoding the IFN-γ/IFN-γ receptor pathway are valid candidate genes for the control of PPF. This led us to evaluate polymorphisms in the genes encoding IFN-γ and its receptor. Various polymorphisms were found in both genes and an ongoing study indicates that three polymorphisms in the IFN-γ gene are associated

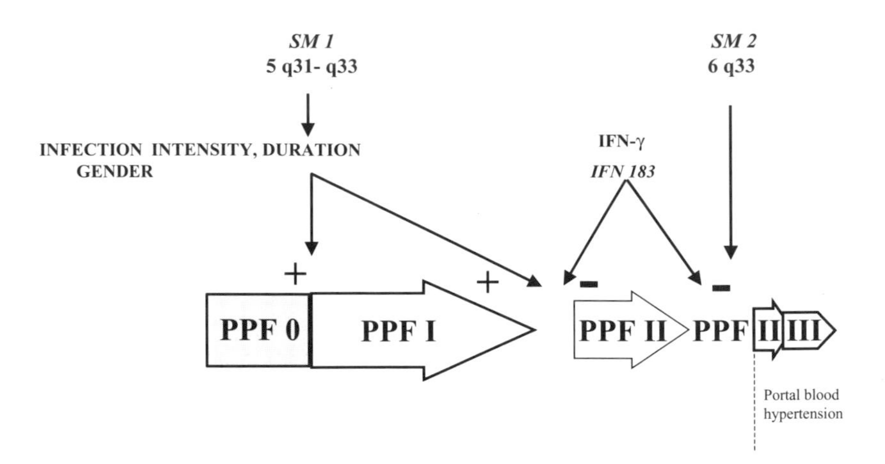

Figure 12.7. Summary of the genetic control of susceptibility to human schistosomiasis. Disease is initiated by eggs trapped in liver sinusoids and amplified by antigens released by adult worms. The first stage is perivascular inflammation associated with the deposition of small amounts of ECMP [grade I (FI) according to the WHO scale]. In some subjects ECMP accumulates around the secondary branches of the portal vein as long stretches of fibrosis (stage II). In some subjects ECMP accumulates further and fibrosis extends to the periphery of the organ, occludes certain secondary branches, and twists major branches; the gall bladder is also affected (stage III). Portal hypertension (P.H), detected by abnormal portal and splenic vein diameters, is observed in some subjects with FII and in all FIII subjects. SM2 accelerates the progression from FII to FIII + P.H or to FII + P.H. The protective role of IFN-γ is indicated by the association between FIII and low IFN-γ production, by the association between a mutation in the *IFN-γ* gene and FIII and by various studies showing that IFN-γ decreases ECMP production and increases ECMP degradation. Other factors important in fibrosis are the duration of infection, the intensity of infection and gender (males are more susceptible than females). The intensity of infection is probably most important during the early stage of disease (FI) and is controlled by SM1, which probably acts on T-cell differentiation.

with advanced PPF (C. Chevillard et al. manuscript in preparation). Further studies should now focus on the IFN-γ receptor.

5. CONCLUSION

These studies on human schistosomiasis have allowed us to describe, for the first time in an infectious disease, the control of infection levels by a major gene (*SM1*) in a population from northeastern Brazil. The existence of this control and the position of the gene involved were confirmed in a second study on a different population from West Africa. Based on the results of immunological studies carried out on infection-resistant and -susceptible subjects, it is likely that *SM1* controls T-helper cell differentiation. Studies on larger populations are required to define this control and to identify the genes and susceptibility alleles involved. Moreover, there is some evidence suggesting that minor loci also affect human susceptibility to schistosomes.

Genetic studies also have enabled us to demonstrate that the control of infection and disease in *S. mansoni*-infected subjects are distinct, illustrating the difference between sterile immunity and clinical immunity. The second locus (*SM2*) is also a major locus, indicating that both infection and disease are influenced considerably by the host's genetic background. Immunological studies and the analysis of gene polymorphisms showed that the IFN-γ pathway is critical for the control of resistance/susceptibility to hepatic fibrosis (Fig. 12.7). Further studies are required to determine whether the same genetic control applies to hepatic fibrosis caused by other pathogens such as hepatitis C virus. Finally, the results presented here show for the first time the usefulness of an integrated epidemiological/immunological/genetic approach to understand the mechanisms of pathogenesis in a complex infectious disease.

ACKNOWLEDGMENTS

This work received financial assistance from the Institut National de la Santé et de la Recherche Médicale, the World Health Organization (ID096546), the European Economic Community (TS3 CT940296, IC18CT970212), the Scientific and Technical Cooperation with Developing Countries (IC18CT980373), the French Ministere de la Recherche et des Techniques (PRFMMIP), the Conseil General Provence Alpes Côte d'Azur and the Conseil Regional Provence Alpes Côte d'Azur. S.H. and C.C. are supported by fellowships from the French Ministere de la Recherche et des Techniques and from the Conseil General PACA, respectively, and from the Fondation pour la Recherche Médicale.

REFERENCES

Abaza, H., Asser, L., el Sawy, M., et al. (1985). HLA antigens in schistosomal hepatic fibrosis patients with haematemesis. *Tissue Antigens* **26**, 307–309.

Abdel-Wahab, M. F., Esmat, G., Narooz, S. I., et al. (1990). Sonographic studies of schoolchildren in a village endemic for *Schistosoma mansoni*. *Trans. R. Soc. Trop. Med. Hyg.* **84**, 69–73.

Abel, L., Demenais, F., Prata, A., et al. (1991). Evidence for the segregation of a major gene in human susceptibility/resistance to infection by *Schistosoma mansoni*. *Am. J. Hum. Genet.* **48**, 959–970.

Abel, L. and Dessein, A. (1991). Genetic predisposition to high infections in an endemic area of *Schistosoma mansoni*. *Rev. Soc. Bras. Med. Trop.* **24**, 1–3.

Amiri, P., Locksley, R. M., Parslow, T. G., et al. (1992). Tumour necrosis factor alpha restores granulomas and induces parasite egg-laying in schistosome-infected SCID mice. *Nature* **356**, 604–607.

Berkow, R. L., Wang, D., Larrick, J. W., et al. (1987). Enhancement of neutrophil superoxide production by preincubation with recombinant human tumor necrosis factor. *J. Immunol.* **139**, 3783–3791.

Butterworth, A. E. (1977). The eosinophil and its role in immunity to helminth infection. *Curr. Top. Microbiol. Immunol.* **77**, 127–168.

Butterworth, A. E., Capron, M., Cordingley, J. S., et al. (1985). Immunity after treatment of human schistosomiasis mansoni. II. Identification of resistant individuals, and analysis of their immune responses. *Trans. R. Soc. Trop. Med. Hyg.* **79**, 393–408.

Capron, A., Dessaint, J. P., Capron, M., et al. (1975). Specific IgE antibodies in immune adherence of normal macrophages to *Schistosoma mansoni* schistosomules. *Nature* **253**, 474–475.

Capron, A., Dessaint, J. P., Capron, M., et al. (1980). Role of anaphylactic antibodies in immunity to schistosomes. *Am. J. Trop. Med. Hyg.* **29**, 849–857.

Capron, M., Capron, A., Dessaint, J. P., et al. (1981). Fc receptors for IgE on human and rat eosinophils. *J. Immunol.* **126**, 2087–2092.

Capron, M., Spiegelberg, H. L., Prin, L., et al. (1984). Role of IgE receptors in effector function of human eosinophils. *J. Immunol.* **132**, 462–468.

Cheever, A. W., Williams, M. E., Wynn, T. A., et al. (1994). Anti-IL-4 treatment of *Schistosoma mansoni*-infected mice inhibits development of T cells and non-B, non-T cells expressing Th2 cytokines while decreasing egg-induced hepatic fibrosis. *J. Immunol.* **153**, 753–759.

Chensue, S. W., Warmington, K. S., Ruth, J., et al. (1994). Cross-regulatory role of interferon-gamma (IFN-gamma), IL-4 and IL-10 in schistosome egg granuloma formation: In vivo regulation of Th activity and inflammation. *Clin. Exp. Immunol.* **98**, 395–400.

Chitsulo, L., Engels, D., Montresor, A. et al. (2000). The global status of schistosomiasis and its control. *Acta Trop.* **77**, 41–51.

Couissinier, P. and Dessein, A. J. (1995). Schistosoma-specific helper T cell from subjects resistant to infection by *Schistosoma mansoni* are ThO/2. *Eur. J. Immunol.* **25**, 2295–2302.

Czaja, M. J., Weiner, F. R., Eghbali, M., et al. (1987). Differential effects of gamma-interferon on collagen and fibronectin gene expression. *J. Biol. Chem.* **262**, 13348–13351.

Czaja, M. J., Weiner, F. R., Takahashi, S., et al. (1989). Gamma-interferon treatment inhibits collagen deposition in murine schistosomiasis. *Hepatology* **10**, 795–800.

Demeure, C. E., Rihet, P., Abel, L., et al. (1993). Resistance to *Schistosoma mansoni* in humans: influence of the IgE/IgG4 balance and IgG2 in immunity to reinfection after chemotherapy. *J. Infect. Dis.* **168**, 1000–1008.

Dessein, A. J., Begley, M., Demeure, C., et al. (1988). Human resistance to *Schistosoma mansoni* is associated with IgG reactivity to a 37-kDa larval surface antigen. *J. Immunol.* **140**, 2727–2736.

Dessein, A. J., Couissinier, P., Demeure, C., et al. (1992). Environmental, genetic and immunological factors in human resistance to *Schistosoma mansoni*. *Immunol. Invest.* **21**, 423–453.

Dessein, A. J., Hillaire, D., Elwali, E. N. M. A., et al. (1999). Severe hepatic fibrosis in *Schistosoma mansoni* infection is controlled by a major locus that is closely linked to the interferon-γ receptor gene. *Am. J. Hum. Genet.* **68**, 709–721.

Duncan, M. R. and Berman, B. (1985). Gamma interferon is the lymphokine and beta interferon the monokine responsible for inhibition of fibroblast collagen production and late but not early fibroblast proliferation. *J. Exp. Med.* **162**, 516–527.

Dunne, D. W., Butterworth, A. E., Fulford, A. J., et al. (1992). Immunity after treatment of human schistosomiasis: association between IgE antibodies to adult worm antigens and resistance to reinfection. *Eur. J. Immunol.* **22**, 1483–1494.

Garcia, A., Marquet, S., Bucheton, B., et al. (1998). Linkage analysis of blood *Plasmodium falciparum* levels: interest of the 5q31-q33 chromosome region. *Am. J. Trop. Med. Hyg.* **58**, 705–709.

Gressner, A. M. and Bachem, M. G. (1995). Molecular mechanisms of liver fibrogenesis – A homage to the role of activated fat-storing cells. *Digestion*, **56**, 335–346.

Grimaud, J. A. and Borojevic, R. (1977). Chronic human *Schistosomiasis mansoni*: Pathology of the Disse's space. *Lab. Invest.* **36**, 268–273.

Hagan, P., Blumenthal, U. J., Dunn, D., et al. (1991). Human IgE, IgG4 and resistance to reinfection with *Schistosoma haematobium*. *Nature* **349**, 243–245.

Henri, S., Chevillard, C., Mergani, A., et al. (2002). Cytokine regulation of periportal fibrosis in humans infected with *Schistosoma mansoni*: IFN-gamma is associated with protection against fibrosis and TNF-alpha with aggravation of disease. *J. Immunol.* **169**, 929–936.

Hoffmann, K. F., Caspar, P., Cheever, A. W., et al. (1998). IFN-gamma, IL-12, and TNF-alpha are required to maintain reduced liver pathology in mice vaccinated with *Schistosoma mansoni* eggs and IL-12. *J. Immunol.* **161**, 4201–4210.

Hoffmann, K. F., Cheever, A. W. and Wynn, T. A. (2000). IL-10 and the dangers of immune polarization: excessive type 1 and type 2 cytokine responses induce distinct forms of lethal immunopathology in murine schistosomiasis. *J. Immunol.* **164**, 6406–6416.

Homeida, M., Ahmed, S., Dafalla, A., et al. (1988a). Morbidity associated with *Schistosoma mansoni* infection as determined by ultrasound: A study in Gezira, Sudan. *Am. J. Trop. Med. Hyg.* **39**, 196–201.

Homeida, M. A., Fenwick, A., DeFalla, A. A., et al. (1988b). Effect of antischistosomal chemotherapy on prevalence of Symmers' periportal fibrosis in Sudanese villages. *Lancet* **ii**, 437–440.

Jimenez, S. A., Freundlich, B. and Rosenbloom, J. (1984). Selective inhibition of human diploid fibroblast collagen synthesis by interferons. *J. Clin. Invest.* **74**, 1112–1116.

Joseph, M., Auriault, C., Capron, A., et al. (1983). A new function for platelets: IgE-dependent killing of schistosomes. *Nature* **303**, 810–812.

Kaplan, M. H., Whitfield, J. R., Boros, D. L., et al. (1998). Th2 cells are required for the *Schistosoma mansoni* egg-induced granulomatous response. *J. Immunol.* **160**, 1850–1856.

Kigoni, E. P., Elsas, P. P., Lenzi, H. L., et al. (1986). IgE antibody and resistance to infection. II. Effect of IgE suppression on the early and late skin reaction and resistance of rats to *Schistosoma mansoni* infection. *Eur. J. Immunol.* **16**, 589–595.

Kimani, G., Chunge, C. N., Butterworth, A. E., et al. (1991). Eosinophilia and eosinophil helminthotoxicity in patients treated for *Schistosoma mansoni* infections. *Trans. R. Soc. Trop. Med. Hyg.* **85**, 489–492.

Mallat, A., Preaux, A. M., Blazejewski, S., et al. (1995). Interferon alfa and gamma inhibit proliferation and collagen synthesis of human Ito cells in culture. *Hepatology* **21**, 1003–1010.

Marquet, S., Abel, L., Hillaire, D., et al. (1998). Full results of the genome-wide scan which localises a locus controlling the intensity of infection by Schistosoma mansoni on chromosome 5q31-q33. *Eur. J. Hum. Genet.* **7**, 88–97.

Marquet, S., Abel, L., Hillaire, D., et al. (1996). Genetic localization of a locus controlling the intensity of infection by Schistosoma mansoni on chromosome 5q31-q33. *Nat. Genet.* **14**, 181–184.

Marsh, D. G., Neely, J. D., Breazeale, D. R., et al. (1994). Linkage analysis of IL4 and other chromosome 5q31.1 markers and total serum immunoglobulin E concentrations. *Science* **264**, 1152–1156.

Martinez, F. D., Solomon, S., Holberg, C. J., et al. (1998). Linkage of circulating eosinophils markers on chromosome 5q. *Am. J. Respir. Crit. Care Med.* **158**, 1739–1744.

Mohamed-Ali, Q., Elwali, N. E., Abdelhameed, A. A., et al. (1999). Susceptibility to periportal (Symmers) fibrosis in human *Schistosoma mansoni* infections: Evidence that intensity and duration of infection, gender, and inherited factors are critical in disease progression. *J. Infect. Dis.* **180**, 1298–1306.

Muller-Myhsok, B., Stelma, F. F., Guisse-Sow, F., et al. (1997). Further evidence suggesting the presence of a locus, on human chromosome 5q31-q33, influencing the intensity of infection with *Schistosoma mansoni. Am. J. Hum. Genet.* **61**, 452–454.

Poli, G. (2000). Pathogenesis of liver fibrosis: role of oxidative stress. *Mol. Aspects Med.* **21**, 49–98.

Postlethwaite, A. E., Raghow, R., Stricklin, G. P., et al. (1988). Modulation of fibroblast functions by interleukin 1: Increased steady- state accumulation of type I procollagen messenger RNAs and stimulation of other functions but not chemotaxis by human recombinant interleukin 1 alpha and beta. *J. Cell Biol.* **106**, 311–318.

Poynard, T., Bedossa, P. and Opolon, P. (1997). Natural history of liver fibrosis progression in patients with chronic hepatitis C. *Lancet* **349**, 825–832.

Prata, A. (1987). Schistosomiasis mansoni in Brazil. *Baillière Clin. Trop. Med. Commun. Dis.* **2**, 349–369.

Rihet, P., Demeure, C. E., Bourgois, A., et al. (1991). Evidence for an association between human resistance to *Schistosoma mansoni* and high anti-larval IgE levels. *Eur. J. Immunol.* **21**, 2679–2686.

Rihet, P., Demeure, C. E., Dessein, A. J., et al. (1992). Strong serum inhibition of specific IgE correlated to competing IgG4, revealed by a new methodology in subjects from a *S. mansoni* endemic area. *Eur. J. Immunol.* **22**, 2063–2070.

Rihet, P., Traore, Y., Abel, L., et al. (1998). Malaria in humans: Plasmodium falciparum blood infection levels are linked to chromosome 5q31-q33. *Am. J. Hum. Genet.* **63**, 498–505.

Rioux, J. D., Stone, V. A., Daly, M. J., et al. (1998). Familial eosinophilia maps to the cytokine gene cluster on human chromosomal region 5q31-q33. *Am. J. Hum. Genet.* **63**, 1086–1094.

Roberts, A. B., Sporn, M. B., Assoian, R. K., et al. (1986). Transforming growth factor type beta: Rapid induction of fibrosis and angiogenesis in vivo and stimulation of collagen formation in vitro. *Proc. Natl. Acad. Sci. USA* **83**, 4167–4171.

Rockey, D. C. and Chung, J. J. (1994). Interferon gamma inhibits lipocyte activation and extracellular matrix mRNA expression during experimental liver injury: Implications for treatment of hepatic fibrosis. *J. Investig. Med.* **42**, 660–670.

Rodriguez, V., Piper, K., Couissinier-Paris, P., et al. (1999). Genetic control of Schistosome infections by SM1 locus of the 5q31-q33 region is linked to differentition of type 2 helper T lymphocytes. *Infect. Immun.* **67**, 4689–4692.

Salam, E. A., Ishaac, S. and Mahmoud, A. A. (1979). Histocompatibilty-linked susceptibility for hepatosplenomegaly in human schistosomiasis mansoni. *J. Immunol.* **123**, 1829–1831.

Secor, W. E., del Corral, H., dos Reis, M. G., et al. (1996). Association of hepatosplenic schistosomiasis with HLA-DQB1*0201. *J. Infect. Dis.* **174**, 1131–1135.

Slungaard, A., Vercellotti, G. M., Walker, G., et al. (1990). Tumor necrosis factor alpha/cachectin stimulates eosinophil oxidant production and toxicity towards human endothelium. *J. Exp. Med.* **171**, 2025–2041.

Tamai, K., Ishikawa, H., Mauviel, A., et al. (1995). Interferon-gamma coordinately upregulates matrix metalloprotease (MMP)-1 and MMP-3, but not tissue inhibitor of metalloproteases (TIMP), expression in cultured keratinocytes. *J. Invest. Dermatol.* **104**, 384–390.

Tiggelman, A. M., Boers, W., Linthorst, C., et al. (1995). Collagen synthesis by human liver (myo)fibroblasts in culture: evidence for a regulatory role of IL-1 beta, IL-4, TGF beta and IFN gamma. *J. Hepatol.* **23**, 307–317.

Von Lichtenberg, F. (1962). Host response to eggs of *S. mansoni*. I. Granuloma formation in the unsensitized laboratory mouse. *Am. J. Pathol.* **41**, 711–731.

Warren, K. S. (1977). Modulation of immunopathology and disease in schistosomiasis. *Am. J. Trop. Med. Hyg.* **26**, 113–119.

Warren, K. S., Domingo, E. O. and Cowan, R. B. (1967). Granuloma formation around schistosome eggs as a manifestation of delayed hypersensitivity. *Am. J. Pathol.* **51**, 735–756.

World Health Organization. (1974). Immunology of schistosomiasis. *Bull. World Health Organ.* **51**, 553–595.

Wynn, T. A., Morawetz, R., Scharton-Kersten, T., et al. (1997). Analysis of granuloma formation in double cytokine-deficient mice reveals a central role for IL-10 in polarizing both T helper cell 1- and T helper cell 2-type cytokine responses in vivo. *J. Immunol.* **159**, 5014–5023.

Zinn-Justin, A., Marquet, S., Hillaire, D., et al. (2001). Genome search for additional human loci controlling infection levels by *Schistosoma mansoni*. *Am. J. Trop. Med. Hyg.* **65**, 754–758.

CHAPTER 13

Genetic susceptibility to prion diseases

Matthew Bishop and J. W. Ironside

Departments of Pathology and Clinical Neurosciences, CJD Surveillance Unit, University of Edinburgh, Western General Hospital, Edinburgh, Scotland

1. WHAT ARE PRION DISEASES?

Prion diseases are otherwise known as transmissible spongiform encephalopathies (TSE), which are fatal neurodegenerative disorders that occur in many species of mammal, including humans. They include the cattle disease bovine spongiform encephalopathy (BSE) and the human form Creutzfeldt–Jakob disease (CJD). The oldest recognised TSE is scrapie in sheep and goats which has been recorded in these animals albeit under different names since the eighteenth century. Other species that have since been diagnosed with a specific form of TSE disease are included in the list shown in Table 13.1. The list has been divided into two groups, those that are host specific and those that have occurred as a consequence of the United Kingdom's highly publicised BSE outbreak of the 1980s–1990s.

The name TSE is derived from the fact that these diseases are transmissible from one animal to another and among species, and they are pathologically defined by the appearance of spongelike vacuolation in the gray matter of the brain.

Experimental animal transmission studies in primates and rodents have been performed using material from cattle diagnosed with BSE. This has indicated, to the acceptance of the majority of the scientific community, that the latest form of TSE in humans, variant Creutzfeldt–Jakob disease (vCJD), is attributable to infection by the bovine prions found in cases of cattle with BSE. The route of infection from contaminated bovine products is as yet undetermined but is most probably through dietary consumption. Since vCJD was first characterised in 1996 (Will et al., 1996) there has been widespread discussion about the implications for the general public who were unknowingly exposed to infectious bovine material before the UK government instigated

Table 13.1. *TSE in non-human species. List of recognised TSE in nonhuman species categorised according to whether they are causally linked to BSE*

BSE-related diseases (recognised since the BSE epidemic)	Host specific
Cattle with bovine spongiform encephalopathy (BSE)	Sheep and goats with Scrapie
Domestic cats with feline spongiform encephalopathy (FSE)	Deer and elk with chronic wasting disease (CWD)
Others thought to have originated from BSE exposure: domestic and captive wild cats, greater kudu, nyala, Arabian oryx, Scimitar horned oryx, eland, gemsbok, bison, ankole, tiger, cheetah, ocelot, and puma	Mink with transmissible mink encephalopathy (TME)

changes to the preparation of meat for human consumption. As the UK exported large quantities of meat produce throughout the world vCJD has become of international concern. For a full report on this topic the UK Government's BSE Enquiry can be viewed online (http://www.bse.org.uk/).

Studies of experimental and natural TSE have indicated that genetic factors are important in influencing susceptibility to the transmissible agent, initially thought to be a virus. However, there is an increasing body of evidence to support the prion hypothesis (following).

2. WHAT IS A PRION?

The term *prion disease* is widely used for these diseases, since TSEs are associated with the accumulation of an abnormal isoform of the prion protein in the brain, discovered through the work of Prusiner. It was Prusiner who coined the term *prion* to describe a proteinaceous infectious agent (Prusiner, 1982). Prion protein (PrP) occurs naturally in mammals, birds, and reptiles, as a membrane-bound, glycosylphophatidylinositol (GPI) anchored glycoprotein with proposed roles in synaptic transmission and signal transduction in the central nervous system (CNS) (Collinge et al., 1994). The normal isoform of prion protein is expressed in a wide range of tissues, including lymphoid tissues, heart, lungs, skeletal muscle, and salivary glands, but the highest levels of expression are found in neurons of the CNS (Brown et al., 1990; McLennan et al., 2001). In TSE this host protein undergoes

Table 13.2. *Comparison of the two isoforms of prion protein*

PrP^C cellular form	PrP^{Sc} disease-associated form
Mainly alpha-helical structure	Mainly beta-sheet structure
Monomeric	Forms aggregates (amyloid)
Soluble	Insoluble
Protease sensitive	Partially protease resistant
Rapidly metabolised	Stable
Present in normal tissue	Only present in prion disease tissue

posttranslation modification of its three-dimensional tertiary structure from an alpha-helix-rich structure to one of a predominantly beta-sheet structure (Prusiner, 1998). Isolation of the two forms is straightforward and has been used to confirm the different structural components (Pan et al., 1993). The disease associated (beta-sheet) form of the prion protein is designated PrP^{Sc} (Sc = Scrapie form) and the normal form is designated PrP^C (C = Cellular). The differences between the two protein isoforms are summarised in Table 13.2.

Through a hypothesised protein–protein interaction, current *in vitro* experimental evidence shows that in the presence of the disease associated form (PrP^{Sc}), the normal form (PrP^C) undergoes structural conversion to PrP^{Sc} (Raymond et al., 1997; Saborio et al., 2001). This process of conversion continues, resulting in abundance of PrP^{Sc} in the central nervous system (CNS). The precise site and mechanism of conversion is unclear (Harris, 1999). Models of scrapie PrP conversion have been used to observe interactions between the more critical components of the protein structure such as the alpha-helical regions. This work has involved the use of synthetic peptides as a means of triggering the conformational transition. (Kaneko et al., 1995; Nguyen et al., 1995). PrP^{Sc} is not degraded by normal cellular processes and is deposited as aggregates, often visible as 'plaques' in histological sections of the CNS. As confirmation that PrP^C is required for these diseases, transgenic mice devoid of PrP^C have been found to be resistant to infection with prions and do not replicate the infectious agent (Bueler et al., 1993).

3. GENETICS OF PRION DISEASE

To understand the aetiology and epidemiology of prion diseases it is necessary to examine the genetic makeup of those with the disease. This will facilitate an assessment of individual risk and potentially identify those who are in preclinical or asymptomatic stages of disease. With TSE, such as BSE, that have been transmitted to another species it will also be advantageous to examine the genetics of the infecting organism.

The current central biochemical element of prion disease is the prion protein itself and therefore comparative analysis of the protein structure together with the prion gene region is of importance when dealing with a suspected case of TSE.

3.1 Experimental Genetics in Prion Disease

The search for a genetic component responsible for causation and familial inheritance of TSE has been undertaken ever since the laboratory techniques became available. Without the obvious presence of a nucleic acid component of infectivity, such as a virus, research was directed towards finding a mechanism by which strain information could be passed on.

One of the most closely observed components of TSE in research animals is the incubation period in infected mice. The first proposal for genetic control in scrapie passaged mice was the description of the hypothetical 'scrapie incubation', or *Sinc* gene (Dickinson et al., 1968). This was described further with a second name, *Prn-i* (Carlson et al., 1986). This murine gene had been found to possess two alleles, s7 and p7, that segregated according to the specific scrapie strains. Each scrapie strain has a highly reproducible pattern of incubation periods and allele dominance in mice with either of the three *Sinc* genotypes (s7/s7, s7/p7, and p7/p7).

Following characterisation of the prion gene (see below) there was further definition of two alleles in mice that segregated with short and long scrapie incubation period (IP). $Prnp^{a}$ (Leu-108, Thr-189) has short IP and $Prnp^{b}$ (Phe-108, Val-189) has a long IP (Carlson et al., 1988; Westaway et al., 1987). Sheep were also found to have two alleles (sA and pA) in their *Sip* gene (for scrapie incubation period) that determined the incubation period of scrapie. Possession of an sA allele lead to increased incidence of disease (Hunter et al., 1989). Evidence from detailed analysis of the prion gene in these species used for experimental analysis soon pointed to a single source for the majority of effects on incubation period. The *Sinc* and *Prn-i* were confirmed as the same genetic loci as *Prnp* (Hunter et al., 1992; Moore et al., 1998).

Lengthy and highly detailed pedigree analysis of human TSE that appeared to show patterns of inheritance further confirmed the central role of the prion protein gene in association with disease. The first *PRNP* mutation associated with a human TSE was a proline to leucine change at codon 102 (P102L) in a case of Gerstmann Sträussler Scheinker Syndrome (GSS) (Hsiao

Table 13.3. *Disease incubation periods for transgenic mice with the human equivalent GSS mutation at murine codon 101*

Inoculum	GSS	GSS from L101L	None
P101P (wildtype)	450	226	>700
L101L (transgenic)	288	148	>700

Numbers are mean incubation period in days. Data adapted from Manson et al. (1999).

et al., 1989). Since then, an ever-increasing number of *PRNP* mutations have been identified in families with GSS and other inherited forms of human TSE (for reviews, Goldfarb et al., 1995; Prusiner et al., 1997). The high lod ('logarithm of the odds') scores for the association of these mutations with the disease indicates that the mutations are of key pathogenic significance in familial TSE.

A method of experimental analysis for determining the effect of *PRNP* mutations is the production of laboratory lines of mice that have the equivalent mutation in their prion gene, or to use transgenic techniques to insert mutated human prion genes into the mouse genome. The initial use of such techniques was in the development of a GSS mouse model with a P101L mutation. This mutation is equivalent to the human codon P102L mutation in GSS (see section on familial CJD below). Primary studies involved production of transgenic mice with the murine prion gene (*Prnp*) replaced by multiple copies of an altered gene with the P101L substitution. This research produced mice that developed spontaneous neurodegeneration with an ataxic phenotype similar to that seen in human cases of GSS with P102L mutation (Hsiao et al., 1990). Subsequent studies on this model indicated that the disease occurring in these transgenic mice was transmissible to other transgenic mice and hamsters (Hsiao et al., 1994). Further studies are required to confirm this observation and to investigate the potential additional effects of genetic mutation and overexpression on this model.

A more recent study has been performed on mice that have been bred with a P101L mutation as a single copy, to more closely model human GSS (Manson et al., 1999). These mice did not develop spontaneous TSE, a proposed consequence of having single rather than multiple copies of the gene. The initial work performed with these mice was to determine the disease incubation period after inoculation with brain material from a case of human GSS with a P102L mutation. The results are summarised in Table 13.3.

The experiments indicated that the mutation was associated with a dramatic alteration in incubation time. Inoculation with the human GSS material produced an almost 50% reduction in incubation time which was reduced even further when brain material from these inoculated mice was used for further inoculations. This has suggested that addition of the human equivalent GSS mutation has removed part of the barrier to cross-species infectivity. It also supports the theory that humans carrying the P102L mutation may be more susceptible to infection with certain 'strains' of TSE.

The most recent technical advances in genetic analysis allow rapid scanning of the whole genome for quantitative trait loci (QTL) linked to a disease state. Such analysis has been carried out to observe differences between lines of mice with short and long scrapie incubation periods, those with *Prnp*^{a} and *Prnp*^{b} genotypes respectively (Lloyd et al., 2001). This work identified three highly significantly linked regions on chromosomes 2, 11, and 12 (with the possibility of multiple QTL at each site) and suggestive evidence for linkage on chromosomes 6 and 7. The *Prnp* locus was within the 95% confidence interval for the peak lod score on chromosome 2 and is therefore likely to be the candidate for that QTL.

A further QTL analysis was carried out on two lines of mice that showed a significant difference in incubation periods when challenged with a BSE inoculum (Manolakou et al., 2001). Four QTL were identified on regions of chromosomes 2, 4, 8, and 15. In this case the *Prnp* locus was outside the 95% confidence interval for the peak lod score on chromosome 2. In addition, it was found that environmental factors were acting on the incubation period. Both the age of the mice at injection and the maternal age at birth were negatively correlated with the length of the incubation period, and there was a significant increase in the incubation period in males compared to females (in the F_1 mice only).

The QTL found in these studies must now be analysed further to pinpoint candidate mouse genes that are proposed to impact on the TSE phenotype. The equivalent human genes can then be identified. Finding QTL in human TSE cases is difficult due to the small numbers available; the two studies detailed above used more than 1,000 mice each.

3.2 The Prion Protein Gene (*PRNP*)

The human *PRNP* gene region is located on the short arm of chromosome 20 (Band: 20p12-pter) and is composed of two exons. The second exon contains the open reading frame (ORF), the complete DNA sequence required for producing the prion protein, which is 762 bp long and translates to a protein of 253 amino acids. Upstream of this exon lies the noncoding exon 1 and then further still is the region containing the promoter sequences. The latter are currently under investigation to identify possible relationships between disease phenotypes and the direct control mechanisms of gene expression (Mahal et al., 2001). With the recent publication of the DNA sequence of chromosome 20 as part of the Human Genome Project there are now further targets available for analysis of the prion gene region.

The human gene *PRNP* was first described in 1986 (Liao et al., 1986) and its DNA sequence is available through the Internet bioinformatics sites (GenBank accession number AL133396). Changes in the DNA sequence of the prion gene can cause alterations in the protein amino acid composition and subsequently the secondary and tertiary structures. The most important reason for obtaining sequence data is in the identification, or confirmation, of familial cases of CJD, caused by mutations. This is of great significance to the other family members as familial TSE occur as Mendelian inherited, autosomal dominant disorders with high levels of penetrance.

3.3 Prion Gene Polymorphisms

Changes in the DNA sequence of the *PRNP* gene fall into two categories: those that are directly associated with a disease phenotype (generally producing changes in the amino acid sequence) and those that may influence clinical features or susceptibility. (For the sake of simplicity in this chapter the former will be referred to as 'mutations' and the latter as 'polymorphisms'.) More than 20 *PRNP* mutations have been characterised. There have also been more than 20 polymorphisms found throughout the gene; these and the mutations are shown on Fig. 13.1. There have also been recent discoveries of polymorphic sites outside of the gene region where variations may determine changes such as gene expression levels or other biochemical properties of the protein (Mahal et al., 2001).

3.4 Codon 129 Polymorphism

A notable polymorphism that is recognised as a key factor when assessing the genotype/phenotype relationship of prion disease cases is the methionine-to-valine change at codon 129. (DNA sequence change ATG-Met to GTG-Val.) The normal population frequencies for the three genotypes for a selection of regions are shown in Table 13.4. The most obvious effect of the codon 129 polymorphism can be seen in the genotype frequencies of the sporadic form of CJD (sCJD) and vCJD cases. In sCJD there is an increase in the homozygote (MM and VV) frequencies with the level of MM at more than 70%, and in vCJD all cases so far analysed have been MM ($n = 97$ of 114). This

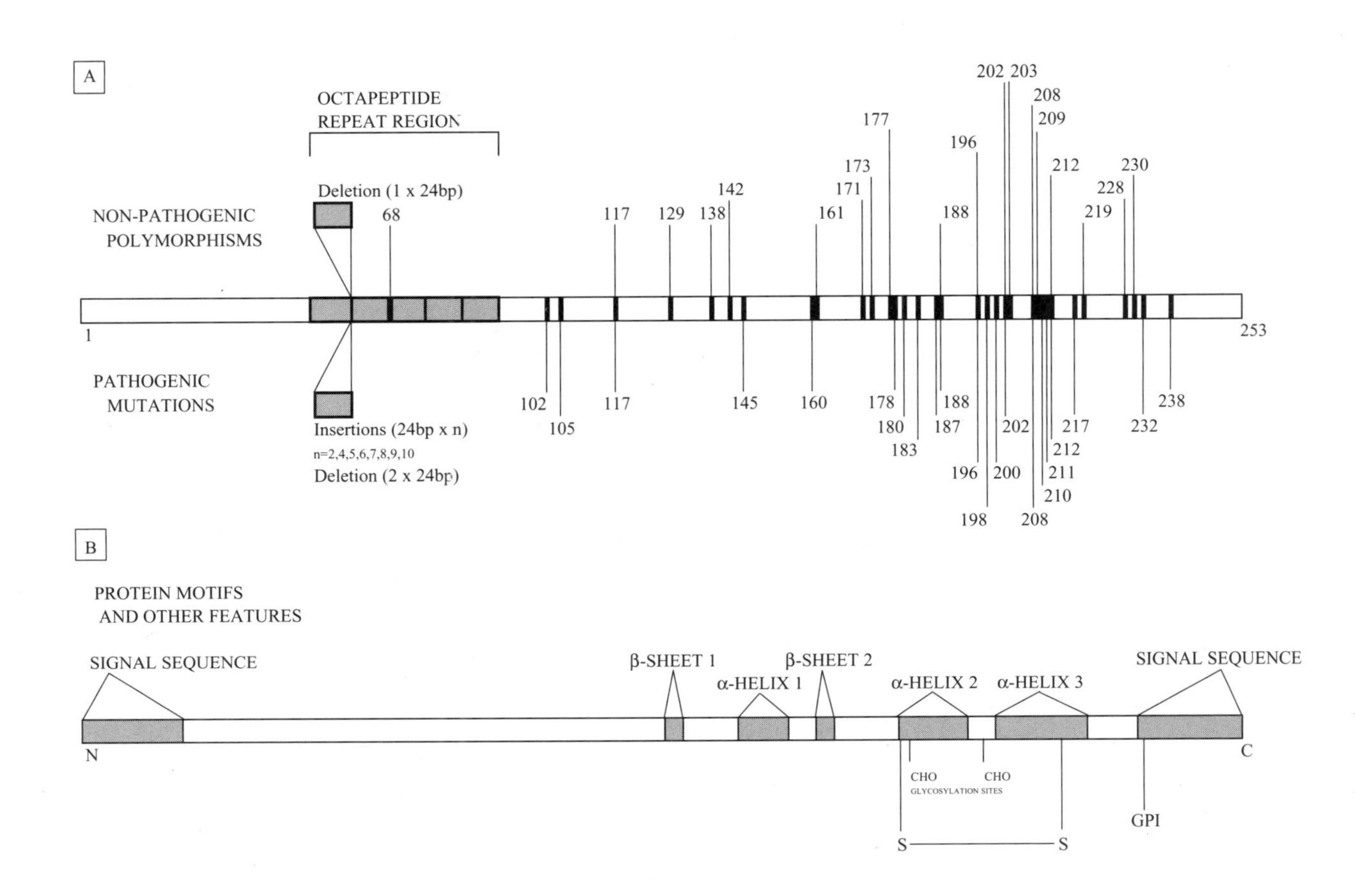
A
OCTAPEPTIDE
REPEAT REGION
Deletion (1 x 24bp)
NON-PATHOGENIC
POLYMORPHISMS
68
117
129
138
142
161
171
173
177
188
196
202 203
208
209
212
219
228
230
1
253
PATHOGENIC
MUTATIONS
Insertions (24bp x n)
n=2,4,5,6,7,8,9,10
Deletion (2 x 24bp)
102
105
117
145
160
178
180
183
187
188
196
198
200
202
208
210
211
212
217
232
238
B
PROTEIN MOTIFS
AND OTHER FEATURES
SIGNAL SEQUENCE
β-SHEET 1
α-HELIX 1
β-SHEET 2
α-HELIX 2
α-HELIX 3
SIGNAL SEQUENCE
N
C
CHO
CHO
GLYCOSYLATION SITES
GPI
S
S

Table 13.4. *Codon 129 genotype frequencies in normal, healthy individuals from three geographically separate locations*

Population	MM	MV	VV
Western Europeans ($n = 106$)[a]	37%	51%	12%
North Americans ($n = 110$)[b]	41%	51%	8%
Japanese ($n = 179$)[c]	92%	8%	0%

[a] Collinge et al. (1991).
[b] Brown et al. (1994).
[c] Doh-ura et al. (1991).

observation indicates that there is probably a level of susceptibility conferred by homozygosity or a degree of protection by the heterozygote genotype. The simplest proposed explanation for this is that the mechanisms involved in prion protein conformational change involve protein–protein interaction. It is therefore suggested that a molecule of PrP^{Sc} will combine only with PrP^{C} if the two have identical amino acid and structural composition, e.g., two codon 129 methionine proteins. If a host is heterozygous then only half the proteins will be accessible for conformational change if the infectious agent is either M or V, resulting in a possible reduction in infectivity (Prusiner et al., 1990). Sequence analysis of the bovine prion gene (gene: *PRNP*; protein: PrP; Genbank: AJ298878) has shown that the equivalent codon to human codon 129 codes for methionine and has not been found to be variable in the populations analysed (Schatzl et al., 1997).

Without knowing the detailed three-dimensional structure of the protein it is difficult to assess the true physicochemical impact of MM at codon 129. We know that this codon is in the second position of four amino acids that make up beta-sheet 1 (see Fig. 13.1B) and therefore could affect elements of the tertiary structure (Petchanikow et al., 2001).

In Japan, codon 129 genotype frequencies are very different, showing that in order to assess the effect of codon 129 genotypes in a disease population

Figure 13.1. (*facing page*). Diagrammatic representations of the prion protein gene (*PRNP*). (A) The position of polymorphisms (above the line) and mutations (below the line) according to codon number. (B) The position of structural motifs that make up the protein tertiary structure as identified by their position on the linearised amino acid sequence.

it is important to identify the underlying population frequency of the M and V alleles. One such study that has highlighted this fact has been undertaken by the Greek CJD surveillance system. They have found that the incidence of sCJD on Crete was fivefold higher than the normal one case per million population per year and so undertook to determine the baseline codon 129 allele frequencies. The results showed that the frequency of MM in the Cretan population was much higher than that expected (57% compared to 37% in Western Europeans) and therefore could be a possible susceptibility factor for the island population (Plaitakis et al., 2001).

The incidence of TSE in Japan has so far shown to be lower than the rates in Western Europe (Nakamura et al., 1999) and therefore other factors as yet to be identified must be responsible for limiting the disease occurrence in this apparently susceptible population. The discovery of a potentially protective polymorphism (codon 219, glutamic acid to lysine, allele frequency = 6%) that has not been found in Japanese sCJD cases highlights this point (Shibuya, 1998). Recent developments in Japanese CJD surveillance systems may show an increase in confirmed cases in the near future.

3.5 Other Gene Regions

As prion diseases are likely to be controlled through multigenic factors there will be other regions of the genome that contain elements with roles in determining the disease phenotype, from the initial susceptibility to infection through to a predisposition to neuronal damage. At the present time there are only candidate regions found from murine studies (Lloyd et al., 2001). One site that has been described in detail is a region 25 kb downstream of *PRNP* where there is a gene that codes for the prionlike doppel protein (gene: *PRND*; protein: Dpl; Genbank: AF106918). This has a high homology to the prion gene with the octapeptide repeat region missing (Moore et al., 1999). The doppel protein has been shown to affect the disease phenotype in transgenic mice (Moore et al., 2001a) but its polymorphic sites have so far shown no association with human disease characteristics (Mead et al., 2000).

As mentioned above the key point of initiation and progression of the disease in human tissues is the conformational change of the prion protein from PrP^{C} to PrP^{Sc} and therefore any destabilising change in the protein structure could be a potential trigger. Unfortunately the three-dimensional structure of PrP^{Sc} cannot currently be determined due to the chemical properties of the protein; however, computer modelling allows a degree of prediction (Zahn et al., 2000). Each of the mutations and polymorphisms could potentially cause changes to the general structure, expression levels, activity, binding site recognition, or stability of the protein. It is proposed that it is because of this that we see highly variable phenotypes arising from subtle DNA sequence changes.

Sequence variations producing different disease phenotypes are categorised into larger groups such as those giving rise to Gerstmann Sträussler Scheinker Disease (GSS), discussed below.

4. HUMAN PRION DISEASE

4.1 History of Human Prion Diseases

Prion diseases in humans were first described in the 1920s by Creutzfeldt and Jakob; the name for the commonest form of human prion disease is Creutzfeldt–Jakob disease. Since that time the specific clinicopathological features of CJD have been defined in great detail from the large numbers of cases subsequently found worldwide in countries where there is active surveillance for such diseases (Parchi et al., 1999; Van Everbroeck et al., 2001; Zeidler et al., 2000). The commonest form of CJD, sporadic CJD (sCJD), occurs at an incidence rate of approximately one case per million population per year (Will et al., 1996).

In addition to the sporadic form of CJD there are also the following additional classifications of CJD and CJD-like TSE:

1. Genetic/inherited CJD
   - 1.1. Familial CJD (fCJD)
   - 1.2. Gerstmann Sträussler Scheinker Syndrome (GSS)
   - 1.3. Fatal Familial Insomnia (FFI)
2. Acquired CJD
   - 2.1. Iatrogenic CJD (iCJD)
   - 2.2. Variant CJD (vCJD)
   - 2.3. Kuru

4.2 Diagnosis of CJD – Clinical

There is significant diversity of clinical and pathological phenotypes in the various forms of CJD and because of the rarity of the disease it can prove difficult to diagnose some of the more atypical cases. With continual publication of clinical presentation data and neuropathology for individual cases and small groups of similar phenotypes, the clinician or pathologist is able to provide a more accurate diagnosis based on the evidence available.

Sporadic CJD has been examined in detail for a large number of cases worldwide and therefore clinical diagnosis has been made easier. The following criteria are used as a general guide to diagnosis:

- Clinical presentation
  - Progressive dementia, myoclonus, and periodic sharpwave electroencephalographic (EEG) activity.
  - Late onset (average age, 55–70 years)
  - Short disease duration (average $<$12 months)
- Pathology
  - Vacuolation of the brain (see Fig. 13.2A)
  - Astrocytosis
  - Nerve cell loss
  - Accumulation of PrP^{Sc} in the brain

Diagnostic criteria for other forms of human prion disease also exist; these similarly depend on close study of both clinical and neuropathological features (Zeidler et al., 1998).

To assist with differential diagnosis of vCJD the UK National CJD Surveillance Unit (Edinburgh) has produced guidelines for the clinical assessment of patients (Will et al., 2000). Only *post mortem* neuropathological examination can truly confirm the diagnosis of vCJD due in part to the specific appearance of 'florid plaques' in the cerebral and cerebellar cortex (Fig. 13.2B) (Ironside et al., 2000).

vCJD is differentially diagnosed from sCJD through the following general criteria:

1. Young age of cases (mean 29 years) and long duration ($>$12 months)
2. Early psychiatric features with progressive ataxia.
3. Presence of 'florid' plaques in the cerebral and cerebellar cortex.

4.3 Diagnosis of TSE – Protein Laboratory

Analysis of the physicochemical properties of the prion protein is undertaken in the diagnosis of TSE. This involves analysis of the biochemical properties of PrP^{Sc} and its protease-resistant core, designated PrP^{res} (res = resistant). Analysis of homogenates of brain tissue, or other tissues, on denaturing polyacrylamide gels (SDS–PAGE) allows separation of proteins according to size. The homogenate is digested with Proteinase K, because PrP^{Sc} (and

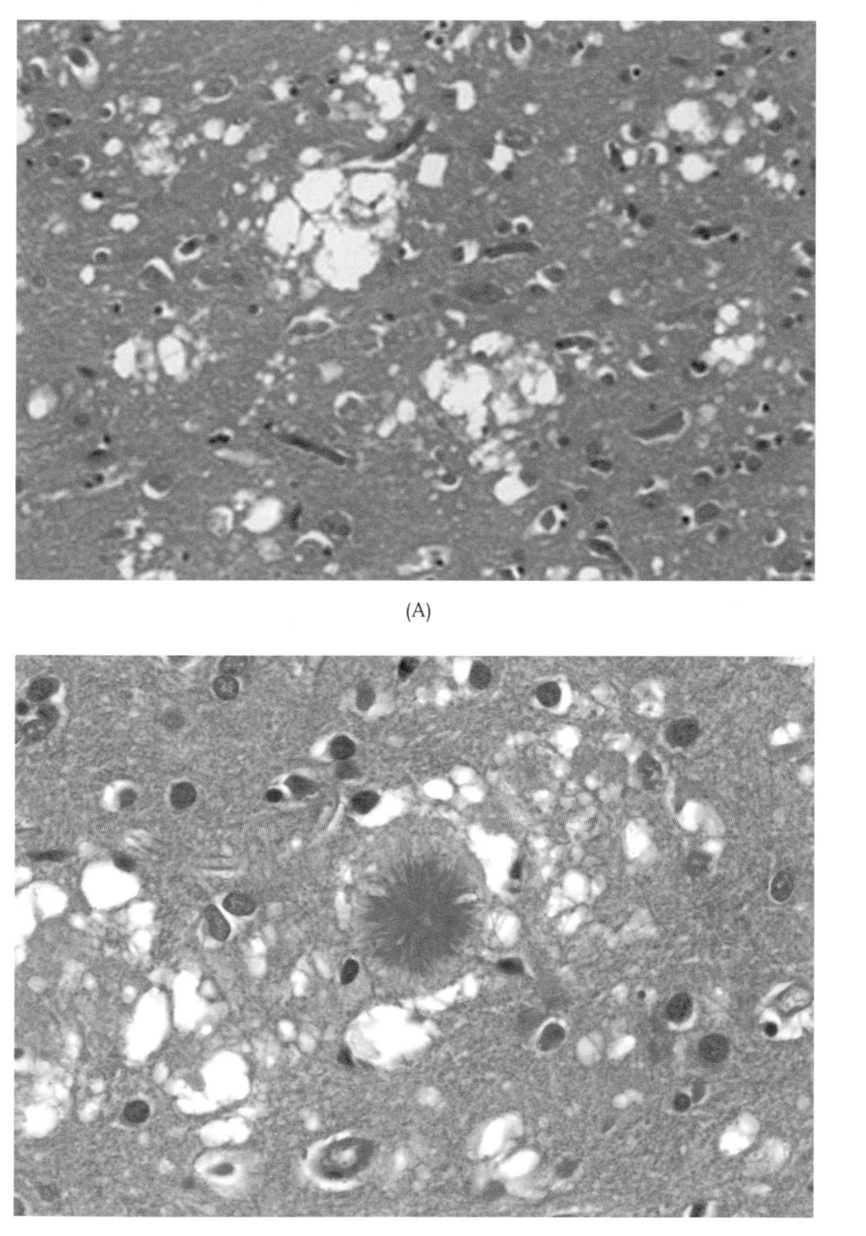

Figure 13.2. (A) Confluent spongiform degeneration in the neuropil of sporadic CJD (haematoxylin and eosin). (B) Florid plaque within the cerebral cortex of a vCJD case (haematoxylin and eosin). (See color plate.)

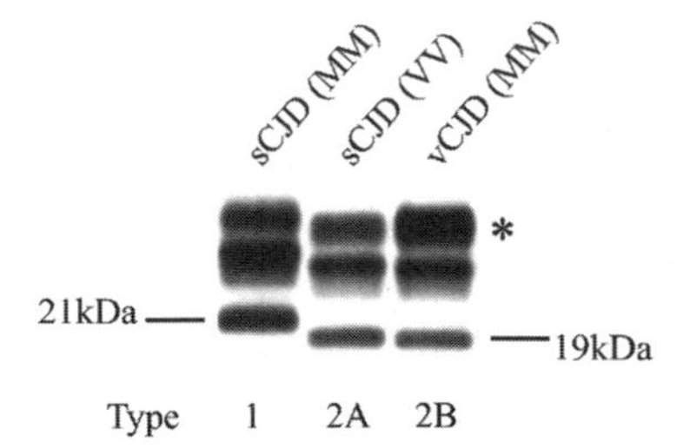

Figure 13.3. Western blot of prion protein (PrP^{res}) isotype in cerebral cortex from cases of sporadic Creutzfeldt–Jakob disease (sCJD) and variant Creutzfeldt–Jakob disease (vCJD). The nonglycosylated PrP^{res} is of either 21 kDa (type 1) or 19 kDa (type 2). PrP^{res} isotype in vCJD is consistently found to be type 2 with a predominance of the diglycosylated form (marked with an asterisk). The *PRNP* codon 129 methionine (M)/valine (V) polymorphism for each individual case is shown in parenthesis.

not PrP^{C}) is partially resistant to its action and will therefore be the only target for the subsequent detection step. Electrophoretic separation is followed by Western blotting of the size-fractionated proteins to a membrane. An antiprion protein antibody is then applied followed by an enzyme-linked secondary antibody. Chemiluminescence or chemifluorescence is then employed and the signal is captured on X-ray film or using a laser scanner. The final image consists of three bands that correspond to diglycosylated, monoglycosylated, and nonglycosylated glycoforms of PrP^{res}. The extent of N-terminal truncation produced by proteolytic degradation results in either a 21-kDa nonglycosylated fragment (Type 1) or a 19-kDa nonglycosylated fragment (Type 2). Further subclassification can be made on the basis of the proportion of the three glycoforms present. Typical examples of the patterns in the two most commonly occurring forms of sCJD and vCJD are shown in Fig. 13.3.

Type 1 and Type 2 isotypes are found in cases of sCJD; however, it is of interest to note that the Type 2B electrophoresis pattern is characteristic of vCJD. This 'glycoform signature' is common to cattle with BSE and other animal species infected with BSE (Collinge et al., 1996).

Using N-terminal sequencing and mass spectrometry the predominant protease cleavage sites for PrP^{Sc} have been determined for both isotypes (Parchi et al., 2000). The Type 1 isotype is a collection of protein fragments that is primarily terminated at the 82nd amino acid (glycine) and the Type 2 isotype is a collection primarily terminated at the 97th amino acid (serine). As a consequence of this analysis it was discovered that variation of the termination point for protease cleavage was dependent on the amino acid at codon 129. Specifically the presence of valine at codon 129 resulted in a more variable

Table 13.5. *Disease phenotype association with codon 129 genotype and prion protein isotype in sporadic and variant CJD*

Disease	Genotype/isotype group	General description
sCJD	MM1 or MV1	"Classical" CJD, myoclonic variant (70% of sCJD cases)
sCJD	VV2	Ataxic variant (16% of sCJD cases)
sCJD	MV2	Kuru plaques variant (9% of sCJD cases)
sCJD	MM2	Thalamic variant, rare (2% of sCJD cases)
sCJD	MM2	Cortical variant, rare (2% of sCJD cases)
sCJD	VV1	Rare (1% of sCJD cases)
vCJD	MM2B	Variant CJD (52 of 114 cases tested)

(Data adapted from Parchi et al., 1999).

size range of protein fragments. This range extends through a sequence of amino acids thought to be associated with tertiary structure elements and could therefore impact significantly on the conversion of PrP^{c} to PrP^{Sc}.

4.4 Diagnosis of TSE – Genetics Laboratory

Molecular genetic analysis of the prion protein gene (*PRNP*) through use of the polymerase chain reaction (PCR) and automated DNA sequencing is a key requirement for the prediction of the TSE phenotype. The main reasons for such analyses are as follows:

1. To distinguish those cases that are not determined by a familial, inherited mutation, e.g., sporadic CJD or variant CJD.
2. To positively identify those cases that have arisen through a familial, inherited mutation or one that has occurred spontaneously in that individual.
3. To identify the presence of polymorphisms that may have effects on the phenotype, particularly at codon 129, an important susceptibility factor and phenotype determinant.

The combination of prion protein isotype and codon 129 genotype in association with disease phenotypes has become an important method of classification of human TSE (see Table 13.5).

Other laboratory analyses are carried out, such as identifying the presence or absence of protein 14-3-3 in the cerebrospinal fluid (CSF), to add support to a diagnosis of TSE. There are many groups around the world that are developing further such tests on a range of tissues and body fluids such as blood. The ultimate goal is to develop a test that can identify the preclinical phenotype in the hope that treatment can then be given to prevent further disease progression.

5. GENETIC SUSCEPTIBILITY FACTORS IN HUMAN PRION DISEASES

5.1 Genetic Susceptibility to Sporadic CJD (sCJD)

Without a specific genetic mutation of the *PRNP* gene determining the onset of sCJD we have to look to other phenotypic components for clues to a possible genetic basis of this TSE. It is by the very nature of a sporadic disease that we are looking for factors that predispose the individual to the clinical phenotype. Codon 129 genotype is the obvious choice for initial investigation of genetic susceptibility as homozygosity for either methionine or valine is overrepresented in the sCJD disease population. Statistical analysis of the relationship between codon 129 and sCJD shows that the association has a high level of significance (P value <0.001) (Parchi et al., 1999). The increase in codon 129 MM frequency is greater than for VV genotypes which suggests that the former genotype in particular is a significant susceptibility factor.

Detailed analysis of the phenotypic impact of codon 129 in large numbers of sCJD cases has led to classification of the disease according to six different groups. These are differentiated by the PrP^{Sc} isotype and the codon 129 genotype and show clear phenotypic variations (Table 13.5).

This strong association among genotype, protein isotype, and phenotype is a highly significant one but can only be defined by brain tissue analysis after disease onset and when the amount of PrP^{Sc} is at a detectable level. Until such time as it is possible to determine which group an individual belongs to at a preclinical or screening stage we will be unaware of their predisposition to a particular phenotype. The future discovery of other genetic markers that segregate with these six groups may allow such analysis to be performed.

If we take the conformational change from PrP^{C} to PrP^{Sc} as being the initial event that could trigger sCJD, then any factor contributing to the instability of the protein may be important. A random somatic mutation of the prion gene in the tissues where PrP is expressed would be an obvious candidate, alongside the spontaneous conformational change from PrP^{C} to

PrP^{Sc}. With regard to possible genetic susceptibility to these two mechanisms we would have to look at factors outwith the prion gene as only codon 129 genotype has been found to show disease specific variation.

Random somatic mutation could occur at any point from embryonic development to adult by a number of means: environmental or biochemical. The chance of a mutation occurring specifically in the prion gene in a cell where that region is transcriptionally active is rare. There could of course be mutations occurring in other genes in any tissue that may be able to act on the modification of the prion protein. With increasing age we become more susceptible to deleterious DNA damage and therefore theoretically the cellular environment of the prion protein could be more biased towards conformational changes to occur. With the average age of onset for sCJD between 60 and 70 years old these factors could possibly play an important role.

It has been found in genetic forms of CJD that the PrP^{Sc} molecules are formed from only those alleles containing the mutated amino acid sequence (Chen et al., 1997). This may suggest that if a somatic mutation has occurred then the resultant PrP^{Sc} molecules should show this, but in practical terms this is difficult to investigate.

The *PRNP* gene itself is not necessarily the only chromosomal region where genetic variation may affect the onset of sCJD. There are other potential sites such as those where direct influences, such as promoter regions (Mahal et al., 2001), or indirect influences, such as genes coding for proteins involved in similar cellular processes. Spontaneous conformational change from PrP^{C} to PrP^{Sc} appears to be the most likely scenario as there are fewer questions that remain unanswered about the hypothesis. It is straightforward to think that all it would take would be for one prion protein molecule to reconfigure to PrP^{Sc} due to a chance collection of factors in the biochemical environment, such as a change in pH (Zanusso et al., 2001). This one molecule may then cause conformational change in its neighbouring proteins and thus the chain-reaction starts.

Apart from the prion protein and its gene there are many other factors such as the clinical phenotypic features of sCJD that may have a degree of genetic-based susceptibility. Two clearly defined diagnostic parameters for the majority of sCJD cases are the late onset and short duration of disease. Rigid time frames for factors such as these may indicate the presence of a mechanism that initiates pathological change and such a mechanism could have important genetic factors. Work on determination of genetic factors for TSE incubation time in mice has shown a number of regions of the genome where such components may exist (Lloyd et al., 2001). Work of a similar nature is under investigation in humans (Mead et al., 2001).

An example of a genetic determinant for disease onset is shown in the case of Alzheimer's disease (AD). It has been found that polymorphisms of the Apolipoprotein E (ApoE) gene segregate according to age of onset for AD, and in particular the $\varepsilon 4$ allele is a significant indicator for late onset sporadic forms (Brousseau et al., 1994). ApoE allele frequency has been determined in a number of sCJD cohorts but has yet to be definitively associated with a phenotypic change. A French study found that the $\varepsilon 4$ allele was a risk factor for the disease (Amouyel et al., 1994), whereas an Italian study found no influence on the disease (Salvatore et al., 1995). No analysis has yet been reported for vCJD.

5.2 Genetic Susceptibility to Familial CJD (fCJD)

Genetic or familial CJD (fCJD), diagnosed through the presence of inherited mutations, is thought to occur because the mutant host protein is more likely to reconfigure to the disease associated form. This is proposed to be due to a destabilisation of the tertiary structure, because of changes in the protein amino acid sequence arising from mutations in the gene. Some specific mutations are associated with particular phenotypes and have thus been classified under alternative names, including GSS and fatal familial insomnia (FFI).

The diagnostic criteria for defining fCJD is a pathogenic mutation present in the prion gene that can be traced through the family pedigree in sufficient detail to show that presence of CJD is linked to inheritance of the mutation. A number of common mutations have been found throughout the world that have been traced back through the generations to where and when the original mutation events are thought to have occurred (Harder et al., 1999; Lee et al., 1999; Nicholl et al., 1995).

It is still a matter of debate as to whether the mutation itself is the actual cause of the ensuing disease or whether it is just an additional level of susceptibility. The majority of mutations that have been well characterised are associated with diseases which show near-complete penetrance and are inherited as autosomal dominant disorders. The discovery of a mutation during screening of the prion gene by DNA sequencing is therefore generally accepted as the causative agent for the disease. Spontaneous mutations of this nature appear to be rare and most cases of fCJD have a family history of neurological disease. However, due to the relatively late onset age for these diseases the mutation may not be 'visible' for a number of generations. Asymptomatic carriers of the mutation could die early of other causes, after producing offspring.

The susceptibility of the prion gene to mutations arises to some degree from the specific DNA sequence of the ORF where there are regions more likely to mutate. This is due to the biochemical nature of DNA bases or effects of secondary structures. For example, CpG 'hotspots', where a methylated form of a cytosine base next to a guanine base can readily cause deamination leaving either TG or CA sequences. Of the potential sites of this nature in the prion gene approximately half have already been found in cases of genetic forms of TSE, including codons 102 and 200.

As well as mutations causing CJD-like phenotypes there are also those that have their own more specific features and have subsequently been given separate names. The following information highlights the most common mutations so far discovered and the specific phenotypes to which they are categorised.

5.2.1 CJD Phenotype Mutations

This group can be recognised because of the following distinct phenotypic characteristics:

- Late age of onset
- Rapidly advancing dementia
- Abnormal EEG
- Widespread spongiform degeneration

The following describes details of the common mutations resulting in this group of phenotypes.

1. **Codon 178**, with valine at codon 129 on the same allele, shows a CJD phenotype and has been found in a number of cases throughout the world, including a large pedigree in Finland (Haltia et al., 1991).
2. **Codon 200** is the most common mutation found causing a CJD phenotype, with clinical features and pathology very similar to those of sCJD. It has been found in many countries with notable clusters in Slovakia, Chile, and Sephardic Jews. In Jews of Libyan origin the incidence of fCJD is 100 times that of sCJD worldwide and accounts for the majority of CJD cases identified by the Israel CJD surveillance system (Lee et al., 1999; Meiner et al., 1997).
3. **Codon 210** found initially in Italian and French families, and now found around the world, shows a phenotype very similar to sCJD (Ripoll et al., 1993).
4. **Insertions** of variable numbers of 24-bp repeat elements in the octapeptide repeat region of the prion gene have been found throughout

the world. Some insertions occur more frequently than others, such as cases with six extra repeats that have been found in UK families. This type of mutation causes a wide spectrum of phenotypes dependent on the number of additional repeats. The disease duration can be between 2 months and 18 years, with phenotypes similar to CJD or GSS. As a general rule the CJD-like histopathological phenotype occurs with up to six extra repeat elements and a duration of less than 1 year. The GSS-like phenotype occurs with a larger number of repeats and a longer duration (Parchi et al., 1998). As these are generalisations from a limited number of cases, there can be difficulties when assessing the clinical features of suspect cases. Subsequent molecular analysis of the prion gene will then assist in definitive diagnosis (Moore et al., 2001b; Nicholl et al., 1995).

5.2.2 Gerstmann Sträussler Scheinker Syndrome (GSS) Phenotype Mutations

The original GSS study involved an Austrian family that now encompasses 221 members in nine generations. The clinical features that are generally found to specifically differentiate GSS from sCJD are earlier age at onset (30–40 years old), longer duration (~5 years) with slowly progressive ataxia, and pathologically identified multicentric amyloid plaques (Hainfellner et al., 1995).

1. **Codon 102** is the commonest GSS phenotype mutation and has been identified throughout the world. It is recognised as the ataxic variant of GSS (due to early cerebellar ataxia) with age at onset approximately 10 years earlier than sCJD (Hainfellner et al., 1995).
2. **Codon 117** has been recently studied in families from Hungary, France, and the UK. It is described as the dementia variant of GSS highlighting the usual occurrence of presenile dementia at onset (Mallucci et al., 1999).
3. **Codon 198**, with valine at codon 129, has been studied in detail for one pedigree commonly referred to as the Indiana kindred. This mutation is recognised by the slow progression of disease and pathologically by the presence of neurofibrillary tangles (NFTs) (Dlouhy et al., 1992).
4. **Codon 217**, with valine at codon 129, found in a Swedish family shows similar clinical features and pathology to codon 198 cases, including the presence of NFTs (Hsaio et al., 1992).

5.2.3 Fatal Familial Insomnia (FFI) Phenotype Mutation

This phenotype is clinically characterised by inattention, sleep loss, dysautonomia, and motor signs and pathologically characterised by a preferential thalamic degeneration. It is associated with a single mutation described below.

1. **Codon 178**, with methionine at codon 129, is the only *PRNP* mutation that is associated with FFI phenotype. This disease is clearly distinct from the fCJD phenotype where the mutation occurs with valine at codon 129. Another example of the important impact that codon 129 has on clinical disease features. This mutation has been found in many countries involving over 20 individual families. It has been recognised since 1986 (Lugaresi et al., 1986) and is characterised clinically by progressive insomnia and pathologically by predominant thalamic degeneration (Cortelli et al., 1999). Variation in this phenotype has been reported in an Australian FFI cohort where individuals were characterised with typical CJD, FFI, and cerebellar ataxia (McLean et al., 1997).

5.3 Genetic Susceptibility to Acquired CJD

This group of TSE arises due to the passage of infectious material from one individual to the next and has so far been limited to the three following causes:

1. Iatrogenic CJD: disease acquired from infected human tissues by medical or surgical procedures.
2. Kuru: ritualistic cannibalism of infected human tissues.
3. Variant CJD: disease acquired from infected bovine tissues.

5.3.1 Iatrogenic CJD (iCJD)

This is an acquired form of the disease caused by the medical or surgical transmission of infectivity from individuals with CJD, e.g., by neurosurgical instruments or dura mater grafts.

iCJD has been found to be transmitted through a variety of means and was first reported in 1974 in the recipient of a corneal graft from a donor who died of unsuspected CJD (Duffy et al., 1974). The current worldwide total for all identified iCJD cases is approximately 250. In the mid-1980s the major risk factors for this acquired disease were identified as contaminated dura mater grafts and human growth hormone, with other routes resulting in only a few cases worldwide (corneal transplant, neurosurgical instruments, and intracerebral EEG needles). There are a few clusters of cases in particular countries where the source of contaminated material has been successfully traced. Particular examples of this are the growth hormone related cases in France, the United States, and the UK and the contaminated dura mater graft

Table 13.6. *Codon 129 genotype for iCJD cases up to 2000*

iCJD Group	Number of cases	MM(%)	MV(%)	VV(%)	Homozygotes(%)
Normal individuals[a]	398	39	50	11	50
Dura mater graft	43	74	19	7	81
Growth hormone	82	48	21	31	79
UK Growth hormone	20	5	40	55	60
Other	3	67	33	0	67
Total iCJD	128	57	20	23	80

(Adapted from Brown et al., 2000).
[a] Alperovitch et al. (1999).

cases in Japan where this type of material is used more frequently than in other countries.

As the diagnosis and general awareness of human prion diseases have improved over the years the production of transplant material and use of surgical instruments has conformed to stricter guidelines to ensure that the risk to recipient patients is at an acceptable level. Improved methods of decontamination and purification of donor material have also decreased significantly the chance of passing on infection. The pharmaceutical industry now uses recombinant bacteria to manufacture human growth hormone free of contamination, although this is currently an expensive option for some countries. Preclinical diagnosis of prion disease is currently unavailable, resulting in the unavoidable use of contaminated material and the proposed continual appearance of small numbers of new cases. The majority of new cases, however, are those where there has been a long incubation period of potentially many decades from the original point of infection.

The pathology and clinical features of iCJD shows similarities to vCJD, which might be predicted, as both are prion diseases initiated by peripheral infection. Following this it is of interest to note the codon 129 genotypes of iCJD cases because the vCJD genotypes may follow the same temporal patterns, e.g., incubation time and the appearance of different codon 129 genotypes at different periods from the point of infection.

The data compiled by Brown (Brown et al., 2000) for codon 129 in iCJD cases up to the year 2000 are shown in Table 13.6.

As for sCJD, homozygosity (MM or VV) is clearly elevated in iCJD cases indicating a degree of host susceptibility. The theory mentioned above where the conformational change of PrP^C to PrP^{Sc} would most easily occur between homologous molecules holds true for iCJD as well as sCJD. The appearance of the disease associated form in the body is, however, different for iCJD as it is an acquired disease not a sporadic one. Therefore the genetic susceptibility to iCJD is a product of not only the recipient's genetic makeup but also of the donor's. If it were simply that the codon 129 genotype had to match between donor and recipient of biological material then this could be assayed prior to any treatment, in a similar way to matching blood groups before transfusion.

It is very difficult to discover what the codon 129 genotype was for the contaminating material in the human growth hormone cases as the source is a pool of many individuals. The UK cases, however, show a marked increase in VV genotypes (see Table 13.6), indicating the possible contamination by solely VV material or a majority of such. The observation that codon 129 genotype frequencies in iCJD closely match that of sCJD cases suggests that the contaminating material has come from a population of sCJD with the expected range of genotypes. It would be of interest to compare the genotype frequencies of the Japanese dura mater cases as the general recipient population has very different codon 129 genotype frequencies to that of Europe, from where the tissue originated. Data from a small number of cases are available and show that 86% were MM, 14% were MV, with no VV cases. This reflects the normal populations frequencies in Japan (Hoshi et al., 2000).

Codon 129 genotype is associated with another effect on the iCJD clinical features and that is the statistically significant increase in the incubation time for heterozygotes, as found in the French growth hormone cases (Hulliard d'Aignaux et al., 1999). Incubation time is very difficult to accurately determine for growth hormone iCJD cases as the contaminating batch could have been administered to the patient at any stage throughout their treatment regime. This observed effect might therefore not be truly representative of all iCJD cases. Homologous protein interaction may add weight to this theory, as only half of the hosts PrP^C molecules could undergo conformational change if the donor was homozygous. This would then lead to a longer time before the levels of host PrP^{Sc} reached a pathological level.

This infectivity scenario is probably too simple and from the finding that the codon 129 genotype frequencies for iCJD and sCJD are quite similar it can be suggested that the same factors, including genetics, are governing the susceptibility of individuals to both types of disease.

5.3.2 Kuru

In addition to the recognised form of iCJD as described above there is also a specific form of 'acquired' CJD that is geographically isolated and does not have any other similar examples anywhere in the world. This is a disease called Kuru, diagnosed in the Fore Tribe of Papua New Guinea, and reported to the scientific community in 1959 (Gajdusek and Zigas 1959). It has been classified as a form of acquired CJD as it is proposed to be a prion disease that arose through ritualistic cannibalistic practices that were performed by the Fore Tribe. Consumption of deceased relatives' brain material was an integral part of the funeral ritual and on reflection was an ideal method of infection of TSE from one generation to the next. All that was required to initiate this disease was for one member of the tribe to fall victim to sCJD and to die in the latter stage of the disease when the brain contained high levels of infectious PrP^{Sc}. The tribesmen and the family of the deceased would then have consumed this material. Unfortunately due to the nature of the protein, even if the brain material had been cooked the prion protein would have remained infectious.

The numbers of Kuru cases currently stands at approximately 2,500 with the initial cases probably occurring between 1900 and 1920 and the last stretching into the 21st century. It is believed that there is unlikely to be many more as there are very few members of the tribe alive today who took part in the cannibalistic practices. The last cases appeared more than 40 years after the last known infectious point which has given us an idea as to the extent of prion disease incubation time. This is of significant relevance to vCJD, another dietary linked acquired prion disease, and the future predictions for expected numbers of such cases.

The genetics of susceptibility to Kuru has been limited to analysis of codon 129 although this has shown some interesting results. It is a relatively complete data set with samples available from cases at the beginning, middle, and end of the disease progression through the generations. The most interesting feature from the codon 129 study is the variation in genotype frequencies of the diseased and survival populations over time. The Kuru survivor population sampled in the late 1950s was found to have a significantly lower level of MM genotypes, indicating that these individuals were the first to die from Kuru. The last group of Kuru cases, who would have been infected at a similar time, was subsequently found to have a high frequency of VV genotypes. These data suggest a variable incubation time between the genotypes with MM being the most susceptible and having the shorter time scale. This may have been due to the 'founder' individual being MM or because of another as yet undetermined factor specific to those people with MM at codon 129.

5.3.3 Variant CJD

Variant CJD (vCJD) is an acquired form of CJD caused by a cross-species transmittance of the BSE infectious agent from cattle to humans. Genetic predisposition to vCJD could, as shown for the other types of TSE, be observed through many factors. The codon 129 genotype is one; all cases so far analysed have been found to be MM, indicating that approximately 40% of the UK population would be at risk. Additionally, all cases so far analysed for PrP isotype have shown a Type 2B pattern that has not been seen in any other forms of disease. These two observations point to a susceptible fraction of the population; however, the fraction appears too large for such a small number of cases so far diagnosed (total to March 2002 = 116).

Two of the more noticeable phenotypic traits of vCJD are the younger patient age (median age at death 28 years) and the longer disease duration (median time 13 months), when compared to sCJD (65 years and <12 months). It is unknown whether these traits could have other genetic components at this stage.

With the large number of infected cattle entering the human food chain at the height of the BSE epidemic the probability of a person consuming contaminated food would theoretically be quite high. Why, then, are there so few cases currently diagnosed? The answer to this is likely to have some elements of personal genetic predisposition and the search for such factors is under investigation in laboratories around the world.

An example of the current research to find other genetic cofactors for vCJD involved typing of vCJD, sCJD, and non-CJD controls for the human leucocyte antigen (HLA) loci. The potential role of major histocompatibility complex (MHC) molecules in immune response and disease pathogenesis makes this an interesting target for analysis. It was observed that there was a statistically significant reduction in frequency of HLA class-II type DQ7 in patients with vCJD (Jackson et al., 2001). The reported protective association between this genetic marker and vCJD is still open to investigation and further typing of cases is required to make a more complete data set.

The internal route of infection after consumption of contaminated material is one of the key pathways along which the involvement of genetic factors is being closely examined. There are many possible targets for this work, such as the following:

1. Mechanisms of uptake of bovine PrP^{Sc} by cells lining the gastrointestinal tract.
2. Breakdown or accumulation of bovine PrP^{Sc} in organs such as the spleen, tonsil, appendix, or lymph nodes.
3. Propagation of bovine PrP^{Sc} into the peripheral nervous system and from there into the central nervous system and the brain.
4. Accumulation and propagation of subsequent host PrP^{Sc} through specific tissue types.

With the current concern over the unknown extent of people with asymptomatic vCJD infection, there is potential for future spread of disease resulting from cross-contamination by other sources such as transplants and blood and blood fractionation products. It is the latter blood related products that have demanded recent attention as there are no man-made alternatives to the majority of the materials required and the demand for supplies is great. However, no link has yet been made between blood and transmission of human prion diseases. To limit the chances of this occurring the UK blood transfusion service deplete the blood of leucocytes, cells that are known to bind PrP^{C} and therefore could be carrying PrP^{Sc}. They also import blood fractions from countries where vCJD has not yet been diagnosed, such as the United States.

For some current opinions on vCJD and the emerging data from recent studies, reviews by Jackson and Collinge (2001) and Brown et al. (2001) are recommended.

ACKNOWLEDGMENTS

We would like to thank Dr. Mark Head for providing the Western blot image and Ms. Diane Ritchie for the pathology figures.

REFERENCES

Alperovitch, A., Zerr, I., Pocchiari, M., et al. (1999). Codon 129 prion protein genotype and sporadic Creutzfeldt–Jakob disease. *Lancet* **353**, 1673–1674.

Amouyel, P., Vidal, O., Launay, J. M., et al. (1994). The apolipoprotein E alleles as major susceptibility factors for Creutzfeldt-Jakob disease. The French Research Group on Epidemiology of Human Spongiform Encephalopathies. *Lancet* **344**, 1315–1318.

Brousseau, T., Legrain, S., Berr, C., et al. (1994). Confirmation of the epsilon 4 allele of the apolipoprotein E gene as a risk factor for late-onset Alzheimer's disease. *Neurology* **44**, 342–344.

Brown, H. R., Goller, N. L., Rudelli, R. D., et al. (1990). The mRNA encoding the scrapie agent protein is present in a variety of non-neuronal cells. *Acta Neuropathol. (Berl.)* **80**, 1–6.

Brown, P., Cervenakova, L., Goldfarb, L. G., et al. (1994). Iatrogenic Creutzfeldt–Jakob disease: an example of the interplay between ancient genes and modern medicine. *Neurology* **44**, 291–293.

Brown, P., Preece, M., Brandel, J. P., et al. (2000). Iatrogenic Creutzfeldt–Jakob disease at the millennium. *Neurology* **55**, 1075–1081.

Brown, P., Will, R. G., Bradley, R., et al. (2001). Bovine spongiform encephalopathy and variant Creutzfeldt–Jakob disease: background, evolution, and current concerns. *Emerg. Infect. Dis.* **7**, 6–16.

Bueler, H., Aguzzi, A., Sailer, A., et al. (1993). Mice devoid of PrP are resistant to scrapie. *Cell* **73**, 1339–1347.

Carlson, G. A., Kingsbury, D. T., Goodman, P. A., et al. (1986). Linkage of prion protein and scrapie incubation time genes. *Cell* **46**, 503–511.

Carlson, G. A., Westaway, D., Goodman, P. A., et al. (1988). Genetic control of prion incubation period in mice. *Ciba Found. Symp.* **135**, 84–99.

Chen, S. G., Parchi, P., Brown, P., et al. (1997). Allelic origin of the abnormal prion protein isoform in familial prion diseases. *Nat. Med.* **3**, 1009–1015.

Collinge, J., Palmer, M. S. and Dryden, A. J. (1991). Genetic predisposition to iatrogenic Creutzfeldt–Jakob disease. *Lancet* **337**, 1441–1442.

Collinge, J., Sidle, K. C., Meads, J., et al. (1996). Molecular analysis of prion strain variation and the aetiology of 'new variant' CJD. *Nature* **383**, 685–690.

Collinge, J., Whittington, M. A., Sidle, K. C., et al. (1994). Prion protein is necessary for normal synaptic function. *Nature* **370**, 295–297.

Cortelli, P., Gambetti, P., Montagna, P., et al. (1999). Fatal familial insomnia: clinical features and molecular genetics. *J. Sleep Res.* **8** (Suppl 1), 23–29.

Dickinson, A. G., Meikle, V. M. and Fraser, H. (1968). Identification of a gene which controls the incubation period of some strains of scrapie agent in mice. *J. Comp. Pathol.* **78**, 293–299.

Dlouhy, S. R., Hsiao, K., Farlow, M. R., et al. (1992). Linkage of the Indiana kindred of Gerstmann-Straussler–Scheinker disease to the prion protein gene. *Nat. Genet.* **1**, 64–67.

Doh-ura, K., Kitamoto, T., Sakaki, Y., et al. (1991). CJD discrepancy. *Nature* **353**, 801–802.

Duffy, P., Wolf, J., Collins, G., et al. (1974). Letter: Possible person-to-person transmission of Creutzfeldt–Jakob disease. *N. Engl. J. Med.* **290**, 692–693.

Gajdusek, D. C. and Zigas, V. (1959). Kuru: Clinical, pathological and epidemiological study of an acute progressive degenerative disease of the central nervous system among natives of the Eastern Highlands of New Guinea. *Am. J. Med.* **26**, 442–469.

Goldfarb, L. G. and Brown, P. (1995). The transmissible spongiform encephalopathies. *Annu. Rev. Med.* **46**, 57–65.

Hainfellner, J. A., Brantner-Inthaler, S., Cervenakova, L., et al. (1995). The original Gerstmann-Straussler-Scheinker family of Austria: divergent clinicopathological phenotypes but constant PrP genotype. *Brain Pathol.* **5**, 201–211.

Haltia, M., Kovanen, J., Goldfarb, L. G., et al. (1991). Familial Creutzfeldt–Jakob disease in Finland: epidemiological, clinical, pathological and molecular genetic studies. *Eur. J. Epidemiol.* **7**, 494–500.

Harder, A., Jendroska, K., Kreuz, F., et al. (1999). Novel twelve-generation kindred of fatal familial insomnia from Germany representing the entire spectrum of disease expression. *Am. J. Med. Genet.* **87**, 311–316.

Harris, D. A. (1999). Cellular biology of prion diseases. *Clin. Microbiol. Rev.* **12**, 429–444.

Hoshi, K., Yoshino, H., Urata, J., et al. (2000). Creutzfeldt–Jakob disease associated with cadaveric dura mater grafts in Japan. *Neurology* **55**, 718–721.

Hsiao, K., Baker, H. F., Crow, T. J., et al. (1989). Linkage of a prion protein missense variant to Gerstmann–Straussler syndrome. *Nature* **338**, 342–345.

Hsiao, K., Dlouhy, S. R., Farlow, M. R., et al. (1992). Mutant prion proteins in Gerstmann–Straussler–Scheinker disease with neurofibrillary tangles. *Nat. Genet.* **1**, 68–71.

Hsiao, K. K., Groth, D., Scott, M., et al. (1994). Serial transmission in rodents of neurodegeneration from transgenic mice expressing mutant prion protein. *Proc. Natl. Acad. Sci. USA* **91**, 9126–9130.

Hsiao, K. K., Scott, M., Foster, D., et al. (1990). Spontaneous neurodegeneration in transgenic mice with mutant prion protein. *Science* **250**, 1587–1590.

Huillard d'Aignaux, J., Costagliola, D., Maccario, J., et al. (1999). Incubation period of Creutzfeldt–Jakob disease in human growth hormone recipients in France. *Neurology* **53**, 1197–1201.

Hunter, N., Dann, J. C., Bennet, A. D., et al. (1992). Are Sinc and the PrP gene congruent? Evidence from PrP gene analysis in Sinc congenic mice. *J. Gen. Virol.* **73**, 2751–2755.

Hunter, N., Foster, J. D., Dickinson, A. G., et al. (1989). Linkage of the gene for the scrapie-associated fibril protein (PrP) to the Sip gene in Cheviot sheep. *Vet. Rec.* **124**, 364–366.

Ironside, J. W., Head, M. W., Bell, J. E., et al. (2000). Laboratory diagnosis of variant Creutzfeldt–Jakob disease. *Histopathology* **37**, 1–9.

Jackson, G. S., Beck, J. A., Navarrete, C., et al. (2001). HLA-DQ7 antigen and resistance to variant CJD. *Nature* **414**, 269–270.

Jackson, G. S. and Collinge, J. The molecular pathology of CJD: old and new variants. (2001). *Mol. Pathol.* **54**, 393–399.

Kaneko, K., Peretz, D., Pan, K. M., et al. (1995). Prion protein (PrP) synthetic peptides induce cellular PrP to acquire properties of the scrapie isoform. *Proc. Natl. Acad. Sci. USA* **92**, 11160–11164.

Lee, H. S., Sambuughin, N., Cervenakova, L., et al. (1999). Ancestral origins and worldwide distribution of the *PRNP* 200K mutation causing familial Creutzfeldt–Jakob disease. *Am. J. Hum. Genet.* **64**, 1063–1070.

Liao, Y. C., Lebo, R. V., Clawson, G. A., et al. (1986). Human prion protein cDNA: Molecular cloning, chromosomal mapping, and biological implications. *Science* **233**, 364–367.

Lloyd, S. E., Onwuazor, O. N., Beck, J. A., et al. (2001). Identification of multiple quantitative trait loci linked to prion disease incubation in mice. *Proc. Natl. Acad. Sci. USA* **98**, 6279–6283.

Lugaresi, E., Medori, R., Montagna, P., et al. (1986). Fatal familial insomnia and dysautonomia with selective degeneration of thalamic nuclei. *N. Engl. J. Med.* **315**, 997–1003.

Mahal, S. P., Asante, E. A., Antoniou, M., et al. (2001). Isolation and functional characterisation of the promoter region of the human prion protein gene. *Gene* **268**, 105–114.

Mallucci, G. R., Campbell, T. A., Dickinson, A., et al. (1999). Inherited prion disease with an alanine to valine mutation at codon 117 in the prion protein gene. *Brain* **122**, 1823–1837.

Manolakou, K., Beaton, J., McConnell, I., et al. (2001). Genetic and environmental factors modify bovine spongiform encephalopathy incubation period in mice. *Proc. Natl. Acad. Sci. USA* **98**, 7402–7407.

Manson, J. C., Jamieson, E., Baybutt, H., et al. (1999). A single amino acid alteration (101L) introduced into murine PrP dramatically incubation time of transmissible spongiform encephalopathy. *EMBO J.* **18**, 6855–6864.

McLean, C. A., Storey, E., Gardner, R. J., et al. (1997). The D178N (cis-129M) "fatal familial insomnia" mutation associated with diverse clinicopathologic phenotypes in an Australian kindred. *Neurology* **49**, 552–558.

McLennan, N. F., Rennison, K. A., Bell, J. E., et al. (2001). *In situ* hybridization analysis of PrP mRNA in human CNS tissues. *Neuropathol. Appl. Neurobiol.* **27**, 373–383.

Mead, S., Beck, J., Dickinson, A., et al. (2000). Examination of the human prion protein-like gene doppel for genetic susceptibility to sporadic and variant Creutzfeldt–Jakob disease. *Neurosci. Lett.* **290**, 117–120.

Mead, S., Mahal, S. P., Beck, J., et al. (2001). Sporadic – but not variant – Creutzfeldt–Jakob disease is associated with polymorphisms upstream of *PRNP* exon 1. *Am. J. Hum. Genet.* **69**, 1225–1235.

Meiner, Z., Gabizon, R. and Prusiner, S. B. (1997). Familial Creutzfeldt–Jakob disease: Codon 200 prion disease in Libyan Jews. *Medicine (Baltimore)* **76**, 227–237.

Moore, R. C., Hope, J., McBride, P. A., et al. (1998). Mice with gene targetted prion protein alterations show that *Prnp*, Sinc and Prni are congruent. *Nat. Genet.* **18**, 118–125.

Moore, R. C., Lee, I. Y., Silverman, G. L., et al. (1999). Ataxia in prion protein (PrP)-deficient mice is associated with upregulation of the novel PrP-like protein doppel. *J. Mol. Biol.* **292**, 797–817.

Moore, R. C., Mastrangelo, P., Bouzamondo, E., et al. (2001a). Doppel-induced cerebellar degeneration in transgenic mice. *Proc. Natl. Acad. Sci. USA* **98**, 15288–15293.

Moore, R. C., Xiang, F., Monaghan, J., et al. (2001b). Huntington disease phenocopy is a familial prion disease. *Am. J. Hum. Genet.* **69**, 1385–1388.

Nakamura, Y., Yanagawa, H., Hoshi, K., Yoshino, H., Urata, J. and Sato, T. (1999). Incidence rate of Creutzfeldt–Jakob disease in Japan. *Int. J. Epidemiol.* **28**, 130–134.

Nguyen, J., Baldwin, M. A., Cohen, F. E., et al. (1995). Prion protein peptides induce alpha-helix to beta-sheet conformational transitions. *Biochemistry* **34**, 4186–4192.

Nicholl, D., Windl, O., de Silva, R., et al. (1995). Inherited Creutzfeldt–Jakob disease in a British family associated with a novel 144 base pair insertion of the prion protein gene. *J. Neurol. Neurosurg. Psychiat.* **58**, 65–69.

Pan, K. M., Baldwin, M., Nguyen, J., et al. (1993). Conversion of alpha-helices into beta-sheets features in the formation of the scrapie prion proteins. *Proc. Natl. Acad. Sci. USA* **90**, 10962–10966.

Parchi, P., Gambetti, P., Piccardo, P., et al. (1998). Human prion diseases. In: *Progress in Pathology*, Kirkham, N., and Lemoine, N. R. (eds.). Edinburgh, UK: Harcourt Publishers Limited, Vol. 4, pp. 39–77.

Parchi, P., Giese, A., Capellari, S., et al. (1999). Classification of sporadic Creutzfeldt–Jakob disease based on molecular and phenotypic analysis of 300 subjects. *Ann. Neurol.* **46**, 224–233.

Parchi, P., Zou, W., Wang, W., et al. (2000). Genetic influence on the structural variations of the abnormal prion protein. *Proc. Natl. Acad. Sci. USA* **97**, 10168–10172.

Petchanikow, C., Saborio, G. P., Anderes, L., et al. (2001). Biochemical and structural studies of the prion protein polymorphism. *FEBS Lett.* **509**, 451–456.

Plaitakis, A., Viskadouraki, A. K., Tzagournissakis, M., et al. (2001). Increased incidence of sporadic Creutzfeldt–Jakob disease on the island of Crete associated with a high rate of *PRNP* 129-methionine homozygosity in the local population. *Ann. Neurol.* **50**, 227–233.

Prusiner, S. B. (1982). Novel proteinaceous infectious particles cause scrapie. *Science* **216**, 136–144.

Prusiner, S. B. (1998). Prions. *Proc. Natl. Acad. Sci. USA.* **95**, 13363–13383.

Prusiner, S. B. and Scott, M. R. (1997). Genetics of prions. *Annu. Rev. Genet.* **31**, 139–175.

Prusiner, S. B., Scott, M., Foster, D., et al. (1990). Transgenetic studies implicate interactions between homologous PrP isoforms in scrapie prion replication. *Cell* **63**, 673–686.

Raymond, G. J., Hope, J., Kocisko, D. A., et al. (1997). Molecular assessment of the potential transmissibilities of BSE and scrapie to humans. *Nature* **388**, 285–288.

Ripoll, L., Laplanche, J. L., Salzmann, M., et al. (1993). A new point mutation in the prion protein gene at codon 210 in Creutzfeldt–Jakob disease. *Neurology* **43**, 1934–1938.

Saborio, G. P., Permanne, B. and Soto, C. (2001). Sensitive detection of pathological prion protein by cyclic amplification of protein misfolding. *Nature* **411**, 810–813.

Salvatore, M., Seeber, A. C., Nacmias, B., et al. (1995). Apolipoprotein E in sporadic and familial Creutzfeldt–Jakob disease. *Neurosci. Lett.* **199**, 95–98.

Schatzl, H. M., Wopfner, F., Gilch, S., et al. (1997). Is codon 129 of prion protein polymorphic in human beings but not in animals? *Lancet* **349**, 1603–1604.

Shibuya, S., Higuchi, J., Shin, R. W., et al. (1998). Protective prion protein polymorphisms against sporadic Creutzfeldt–Jakob disease. *Lancet* **351**, 419.

Van Everbroeck, B., Pals, P., Quoilin, S., et al. (2001). The many faces of human prion diseases in Belgium and the world. *Acta. Neurol. Belg.* **101**, 81–87.

Westaway, D., Goodman, P. A., Mirenda, C. A., et al. (1987). Distinct prion proteins in short and long scrapie incubation period mice. *Cell* **51**, 651–662.

Will, R. G., Ironside, J. W., Zeidler, M., et al. (1996). A new variant of Creutzfeldt–Jakob disease in the UK. *Lancet* **347**, 921–925.

Will R. G., Ironside J. W., Hornlimann B., et al. (1996). Creutzfeldt–Jakob disease. *Lancet* **347**, 65–66.

Will, R. G., Zeidler, M., Stewart, G. E., et al. (2000). Diagnosis of new variant Creutzfeldt–Jakob disease. *Ann. Neurol.* **47**, 575–582.

Zahn, R., Liu, A., Luhrs, T., et al. (2000). NMR solution structure of the human prion protein. *Proc. Natl. Acad. Sci. USA* **97**, 145–150.

Zanusso, G., Farinazzo, A., Fiorini, M., et al. (2001). pH-dependent prion protein conformation in classical Creutzfeldt–Jakob disease. *J. Biol. Chem.* **276**, 40377–40380.

Zeidler, M. and Gibbs, C. J. Jnr. (1998). *WHO Manual for Strengthening Diagnosis and Surveillance of Creutzfeldt–Jakob Disease.* Geneva: World Health Organisation.

Zeidler, M. and Ironside, J. W. (2000). The new variant of Creutzfeldt–Jakob disease. *Rev. Sci. Tech.* **19**, 98–120.

Index